AF600677

NORTHERN STAR

J.S. Plaskett

Northern Star

J.S. Plaskett

R. PETER BROUGHTON

With a foreword by James E. Hesser

UNIVERSITY OF TORONTO PRESS
Toronto Buffalo London

Toronto Buffalo London
www.utppublishing.com

ISBN 978-1-4426-3017-8

Library and Archives Canada Cataloguing in Publication

Broughton, R. Peter, 1940–, author
Northern star : J.S. Plaskett / R. Peter Broughton ; with a foreword by James E. Hesser.

Includes bibliographical references and index.
ISBN 978-1-4426-3017-8 (hardcover)

1. Plaskett, J. S. (John Stanley), 1865–1941. 2. Astronomers – Canada – Biography. 3. Biographies. I. Title. II. Title: J.S. Plaskett.

QB36.P63B76 2018 520.92 C2017-907026-6

University of Toronto Press acknowledges the financial assistance to its publishing program of the Canada Council for the Arts and the Ontario Arts Council, an agency of the Government of Ontario.

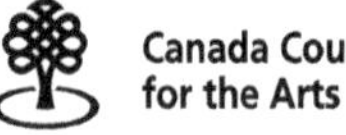

Contents

Part Four: The Fruits of His Labour

Illustrations are throughout and in a section following page 114.

Foreword

The establishment of a major observatory near Victoria, British Columbia, a century ago can be argued to represent Canada's first foray into what today would be termed "big science." That the federal government agreed to such a challenging development a continent away from the capital and adhered to its decision throughout the First World War remains powerful testimony to Canada's determination to begin to play a strong role in international science and technology. The creation by the Canadian government of the Dominion Astrophysical Observatory (DAO) only some fifty years after Confederation stands in contrast to the situation in the United States, where private funding stimulated initial development and operation of their famous western optical observatories.

The scientific, technical, and administrative accomplishments of John Stanley Plaskett, the founder of the observatory, make for compelling reading as we anticipate with twenty-first-century perceptions the "first-light" centenary, on 6 May 2018, of the telescope that now bears his name. His was a well-lived life. His journey from childhood on an Ontario farm, through challenging early life experiences following the premature death of his father, and subsequent employment in machining and electrical trades nourished his propensity for technical matters. The accumulated experiences formed rock-solid principles that enabled him to lead the development of Canada's first scientific facility designed to compete on the world stage. His hard-earned knowledge of building things led through assiduous research to his becoming expert in the design of scientific instrumentation (chiefly spectrographs) second to none in the world, and to being consulted far and wide for advice. Family members and contemporaries in Canada, the United

States, England, and continental Europe played key parts in Plaskett's remarkable story, which includes personal shortcomings, professional disagreements, and determined boldness in the face of opportunities and their challenges.

Plaskett's vision, determination, style, and ingenuity continue to be well reflected in contemporary Canadian astronomy and, indeed, in Canadian science generally. His quest to build what became for almost a year the world's largest operational telescope was driven by carefully articulated scientific goals arising from extensive international consultation that Plaskett initiated with the leading astronomers of the day. In everything, he insisted that the science be kept front and centre, which is a *sine qua non* for scientific success. In designing and overseeing the implementation of the initial research programs for the DAO, his plans were ambitious and farsighted, characteristics we associate with "key" or "large" projects in international astronomical facilities of the late twentieth and early twenty-first centuries. Research that Plaskett began in 1918 on the properties of the most luminous stars in the Milky Way culminated (through the crooked path that research typically follows) some fifteen years later in Plaskett and his younger colleague J.A. Pearce determining the size, mass, and general structural picture of our home galaxy, the Milky Way – determinations that have stood the test of time remarkably well. Similarly, DAO astronomers' work on spectroscopic binary stars played key roles in early-twentieth-century understanding of basic properties of stars and their evolution that underpinned the transition to extensive results on X-ray binaries and black holes in the 1980s and 1990s.

In this biography, the reader comes to appreciate that Plaskett, through his extensive, high-quality research publications and frequent presentations at scientific conferences in Canada, the United States, and Europe, became *the* face of professional astronomy in Canada and internationally. He received national and international award upon award throughout his astronomy career, remaining without equal today. Moreover Plaskett felt deep responsibility for disseminating astronomical discoveries to the public, and encouraged others to do likewise. From day one, he ensured that the DAO was open to the public for Saturday evening viewing. Plaskett and his small staff frequently shared astronomical discoveries and associated developments at venues in Victoria and across the country. During his life, he contributed much to the Royal Astronomical Society of Canada (RASC), publishing numerous popular and professional research articles in its *Journal.* (I speculate that, were he to return today, he would not only be pleased with the impact of the RASC in Canadian society, but equally delighted with the existence of an effective body of professional astronomy founded in 1971, the Canadian Astronomical Society.)

To this day, the DAO remains federally funded, so that many of the stories of Plaskett's experiences with Ottawa regarding such things as budgets, travel approvals, and hiring have an all-too-familiar ring. With the observatory being, since 1970, a part of the National Research Council (founded in 1916 – that is, in the same timeframe in which DAO was being built), the thought that the present-day DAO director, or that person's superiors in the NRC organization, could communicate directly with the prime minister on an issue of operational importance is unfathomable, yet, as we learn in this book, Plaskett did so as a last resort on more than one occasion.

Perusal of volume 1 of the *Publications of the Dominion Astrophysical Observatory* reveals Plaskett's involvement in, and deep appreciation for, every aspect of the successful development of the telescope now bearing his name and its ancillary instrumentation. In modern parlance it is clear that he exemplified the precepts of what has come to be known as "systems engineering," which begins with clear specifications of the desired attributes. Over the years he generously shared with others what he learned from his meticulous research. Through mentoring of younger DAO staff, some of whom subsequently joined the staff of the fledgling David Dunlap Observatory of the University of Toronto, his knowledge of stellar astronomy and of how to build efficient spectrographs capable of yielding accurate stellar radial velocities was passed on to subsequent generations of students.

Since initiation of the DAO under Plaskett's inspired scientific leadership, Canadian astronomy has continually aspired to and succeeded at the highest international standards. Consolidation of federal astronomy in the NRC with a mandate from Parliament to operate and administer the observatories established by the government of Canada provided a basis for exploring opportunities that have dramatically altered the landscape in the past four decades. The creation in 1974 of Canada's first international astronomical facility partnership (to build and operate the Canada-France-Hawaii Telescope at 4,000 m on Maunakea, Hawaii), led to the DAO staff – and the Herzberg Institute of Astrophysics of which the observatory became a founding element – evolving to focus increasingly on providing access to facilities needed by astronomy community members at large for their observational research. Today most community members work in the university system, which underwent marked growth after the Second World War. Effective interaction with that broader community regarding development and implementation of plans to remain internationally competitive is the focus for the scientists and engineers at the NRC Herzberg, including the DAO. Through continued research excellence, they are strong contributors at national and international venues developing or operating major observational facilities.

As I reflect on Plaskett's story, I am struck by how strongly the guidelines he lived by are mirrored in the observatory's highly successful twenty-first-century support of Canada's national and international astronomy activities. The community is strongly science-driven to ensure that Canadian resources are addressing the most pressing astronomical challenges of the present era. NRC Herzberg staff today provide unique instrumentation and data management tools, as well as project management expertise, for the Canadian and international communities to tackle scientific problems unknown during Plaskett's lifetime. I'm fond of pointing out to visitors that DAO's current renown for extremely efficient, reliable spectrograph designs reflects directly Plaskett's heritage and his influence on Canadian astronomy. DAO astronomer David Crampton led the development of major spectrographs for both the Canada-France-Hawaii Telescope and the twin Gemini 8-m telescopes. His career began by determining accurate stellar radial velocities at the David Dunlap Observatory, whose spectrograph reflected Plaskett's design precepts, particularly in terms of the mechanical rigidity that facilitates radial velocity determinations on Cassegrain-focus instruments. With the introduction of digital detectors in astronomy, Crampton's insistence on, among other elements, integrity of mechanical structural design for spectrographs destined for the international facilities substantially reduced the time required to calibrate the data, thereby greatly increasing operational efficiency of heavily oversubscribed large telescopes and thus their enhanced scientific productivity.

A century after the founding of the DAO, Plaskett's achievements merit deep respect and wonder, and Peter Broughton's telling of his life and career will help readers – particularly Canadians – appreciate why. I am confident that were J.S. Plaskett to visit DAO today, he would feel that the scientific, instrumentation, data management, and community leadership of the DAO staff in the past half century remains true to the spirit he fostered in the first three decades of the observatory's existence. However, the telling of the story of the past fifty years awaits another day. May such an effort be encouraged by this inspirational biography of John Stanley Plaskett, a Canadian of immense foresight, ability, and perseverance who had an enormous impact on the development of twentieth-century astrophysics.

James E. Hesser
DAO Director (1986–2013)
Strategic Advisor
NRC Herzberg Astronomy and Astrophysics Program
National Research Council of Canada

Preface

John Stanley Plaskett saw first light on 17 November 1865 in a provincial corner of the British Empire known as Canada West. There, on the family farm in Oxford County, his parents welcomed their firstborn. A lifetime later he stood before a crowd of luminaries at Oxford University to deliver the prestigious 1935 Halley lecture. He was by then a Companion of the British Empire and Fellow of the Royal Society, winner of the Royal Astronomical Society's Gold Medal and many North American scientific awards – more, in fact, than any other Canadian in the first half of the twentieth century.[1] This book tells the story of his accomplishments from the plough to the podium. It was quite a journey. Like the North Star, Plaskett pointed the way for future exploration, enabling him and his scientific heirs to build an international legacy for themselves and the nation. In many ways, Canadian astronomy seemed to revolve around him, though some of his contemporaries fought against that perception.

Why should a biography of an astronomer, albeit a celebrated one, arouse much interest? In the first place, astronomy holds a unique place in the human psyche. We all question where we fit in. The answer, of course, is always evolving, but the story of the intellectual trip that has brought us to our present understanding is a superb adventure. Many of the great advances in knowledge have their roots in astronomy – the oldest of the sciences. There is plenty of evidence of a deep attraction to astronomy. The most-watched show in public television history was Carl Sagan's *Cosmos* series.[2] It has been seen by more than 600 million people in 60 different countries since its debut in 1980 and has recently been revived as a sequel hosted by Neil deGrasse Tyson. Another TV documentary, *The Sky at Night* with Patrick Moore, must hold the record for enduring popularity, as it was aired for over fifty years (1957–2012). From widely-read sky guides such as Canadian Terence Dickinson's *Night Watch* to more esoteric writing like Stephen Hawking's *A Brief History of Time*, there is an avid appetite for books about the universe.[3] Each year, conventions of the American Astronomical Society attract nearly a hundred journalists and associated media types, eager to hear about the latest discoveries and theories. Why all the fuss? Canada's grand dame of astronomy, the late Helen Sawyer Hogg, put her finger on it, when she said, "Not to know what's beyond is like spending your life in the cellar, being completely oblivious of all the wonderful things around us."[4] Or, as Canadian science historian Richard Jarrell argued in a lecture he gave on the value of astronomy for a civilized society, life without astronomy is like life without art or music.[5] Yes, we would probably have most of our marvellous technology, but what poor, impoverished lives we would lead.

In any narrative, people are always the most important ingredient. Biography opens a window for us to see through others how families and education in addition to political, social, and economic factors affect each of us. Plaskett's life illustrates how his career was shaped by these agents and by the scientific milieu in which he worked; his story parallels the transition of the nation of his birth from its rural roots to an important international power. By reviewing his life, we are invited to question the role of nature and nurture, luck and pluck, challenges and opportunities in life's journey. As we consider the influences on his work, we may confront our own endeavours and question the extent to which we are each masters of our own destiny.

By tracing Plaskett's active career, we are actually sketching the first thirty years of astrophysics in Canada, as he and his assistants had the field pretty much to themselves. In contrast to the traditional astronomers, who were almost exclusively concerned with the apparent position and motion of the stars and planets, astrophysicists analysed the light of stars to determine their physical properties such as size, mass, temperature, and variability. Plaskett's biography can give a good idea of what occupied his profession, especially in the highly productive period between the world wars. I hope that readers without a background in science will find my approach friendly; I have striven to explain any essential technicalities as they arise. However, for anyone feeling the need for a fuller appreciation of Plaskett's research interests – the astrophysics of the stars and the Sun – I would highly recommend

James B. Kaler's *Stars and Their Spectra*.[6] It successfully bridges the gap between popular and professional needs.

The names of a few famous astronomers – Copernicus, Galileo, and Hubble – are very familiar, helped by their eponymous space probes and substantial biographies. For most of us, naming any Canadian astronomers could be more problematic. Surely if we are fascinated by astronomy, we would enjoy knowing more about those who devoted their lives to it. If we Canadians know our artists, such as the Group of Seven, or authors such as Stephen Leacock and Lucy Maud Montgomery, who all flourished at the same time as Plaskett, should we not be informed about this astronomer who also put the nation on the world stage? Science has not always played second fiddle to the arts: at least one commentator in the 1920s saw that decade as a Canadian Elizabethan age, not for its writers and artists, but for its scientists.[7]

Canadian astronomers have certainly done their best to perpetuate Plaskett's memory, but their efforts probably have not percolated down to the public at large.[8] The problem is partly a "two cultures" divide, but not entirely so. The best of art has an eternal appeal but science moves on. Ideas that were indispensable at one time become mere building blocks in the magnificent structure of understanding that is continuously being erected. Medical discoveries can resonate more personally, as they may continue to save countless lives. Canadian historian Michael Bliss knew this intuitively when he wrote his acclaimed biography of Plaskett's contemporary, Nobel laureate Frederick Banting, the co-discoverer of insulin.

There is no Nobel Prize for astronomy – the closest link between Canadian astronomy and the Nobel Prize came with the 1971 award in chemistry to Gerhard Herzberg. Though only a dozen papers in his output of a couple of hundred are directly on astronomical topics, a lot of his lifelong spectroscopic research has a bearing on astronomy. Canadian astronomers are happy to call him one of their own – the National Research Council of Canada's Herzberg Astronomy and Astrophysics Programs, which now oversees the observatory that Plaskett built, is evidence of their pride in Herzberg's work. His colleague, Boris Stoicheff, wrote a masterful biography: *Gerhard Herzberg: An Illustrious Life in Science* and Herzberg's son, Paul, penned a personal memoir of his mother Luise, an astrophysicist in her own right. For me, these books have set a stimulating standard.

As appealing as it might be for a colleague or family member to write a biography, Plaskett has now been dead for seventy-five years, so no one who knew him will be coming forward to do so. On the other hand, the lifetime that separates us from Plaskett coupled with the objectivity of a stranger may lead to a more balanced assessment. Beyond any transitory pleasure this biography may bring, I trust that it will result in a lasting understanding of Plaskett's achievements as well as the forces and relationships that made them possible.

Within my own adult life, the academic discipline of the history of science has emerged as a field in its own right. Its practitioners sometimes use biography as a vehicle for their studies but also approach their interests in many other ways – for instance by basing their work on institutions, instrumental developments, intellectual advancement, or public perceptions. While a major biography of Plaskett is without precedent, I have had the benefit of previous relevant research. Over twenty-five years ago, John Hodgson, who knew the Dominion Observatory from the inside, wrote an insightful but poorly publicized history of the two institutions that preoccupied Plaskett during his career, the Dominion Observatory (DO) in Ottawa and the Dominion Astrophysical Observatory (DAO) near Victoria.[9] About the same time, York University historian Richard Jarrell composed a unique general history of Canadian astronomy, in which Plaskett's achievements play a vital part.[10] More recently, Jarrell contributed a number of articles on some aspects of Plaskett's career, highlighting his role in the development and spread of large reflector telescopes so crucial in the advancement of twentieth-century astronomy. Tom Plaskett, the astronomer's brother, wrote an unpublished family history, which gives invaluable clues to their formative years.[11] John Lankford studied the structures and relationships that defined American astronomy during the period 1859–1940, coincidentally close to Plaskett's lifespan.[12] This astronomical community, of which Plaskett was an essential part, was integrated through shared scientific concerns, organizations, journals, and higher education. I have found the work of these four authors to be very valuable.

Lankford dealt with the situation in the United States. Canada – even English-speaking Canada, which was Plaskett's milieu – is, of course, a distinct society.[13] Neither British nor American, the country feels pulls in the directions of those powers, and yet there are features that set it apart from both. In Plaskett's time, the great majority of Canadians, other than Quebeckers, were of British stock, including the Loyalists who had come north after the American Revolution. So ties to the "old country" were strongly felt. Perhaps one manifestation of this was an emphasis on peace, order, and good government, in contrast to the American pursuit of life, liberty, and happiness, abetted by a strong streak of individualism.

Though Plaskett had the greatest admiration for his British colleagues, geography dictated that he was more closely integrated into the American community. Outside Canada, his closest friends and the society meetings he most frequently attended were American. The leaders in these circles unconsciously served him well as role models. Though he shared their "unbridled ambition, vast egos, and great vision," his situation served to temper those characteristics.[14] The fact that he was, after all, a civil servant, and that Canadian astronomers were few in number, meant that he never exercised the same power as the leading observatory directors in the United States. Yet, on both sides of the Atlantic, he was considered as an equal and was recognized with the highest rewards that his profession could provide.

Because of Canada's relatively small population (about a tenth that of its neighbour to the south and a half that of the United Kingdom), and because it moved away from colonial status nearly a century after the United States, its national cultural characteristics, like support for the arts and specialized sciences, made a relatively late appearance. Not only was government funding slow to materialize, but there were almost no private donors or philanthropic foundations to encourage development in these areas. Against this background, it is all the more remarkable that Plaskett managed to secure for Canada an astronomical observatory housing the largest telescope in the world. On the other hand, the smallness of the spheres in which he moved, whether academic, scientific, governmental, or political, meant that almost all those involved knew one another. He understood that, with proper tending, these personal ties could be very fruitful. I also think it is fair to recognize that Plaskett was "a big frog in a small pond" and that he might not have received such exceptional recognition if he had been working in the United States or in Britain.

In doing the research for this book, I have used primary sources as fully as possible. Through Plaskett's extensive professional correspondence and publications, I have examined his substantial contributions to the key new ideas of astronomy evolving during his era. Such an analysis, I believe, should be the main purpose of his biography. Though Plaskett unfortunately left no family letters or diaries, I have been able to use the diaries of some of his contemporaries to create an account of the more private side of his life. While it is inspiring to see how Plaskett rose from ordinary circumstances to gain fame for himself and his country, I am not trying to make a superman of him. By being frank about his few foibles, I trust that his achievements will seem all the more real.

For the many colleagues, correspondents, and contemporaries of Plaskett who appear in this biography, I usually give only enough information for the reader to appreciate who they were and how they interacted with Plaskett. Providing too many details in the text would divert attention from the main narrative; however, I have identified them by their full names and dates in the index. Further information on individual astronomers (at least the better-known ones) can usually be found in the *Biographical Encyclopedia of Astronomers*, while prominent Canadians (and some lesser lights) appear in the *Dictionary of Canadian Biography*.

A few words need to be said here about another remarkable Canadian astronomer born in the same year as Plaskett – Clarence Augustus Chant. Both men are considered fathers of Canadian astronomy, so Chant will appear often in the pages of this book. He played a formative role in setting up an honours course in astronomy at the University of Toronto and in strengthening the Royal Astronomical Society of Canada from a local group into a national organization respected around the world. After decades of effort, he persuaded a generous benefactor to establish a major observatory with a telescope even larger than Plaskett's, but it came about too late in his life for him to do any research with it. In many ways, Chant's life was complementary to Plaskett's. Though his achievements also had great and lasting consequences, the title he chose for his personal story, *The Quiet Life*, succinctly sums it up – quite a contrast from Plaskett's bold adventure.[15]

I conclude with two notes of explanation. First, I often refer to John Stanley Plaskett as JSP. Though his family knew him as Jack, no one else did, and it seemed presumptuous of me to do so once he reached adulthood. This use of initials allowed me to reduce repetitious references to "Plaskett" and to avoid confusion with other members of his family.

Second, except for direct quotations where old units like inches, feet, and miles are used, I have employed metric units. This is a bit awkward, as Plaskett always referred, for example, to the 72 inch reflector in Victoria, whereas it is now described as a 1.83 metre telescope. I trust the reader will make the transition easily.

Acknowledgments

First and foremost, I turn my attention to the Royal Astronomical Society of Canada – my extended family for the past five decades. I am very grateful to the society for its consistent generosity and undue praise, especially now for its special projects grant in aid of this biography of a famous former president.

In the course of researching this book, I have had the cooperation of over a hundred kind and dedicated people, mostly librarians and archivists. Perhaps, as the nature of their work is to serve the public, they will forgive me for not naming them here. Included would be a number of members of staff at the Dominion Astrophysical Observatory who were exceedingly generous with their time. Elizabeth Griffin, who works tirelessly as a volunteer to maintain the DAO archives, deserves special mention for her successful efforts to provide me with a scan of the first spectrogram as well as one of JSP's photographs showing the great globular star cluster in Hercules. Jack Newton, one of Canada's outstanding astrophotographers, provided beautiful images of the Milky Way and of our neighbouring galaxy in Andromeda. RASC's awesome archivist, Randall Rosenfeld, used his artistic talent to produce a map and two diagrams.

It is a pleasure to acknowledge correspondence and conversations with John Stanley Plaskett's descendants – his grandson and namesake in England, his great-granddaughter Deborah Tilley in Halifax, and his niece, Mary Dyson, and her daughters in Victoria.

I am greatly indebted to seven people who read the entire draft of the manuscript at various stages – two anonymous readers for the University of Toronto Press; Dr Jim Hesser, retired director of the DAO, who also has enhanced this book with his foreword; Dr Alan Batten, a renowned expert on binary stars (also retired from the DAO); the aforementioned Randall Rosenfeld; Andrew Oakes, a doctoral student at the University of Toronto's Institute for the History and Philosophy of Science and Technology; and especially my dear one, Marilyn, who also helped with the research and endured (or perhaps enjoyed) times of solitude when I was preoccupied with Plaskett. All these readers made invaluable suggestions and offered much-needed moral support for this project. It has been a long road from the original conception of this book as a joint project with David Smillie, through numerous drafts. I would have given up long ago without the encouragement of all of the above plus the unstinting guidance of historian Ted Binnema, author of *Enlightened Zeal: The Hudson's Bay Company and Scientific Networks, 1670–1870*, recently published to wide acclaim by the University of Toronto Press.

Finally, I express my gratitude to UTP and editor Len Husband for agreeing to publish this book. At times I despaired that they would ever do so, but the delays, so annoying at the time, kept me striving for higher standards. With Frances Mundy's patient care and copy editor Barbara Tessman's substantial improvements in hand, I can only hope that we have reached some sort of plateau from which the lofty summits of perfection can be glimpsed in the distance.

Abbreviations

AAAS	American Association for the Advancement of Science
AAS	American Astronomical Society
ASP	Astronomical Society of the Pacific
BA / BAAS	British Association for the Advancement of Science
BCHA	British Columbia Historical Association
CADC	Canadian Astronomy Data Centre
CASCA	Canadian Astronomical Society / Société Canadienne d'Astronomie
CFHT	Canada-France-Hawaii Telescope
DAO	Dominion Astrophysical Observatory (Victoria, BC)
DO	Dominion Observatory (Ottawa)
IAU	International Astronomical Union
NAS	National Academy of Sciences (USA)
NRC	National Research Council (Canada)
NRC Herzberg	NRC Herzberg Astronomy and Astrophysics
RAS	Royal Astronomical Society (London, England)
RASC	Royal Astronomical Society of Canada (headquartered in Toronto, but with local centres in many Canadian cities)
RSC	Royal Society of Canada (Ottawa)
SPS	School of Practical Science (engineering school affiliated with U of T)
U of T / UT	University of Toronto

NORTHERN STAR

J.S. Plaskett

PART ONE

Preparing the Ground

1
Rural Roots, to 1890

When the great Laurentide ice sheet receded twelve thousand years ago, its last bastion was the vast northern plateau of Precambrian rock known as the Canadian Shield. Around the perimeter of its immense icy tongue, it left behind great lakes and fields of rich glacial till. Eventually this area became attractive and fertile farmland in the southernmost part of Canada, nestling between Lake Huron to the northwest and Lakes Erie and Ontario to the south and east. It has nourished the bodies, minds, and spirits of some remarkable people.

In the heart of this good land lies Oxford County and, at its core, the town of Woodstock. European settlement in this part of Canada began in the 1790s, when the first British lieutenant governor, John Graves Simcoe, imprinted the land with familiar names from home. He even intended that the colonial capital would be at London in Middlesex County, on the River Thames, downstream from Oxford County and the town of Woodstock – just as it was in England. That tradition would have been quite acceptable to the settlers who received their land grants as a reward for their allegiance to Britain during the American Revolution and again following the War of 1812. Chief Joseph Brant and the Six Nations were among those loyal to the Crown, and Brant County was named for him when it was carved out of the eastern part of Oxford County in 1853. It was on the Six Nations reserve where Pauline Johnson, the renowned English-Mohawk poet, was born in 1861. About the same time Oronhyatekha, from the same reserve, was studying in England at the University of Oxford, the first indigenous North American to do so.

Oxford County is presently divided into five townships, though that has changed several times over the years. Immediately to the north of

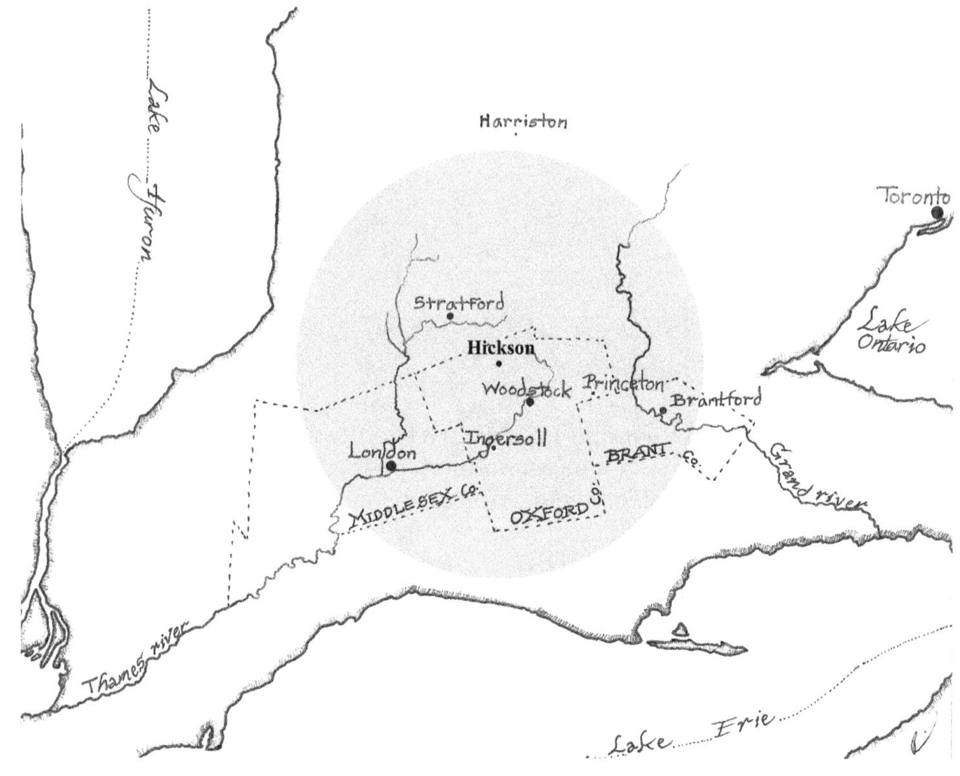

1.1 Map of JSP's southern Ontario, around Oxford County (Map by Randall Rosenfeld)

Woodstock is the township of East Zorra, the home of the Plasketts from 1859.[1] Their farm, called March Hill, was adjacent to the hamlet of Strathallan, originally named by the highland Scots pioneers who, about 1830, had hewn the trees, broken the soil, built houses, and laid out roads.

Many people who rose to prominence had roots similar to those of John Stanley Plaskett (JSP). Within ten years of his birth and fifty kilometres of his birthplace, several such striking coincidences can be found, in the scientific realm alone.[2] John Patterson, one of Canada's most brilliant meteorologists, was born in Oxford County in 1872. Thomas "Carbide" Willson, promoter of hydroelectric development and inventor of an industrial process for producing calcium carbide used in the production of acetylene gas, was born in 1860 at Princeton, twenty kilometres east of Woodstock and later in life lived in town at 210 Vansittart Street, the beginning of the Twelfth Concession running north to the Plasketts' farm. Physicist Sir John McLennan, with whom JSP would work at the University of Toronto, was born in 1867 in Ingersoll, just fifteen kilometres southwest of Woodstock. Up the rail line twenty-five kilometres north of the March Hill farm, Thomas Edison was working at Stratford as a telegrapher in the early 1860s, and within a few years Melville Bell and his family, including his son Alexander Graham Bell, arrived from Scotland to settle near Brantford, forty-eight kilometres to the east. It was here in 1874 that Bell first described the principle of the telephone and made the world's first long-distance call two years later. Very near the Bell homestead was the birthplace of Augusta Stowe-Gullen, the first woman to gain a medical degree in Canada and a strong advocate for women's rights. From Brantford itself came Robert Kennedy Duncan, who pioneered research in industrial chemistry and became the first director of Pittsburgh's Mellon Institute. What was it about this time and area that started so many on the road to fame? Was there something in the water? More likely the healthy, provident environment gave these residents and others like the Plasketts the confidence that hard work could lead to an abundant future. We should also look to JSP's forebears for possible evidence that their characteristics, beliefs, and occupations rubbed off on him.

Ancestors

Both sides of John Stanley Plaskett's family had roots in England's northern Lake District. His maternal grandparents, John Stanley (a miller) and Mary Thwaites, were married in the parish of Crosthwaite, Cumberland. They had ten children; the youngest was Annie Stanley (JSP's mother), born on 12 March 1840 at Dalston.[3] In 1848, after John's death, Mary Stanley and her five surviving children moved to Canada, where she kept house for a bachelor brother, Joseph Thwaites, a former planter on Santa Cruz (now Saint Croix, U.S. Virgin Islands) who had just bought a forty hectare (100 acre) farm north of Woodstock.[4] Likely Mary and the children reached Canada via New York and the Erie Canal; we do know they travelled on Lake Erie by sail boat to Port Maitland, then by scow up the Grand River to Brantford, and finally by ox-cart to the Twelfth Concession, East Zorra, to Thwaites' farm (Maple Grove).[5] Annie was eight when they came. She and her brothers and

1.2 John Stanley Plaskett as an infant, holding an alphabet book (Photograph donated by Mrs. J.S. Plaskett to BCA, Image I-76925 courtesy of the Royal BC Museum and Archives)

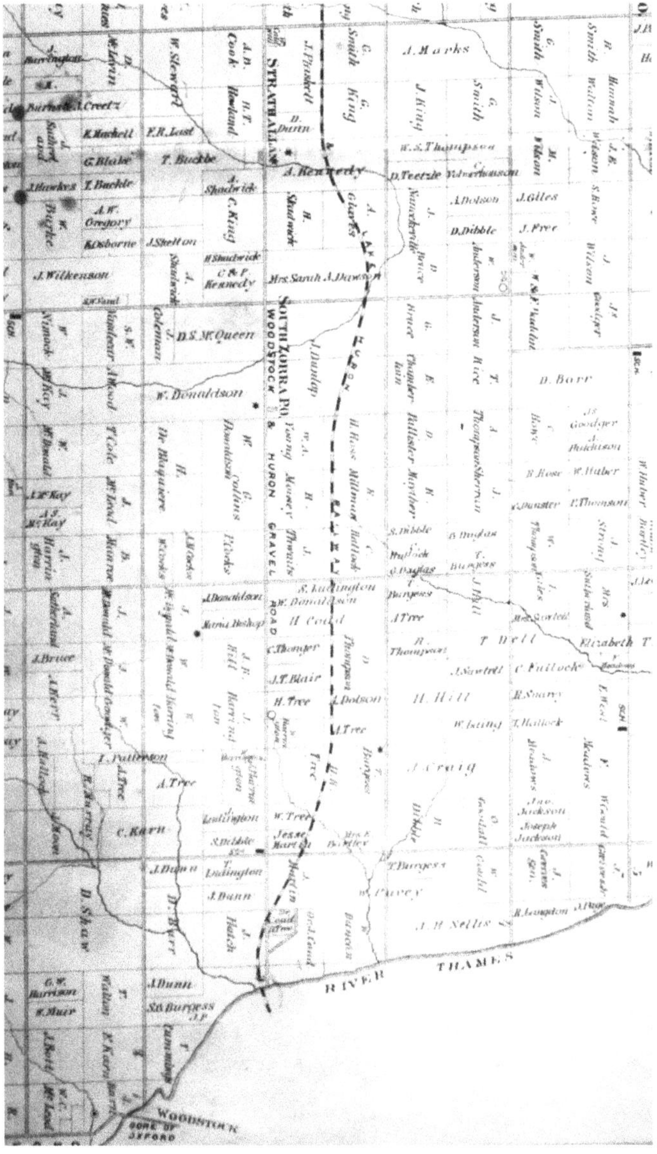

1.3 Part of an 1876 map of the township of East Zorra. The Plaskett farm is shown near the top of the portion reproduced here, on the east side of the main road running north from Woodstock. The church that the family attended is on the west side of the road at the property marked "W. Donaldson," and the primary school that JSP would have attended was, until 1874, in the churchyard. After that date, the school was located on the Dunlop property. The property of "J. Thwaites," JSP's great uncle, is also nearby. (Canadian County Atlas Digital Project, Rare Books and Special Collections, McGill University Library)

1.4 Plaskett memorial tablet and Church of St Mary, Threlkeld, Cumbria, England. The top line of the inscription shows the names of JSP's great-great grandparents John and Isabella. (Image courtesy of Stuart Cresswell)

sisters all shared in the farm chores. After labouring on the farm for fifteen years, Annie married Joseph Plaskett at Christ Church in East Zorra on 30 December 1863.[6] They made their home at the Plaskett farm, not far up the Twelfth Concession from Maple Grove; there at March Hill their first child, John Stanley Plaskett, was born on 17 November 1865.[7] He soon would be nicknamed Jack by his family.

On the paternal side of the family, JSP's known ancestry begins with his great-great-grandfather, John Plaskett and his wife Isabella. As a memorial tablet at St Mary's Church, Threlkeld, England, makes clear, this side of the family also had strong ties to Saint Croix. One of John and Isabella's sons was Wilfred, who married Ann Mounsey in 1797.[8] Over the next ten years they had seven children, including Timothy (JSP's grandfather). Wilfred died at age thirty-four while travelling to the West Indies, leaving Timothy and two other sons and one daughter to be brought up by relatives. The three boys went to live with their Uncle Joseph in Saint Croix. This uncle – Major Plaskett, to give him his proper rank in the island militia – had settled there before 1815, when Britain ceded the islands to Denmark after the Napoleonic wars. He held a large sugar plantation and became very wealthy. Timothy also became a planter on Saint Croix, where he and his wife, Sarah, had five children between 1835 and 1844 and where they remained until 1848.[9]

We have already seen the circumstances that brought the maternal side of JSP's family to Canada, but what motivated his paternal grandparents to immigrate? When the slaves of Saint Croix were promised freedom in 1847, most of their owners got out, knowing that they could not make a profit without slave labour. One of these planters, John Dawson, evidently a friend of the Plasketts, came to Canada and wrote glowingly about his new home. Timothy Plaskett at first returned with his family to the north of England. He had his three daughters educated in art, French, music, and deportment and sent his two sons, John and Joseph, to the Dumfries Academy in Scotland for their education.[10] Joseph (JSP's father) excelled in mathematics and Latin. After graduation he was delegated to come to Canada to see first-hand the conditions in Oxford County that Dawson had extolled. Apparently Joseph's report was favourable for, in 1859, Timothy brought his family out to Woodstock, where several English families of means had taken up residence. JSP's brother Tom wrote later in life that his grandparents and their family

> stayed in Woodstock a short time, entertaining, driving around in a carriage with a coachman, and, as we would say nowadays, putting on the dog. My grandfather bought a farm on the twelfth line, East Zorra,

1.5 A photograph, taken in 1982, of the large brick house located on the former Plaskett property in East Zorra, Ontario, where JSP was born and raised. (Photo by author, courtesy of RASCA)

> near two other quite wealthy West Indies planters – Mr. Thwaites and Mr. Dawson.
>
> [Grandfather] built a large brick house, laid out orchards and a winding evergreen driveway in the English mode, and then started to farm and to entertain in a lavish way. [11]

The farmhouse stood on the east side of what is now Highway 59, just south of what is now Loveys Street South in the amalgamated municipality of East Zorra Tavistock. The house burned down in 1983, but part of the winding driveway, with what are now very large evergreen trees is still there.[12] Most of the Plasketts' original fifty-six hectares is still farmland, but the northern edge along Loveys Street has been subdivided and about twenty family homes now line it. They are part of the village of Hickson. The western boundary of the old farm, along the highway where the settlement of Strathallan originally was located, is now mainly an agricultural supply company.

As Tom tells it, his grandfather Timothy and most of the other wealthy Englishmen who farmed in that district depended completely on hired help and spent their days riding around on horseback, sporting, shooting, and so on.[13] By the time of Timothy's death (24 May 1872), his estate was valued at $12 000 – a moderately large sum for the times.[14]

Family Life after 1865

On the death of Timothy Plaskett his son Joseph, by then married to Annie and the father of JSP (Jack) and Robert (Bert), took over March Hill by paying off his siblings' share of the inheritance and purchasing an adjoining twenty hectares, making the property considerably larger than average. Joseph was active in the East Zorra Agricultural Society, serving as its auditor and taking part in its fall fairs. When the fair was held in Strathallan in 1874, it attracted 430 entries of all sorts and Joseph won three prizes for cattle and one for barley, while his younger sister Sarah won a prize for needlework. Well regarded within the community, Joseph served as a lieutenant in the reserve militia in 1869, deputy reeve in 1875, and later as president of the East Zorra Rifle Association. Joseph was very proud of his children, eventually ten in number, and would line them up by height for visitors.[15] He delighted in playing cricket with his boys and enjoyed life – hosting dinners in the family dining room and holding square dances in their large parlour.

Tom painted a picture of a happy home life. He wrote of the big bedrooms with two double beds, pillow fights, a love of the open spaces, hunting, fishing, swimming, and games. He recalled the joy of getting Christmas presents – hampers of good things from Aunt Esther and Uncle Duncan in Montreal and adventure stories and the *Boy's Own Annual* from his Uncle John and Aunt Eliza in Toronto.

The *Boy's Own Annual* was a natural choice for a gift from Uncle John S. Plaskett, who owned a bookstore on Toronto's main thoroughfare, Yonge Street. The boys loved getting the *Annual*, which was the year's collection of the weekly *Boy's Own Paper*, all bound together. From its inception in 1879, this remarkable publication was profusely illustrated and had plenty of tales of battles, sports, explorers, school, and nature. There were arithmetic puzzles, games, conjuring tricks, and other amusements. Jack was already exploring the "How To" articles; the first volume alone included "The Telephone and How to Make It," "Pleasant Hours with the Magic Lantern," and "My Boat and How I Made It." Regular features were "Boys of English History" and "Boys

Who Became Famous." The article under this heading for 10 May 1879 featured Michael Faraday, who came from a working-class family and learned everything from reading voraciously and experimenting inquisitively. The article, illustrated with a picture of him lecturing to the Royal Institution, began with these words: "One of the most treasured curiosities of the Royal Institution is an old electrical machine, which was made about the beginning of the present century by a boy fourteen years old." JSP, who was also fourteen when he read those words, must have been impressed. As we shall see, he built his own electrical machine.[16] Perhaps a bit of doggerel in the same volume also registered with him. Entitled "An Unrecognized Genius" it began,

I am a great philosopher
Of wondrous penetration
I *think* I am the greatest man
In this or any nation …
While dressing, all my thoughts were bent
On certain calculations
About the distances of stars
In various constellations.
When the result was nearly gained
By process long and slow,
I tumbled from the Milky Way
Brought down by "Milk Below."[17]

The late 1870s were hard economically, and Joseph and Annie sold part of their farm, arranging a mortgage for $4000 on the remainder with Joseph's sister, Mary Ann.[18] There may have been other reasons for the family's difficulties. Joseph, known to be fond of his drink, died of intemperance on 17 November 1881 at the young age of forty-three. His estate, comprising household goods and furnishings, farming implements, horses, horned cattle, sheep and swine, and farm produce of all kinds, was valued at only $2025.[19]

Joseph was eulogized for his excellent ability and for his "free, generous disposition. Politically an ardent Conservative, he never hesitated to announce his views and stand by them, but in all his contests, political, municipal, or personal, and to friend and opponent alike, his sturdy, manly English nature asserted itself and left him on the best of terms with everybody."[20] JSP inherited many of these characteristics, including his father's political preferences.[21]

JSP's mother Annie, unlike her husband, had been used to hard work all her life and was the mainstay of the family. She was a noble figure like "the little mother" portrayed by JSP's contemporary Ralph Connor, in his wildly popular novel, *Glengarry School Days*, "ever on guard, [so that] all the machinery of house and farm moved smoothly and to purpose because of that unsleeping care."[22] After her husband's death, Annie had to fend for herself and her family just as her mother had had to do thirty-three years earlier. Tom said she was the beauty of the community in her younger days, and he wondered where he and his siblings acquired their heavy features. Her husband's death left her with ten children to bring up – nine boys and Josephine, the youngest, two years old. Tom paid his mother a nice tribute: "I never heard [her] utter one discouraging or despairing word, and she never complained of her work. We boys must have caused her many anxious moments when we would go off to the bush with a gun – some of us not old enough to hold it properly ... Fortunately, my mother was blessed with good health, a good background of common sense and her mother's determination."[23]

Her children all turned out to be a credit to her. Though JSP may have become the most famous of his siblings, three of his brothers graduated from university, a very unusual achievement in those days.

School

Like many of their neighbours, the Plaskett boys could attend school only during the winter months because of farm chores. This was officially condoned – the Provincial School Act of 1871, bringing in compulsory education for children seven to twelve years of age, insisted on attendance for only four months of the year.[24]

JSP's siblings all went to the one-room elementary school in Strathallan, known as S.S. No. 6, East Zorra, which operated from 1855 to 1886.[25] It was a couple of kilometres down the road from March Hill. Likely Jack went there too, though the single reference found about his elementary school education lists his name, perhaps mistakenly, among pupils of S.S. No. 7, five concession lines away. At any rate, these rural schools had almost exactly the same outward appearance and would have had very similar methods.

A competitive atmosphere reigned in rural elementary schools of that era, with spelling bees, history and geography matches, and debates.[26] Even snowball fights on the way to and from school pitted one concession against the next. A public examination, with recitations and prizes, was a highlight in the community. On 20 December 1877, the walls of School No. 7 were decorated for the occasion with evergreen boughs and students' drawings.[27] The teacher, George Oliver, was assisted by others from neighbouring schools; parents, trustees, and even the local member of Parliament took great interest in the recitations and demonstrations of reading, arithmetic, and geography. It was probably Jack's final day in primary school before moving on to high school. To take this important step, pupils had to write entrance examinations at the school they hoped to attend. Only half of those attempting the two-day ordeal succeeded. JSP wrote his exams in December 1877, when seventeen boys and an equal number of girls gained admittance to the Woodstock High School.[28] Only ten were from outside the town, and Jack and one other were the only two from East Zorra. Though the odds may seem to have been against farm children, JSP stood third among the thirty-four candidates, earning 406 (out of 444[29]) marks. Once admitted, most of the rural pupils, including JSP's brothers Charlie and Tom, had to board in town. But Jack, because of his responsibilities on the farm, travelled by train to school in Woodstock, about thirteen kilometres away. It was a challenge for him to do farm chores, which he really did not like, while keeping up with school work, so his way of getting more time for studying was to pay his brother Bert one cent per night to cover the chores that were supposed to be his responsibility.[30] The deal suited Bert, who was trying to save money to purchase a rifle. Taking the train to Woodstock was certainly convenient, as the Port Dover and Lake Huron Railway had opened a line down the east side of the Plaskett farm in 1875 and established a station at Hickson. It was this new station, in fact, that spelled the decline of the older settlement of Strathallan. In Woodstock, JSP would have attended the brick grammar school on the corner of Graham and Hunter Streets, where Principal Strauchon was headmaster. By 1881 the staff had increased to four to keep up with burgeoning enrolment. The old school was now overcrowded and in a deplorable state and had to be replaced by a new building on Riddell Street, known as Woodstock Collegiate Institute.[31] JSP barely got a chance to enjoy the new facilities.

JSP after His Father's Death

Jack's sixteenth birthday was marked by the death of his father. Suddenly thrust into manhood, he had to drop out of high school and take up responsibility for the farm, including two labourers, William Budd age sixty-five and Eugene Clendinan, age thirty-three.[32] Some of JSP's brothers helped him, while the youngest boys lent their mother a hand with the housework and minded their little sister. All was not work and drudgery, however. Even after their father's death, there were lots of occasions when, as Tom put it, they should have been farming but were "miles away with a gun on our shoulder or a fishing-pole in our hands."[33]

Jack pursued other interests. Tom wrote:

> Instead of spending his spare time fishing, shooting, or at some other frolic with the rest of us, he would be delving into books or working on some hobby. He was always making something that to the rest of us was both wonderful and uncanny.
>
> Among his many achievements was the construction of a static electrical machine for generating frictional electricity. Money being scarce and mother economical, he had quite a struggle getting the circular glass eighteen inches in diameter. Finally he managed to get a glass plate home from Woodstock, and prepared to drill a large hole in the centre for a shaft, surrounded by a lot of mystified but admiring youngsters. The glass plate was placed on a chair, but being transparent, Bert did not notice it and sat on it. Was there a disappointed Jack and, I'll bet, a crestfallen Bert! Jack who could always get more from mother than the rest of us, finally secured another glass plate and finished the machine. With it we had many happy evenings. He built some Leyden jars for capacities, and around the winter fire, when the air was dry, we would join hands, the end one touching the ball on the Leyden jar, discharging the current through our bodies and all getting a heavy jolt. The best fun was to get our school chums joined up and to hear them scream when they received the shock. Once we gave Collie, the dog, a shock, but after that he would not go near the apparatus.
>
> Jack built a wood-turning lathe, a fret-saw, cross bows for us kids, and started to build a violin. However he could not wait a year for the maple to season, so mother bought him a violin (to keep peace in the family as Jack was very persistent), and he taught himself to play. He dabbled in shorthand, trying to master it by the book method and he also studied chess.[34]

JSP's fascination with electricity was a sign of the times. Arc lighting, where illumination came from a spark bridging a gap between electrodes, was the first form of electric light, but it soon gave way to filament lamps. The first public plant in Canada that produced power for incandescent lighting was built in 1887, just five years after Edison's first central generating station in New York City.[35]

So far as we know, at this stage of his life JSP paid no particular attention to astronomy, but it is worth pointing out that the opportunity was there if he had been so inclined. In an age before electric lighting dotted the landscape, the panorama of the Milky Way would have been spectacular from March Hill and could have aroused his curiosity. Books on astronomy were included in the Woodstock Mechanics Institute, and articles on the subject appeared from time to time in the Woodstock newspapers. One in particular from 1881 might have caught his eye:

> In a recent lecture by the eminent English astronomer, William Huggins, as to the results of spectrum analysis as applied to the heavenly bodies, this striking statement was made:
>
> If you suppose a railway from the earth to the nearest fixed star, which is supposed to be twenty billions of miles from us; and if suppose the price of the fare to be one penny for every one hundred miles – not, mind, a penny per mile – then, if you take a mass of gold to the ticket office equal to the national debt ($3,800,000,000) it would not be sufficient to pay for a ticket to the nearest fixed star. And I think I should not be wrong in saying that there are stars so far off that at the price of one penny for every hundred miles, the whole treasure of the earth would not be sufficient to pay for a ticket.[36]

Or perhaps JSP read about, or maybe even glimpsed, the 1882 transit of Venus, visible throughout North America. This rare passage of the planet across the face of the Sun had been noted on only four occasions in recorded history, and everywhere the press gave wide coverage to it. Moreover, one of the few schools in the country equipped to observe the transit was the Canadian Literary Institute in Woodstock, where astronomy was on the curriculum. Its observatory had been opened in 1879 and housed the second largest telescope in Canada – a 20 cm refractor.[37] (Telescopes are usually described by the diameter of their main lenses when they are refractors or mirrors when they are reflectors.) It seems likely that, on some occasion, an intelligent and curious teenager would have taken advantage of twice-weekly public invitations to come and view the heavens.We can only speculate on the extent to which any of this made a lasting impression on the young Plaskett but it is clear that astronomy was not completely absent from the Woodstock of his youth.

1.6 Whitelaw's foundry, where JSP worked in 1885–8. (From the 1876 Oxford County Atlas, Canadian County Atlas Digital Project, Rare Books and Special Collections, McGill University Library)

Astronomy aside, we do know that JSP had an inquisitive mind. With his aptitude for mechanical things, it is not surprising that, at age nineteen, he left the farm to learn the machinist trade as an apprentice at Robert Whitelaw's firm, the Oxford Foundry and Machine Shop, in Woodstock.

JSP's Family after 1884

After JSP took on his apprenticeship, his next younger brother, Robert (Bert), assumed responsibility for the farm.[38] Of all the remaining boys, he seemed to be the only one suited to the task. Whatever chance he had for education beyond elementary school slipped away as his younger brothers Wilfred ("Wiff"), Joe, Tom, and Charlie soon left for greener pastures. Frank, who was fourteen when JSP went to Whitelaw's, did stay to help Bert for a while.

The time came when it became clear that the farm should be sold. Even Bert wanted to go west to become a cowboy. So the March Hill property was sold in 1889 for $12 000, and an auctioneer was engaged to sell the animals and machinery. No doubt the family pitched in, displaying and arranging the implements, grooming the stock, and getting refreshments ready for the big day. Farmers would have come from miles around looking for bargains, food, and fun. Annie then moved to Woodstock with the remaining children, Frank, Harry, Fritz, and Josephine. Annie's sisters-in-law, Sarah Plaskett and Mary-Ann Little (widowed) both of whom had also lived at March Hill, set up a guest house in town.[39]

Bert got his opportunity to become a cowboy, thanks to his uncle, Duncan McEachran, who was the general manager of the Walrond Cattle Ranch in Alberta – an immense spread of over 200 square kilometers. Years later Tom had this to say about his brother: "Bert, in spite of all the many years among rough 'wild and wooly' cowboys, is more of a gentleman than most of his more cultured brothers … All the fine clothes, exalted positions in society, or University educations do not indicate the gentleman."[40] Was Tom thinking of JSP when he wrote that?

Two other Plaskett boys also owed their careers to the McEachrans. Duncan McEachran, as a young man, had studied veterinary medicine in Scotland before immigrating to Canada. He had come to Woodstock, set up a practice, and then met and married JSP's aunt, Esther Plaskett, in 1868.[41] The couple made their home in Montreal, where Duncan had already established a veterinary college two years earlier. Canada's largest city had a large horse population in a relatively small, urban area, and so was a very appropriate place for such a college. McEachran got the support of McGill University president John William Dawson and dean of medicine George William Campbell to offer courses at the university. It was with the encouragement of Aunt Esther and Uncle Duncan that Wiff and Joe entered and graduated from the veterinary medicine program at McGill. Wiff subsequently built up a large practice in Clinton, Massachusetts. Joe, in the course of his career, handled mules for the British forces during the Boer War and bought and shipped horses to Britain in the First World War.

Tom, to whom we owe much of this family history, admired his oldest brother and followed in his footsteps by entering the machinist trade. Frank, after trying some unrewarding routine jobs, entered the ministry with a master's degree from Bishop's College in Lennoxville, Quebec. We will encounter Tom and Frank again in later chapters. Charlie and Harry died at an early age – Charlie of tuberculosis when he was nineteen, and Harry in his early thirties from an operation to remove a bile duct blockage.

1.7 Spaying at the Upper Walrond Ranch corrals, Alberta, spring 1894. JSP's brother Robert is sitting in the foreground. (Glenbow Archives, NA-237-12)

Frederick, whose nickname Fritz would be discarded during the First World War, graduated from high school with first class honours but, to JSP's disgust, did not pursue further studies and went into business instead. He made a fortune in the seed business but lost most of it in the 1930s. Finally, the baby of the family, and the only girl, Josephine, Tom described as flawless. "Every person who comes in contact with her holds her in the highest esteem. She has a sweet disposition, clever, unselfish, kind to everyone, a leader in the community." We will meet her again later on too.

Church

Religious practice was an essential part of life for the Plasketts, as it was for most families at the time. They worshipped at Christ Church (also known as the Huntingford Church after its donor), about halfway between March Hill and Woodstock. It was a high Anglican parish where JSP's forebears, Joseph Thwaites and Timothy Stanley, had been wardens.[42] The church cemetery holds the mortal remains of JSP's brother Charlie, their father, and three grandparents.[43]

The rector there from 1845 to 1873 was Frederick Fauquier, originally coming from England to farm in East Zorra but then training for the ministry in Canada and later becoming Bishop of Algoma. He was the Plasketts' minister during JSP's early childhood and officiated at the marriage of his parents and several other relatives.[44] During his tenure as priest at Huntingford, his congregation made many improvements to the church – a stone foundation, a bell in the tower, and a barrel organ; the addition of a proper chancel in 1868 enlarged the capacity of the sanctuary to seventy. To pay for these improvements, two-thirds of the seating was rented at a rate of $8 per pew, payable twice a year. In 1875 a substantial rectory was built for Fauquier's successor, W.A. Young, at a cost of $3568.60. Joseph Plaskett was the treasurer of the building committee, and he probably deserves much of the credit for the good management that allowed for the success of such a project. Besides his name, the list of donors includes his wife's family – the Thwaites and the Stanleys – and Robert Millman, another farmer living close to the church.[45] One of his sons, Thomas Millman, graduated from medical school in Toronto in 1873 and immediately joined the British North America Boundary Commission as assistant physician, working on the 49th parallel alongside William F. King, later Canada's chief astronomer. Another son, Robert Malcolm Millman, like Joseph Plaskett's son, Frank, became an Anglican clergyman. What is even a more amazing coincidence, the priest's son, Peter, became another outstanding astronomer, though a generation after JSP.

After the farm was sold and Annie moved with the younger children to Woodstock, she attended St Paul's Church in town, where a new, young minister, John Farthing, was installed in 1889. Over fifty years later he still recalled a Christmas service about this time, when the Plaskett family was reunited.[46] He described the inspiring sight of Mrs Plaskett, "a little widow," leading her ten children up the aisle for communion. As they all knelt in a row to receive the elements, Farthing said he was so touched and overcome that he could hardly minister to them. In the long run, he did much more, encouraging Frank to become a priest. JSP, for his part, remained a devoted Anglican for the rest of his life.

The Path to University

JSP was an apprentice at Whitelaw's foundry from 1885 to 1888, along with school chum Robert A. Ross. At the end of this period, JSP went off to Schenectady, New York, to apprentice for a further year at the rapidly expanding Edison Electric Company, though he may have gained electrical experience prior to that. There are a couple of letters dating from November and December 1885, signed "J.S.P." in the *English Mechanic*.[47] This mass-circulation weekly devoted a great deal of space to queries and responses from readers, and in this case the correspondent was asking why a small dynamo he had made did not seem to fully power ten incandescent lights. There is no way to know if these letters were written by our JSP, but the timing and the nature of his enquiries seem very suggestive. In the summer of 1889 he returned to Canada and worked briefly at the Edison General Electric plant in Sherbrooke, Quebec. Ross, meanwhile, enrolled in engineering at the University of Toronto, graduated from its School of Practical Science in 1890, and then became engineer in charge of the Sherbrooke plant.[48]

It happened that Plaskett and Ross were together in Montreal in the fall of 1889, when they ran into William J. Loudon, a demonstrator in the University of Toronto's Physics Department and one of Ross's instructors.[49] Ross mentioned that his friend Plaskett was looking for a position, and Loudon suggested he contact his uncle, James Loudon, professor of physics at Toronto. The senior Loudon needed someone to set up experiments and assist in lecture demonstrations. He recognized that JSP was well qualified and so he hired him as a "mechanician" (*technician* would be a modern equivalent) starting in January 1890.

Clarence Augustus Chant, the recorder of this story and Plaskett's contemporary, was within months of completing his BA. At this point, a career in astronomy was far in the future for these young men who, each in his own way, would sire the science of the stars in Canada.

2
Toronto, 1890–1903

When John Plaskett began work in the Physics Department at the University of Toronto on 1 January 1890, the university was evolving into a modern institution of higher learning.[1] Until recently it had been a very small, elite place, proudly claiming to be modelled on traditional British institutions, the source of most of its few faculty members. The main building, known then and now as University College, had been completed in 1859 and *was* the University of Toronto until 1889. Bolstered in part by Victoria University's acquiescence to federation and by the absorption of the School of Medicine, enrolment grew from 495 in 1882 to 1095 a decade later.

There were signs of modernization. The admission of women in 1884 was one enlightened change, though forced by provincial legislation. Another was development in the sciences, including the establishment of an engineering school, known as the School of Practical Science (SPS), which opened on campus in 1878. Though the school was technically independent until 1889, its students took some of their courses from university faculty and it housed some of the university's science departments for a time.[2] Adjacent to it, the university erected its only other building devoted to science – the biology building – in 1889. Another step towards specialization occurred when the old arts discipline of natural philosophy was split in 1887 into two departments – Mathematics headed by Alfred Baker and Physics under James Loudon.

Loudon was the first Canadian, and the first graduate of the University of Toronto, to be appointed to the university's teaching staff. Throughout his career, most significantly when he was university president, he favoured filling faculty positions with Canadians, and preferably U of T graduates. This was a radical departure from the normal practice of Canadian universities at the time. He also reformed science teaching at Toronto and understood the vital role of university laboratories, not only as centres of basic investigation and scientific discovery, but as essential foci for learning science. The research model, originally espoused by Wilhelm von Humboldt at the University of Berlin in 1810, took hold in British universities in the early 1870s.[3] Inspired by this "hands-on" approach, Loudon set up the first undergraduate physics laboratory in Canada in 1878.[4] He went to Europe that summer to purchase apparatus; when he returned he was chagrined to find that other departments had completely occupied all the available places in the new SPS building. So, Physics had to make do with the old "round house" (now known as the Croft chapter house) and rooms in the west wing of University College. There the department stayed until a splendid new physics building opened in 1907. (It is now occupied by the Engineering Faculty and is known as the Sir Sandford Fleming Building.)

At first, practical work was optional but by 1885 all honours students were required to spend time in the lab. It took years before a technician could be hired to look after the instruments, but when Plaskett came along in 1890, he was a perfect fit, "a skilled machinist and a master of the electrician's art which was then rising into prominence."[5] He was put in charge of a large range of equipment and would have had to do photographic work, blow glassware, make repairs, and even construct new instruments.[6] Tangible evidence of this aspect of his work recently surfaced when a history of science doctoral student was preparing a display of old electrical apparatus for the University of Toronto's Scientific Instrument Collection.[7] Amazingly he found two resistance boxes that Plaskett had made and signed and an ammeter that he had calibrated. These would have been used by physics or engineering students in the 1890s and probably well into the twentieth century.

Though JSP was attached to the Physics Department, he sometimes was called upon to do other work for the university.[8] Lecturers liked to use slides to enliven the old chalk and talk, but the projector or lantern was a difficult contraption to operate. The source of illumination had to be bright enough to throw an image onto a screen at the front of the classroom. In the old days, limelight – a flame of combined hydrogen and oxygen gas playing on a cylinder of lime (calcium oxide) – was used.[9] The gases, awkwardly stored in bags, had to be combined in the right proportion, and the position of the cylinder had to be frequently adjusted. An improvement in the form of an arc lamp was powered by a large battery of cells consisting of a platinum cathode in concentrated

2.1 Ammeter used in the Physics Department, University of Toronto. It was made by J. Carpentier of Paris about 1880 and recalibrated by JSP in 1893. (University of Toronto Scientific Instrument Collection)

nitric acid and a zinc anode in concentrated sulfuric acid, the two separated by a porous ceramic pot. The cells had to be prepared anew whenever the projector was used, a messy and costly process subjecting the operator to dangerous fumes. Plaskett knew how to handle such contraptions.

Within weeks of starting his new job in 1890, JSP was asked to prepare an electrical demonstration to entertain students and guests at the Valentine's Day conversazione organized by the University College Literary and Scientific Society. It was the big social event of the year, attracting thousands to readings, refreshments, concerts, and promenading bands (but no dancing). JSP and his friend Robert Ross, then in his senior year, set up a showy electrical experiment and attached the railing around it to an induction coil so that visitors would receive a jolt when they touched it. Apparently Plaskett remembered the days at the farm when he delighted in shocking his family and decided it would be fun to try a similar prank on the university students. How many actually got to enjoy the thrill of being electrocuted we cannot say, for a disastrous accident interrupted the evening's pleasures.[10]

Aside from such novelties as JSP's demonstration, electricity had not yet reached the university. The gloomy interior of University College was normally illuminated by gas light, but for the conversazione somebody decided that portable kerosene lamps would add a brighter glow. In the early evening of 14 February, a servant carrying a tray of these lamps up the southeast stairs from the basement lost his balance; the lamps slid off the tray and within minutes the burning oil ignited the wooden staircase. Before long the fire had engulfed the entire east wing and part of the main building, including the massive front doors and the interior of the tower. No lives were lost, but nearly all the library books were consumed. The Physics Department, in the west wing, was unaffected.

Progress sometimes proceeds out of disaster, as a phoenix rises from the ashes. William Loudon, nephew of James, went so far as to say later that "it was the best thing that could have happened [to] the University, as it created an entirely new atmosphere and brought about new conditions ... From the time of the fire to the great rebellion [of 1895], whether by chance or luck or by some predestined plan, the students of the University seemed, during that short five years, to have more intelligence, more initiative, more desire for real knowledge, more ambition, than those of any other time."[11] The college, as part of the rebuilding, got a new heating plant. Attached to it was a steam engine that drove a dynamo supplying direct current for a new lighting system and some laboratory experiments. JSP, with his Edison Electric experience, "designed the dynamo, made the patterns and built the machine."[12] He was considered the superintendent of the university's power plant and was paid $172 out of the building restoration fund for "services as electrical expert re wiring."[13] This was in addition to his $800 annual salary. The dynamo also spelled the end of the cumbersome, old, battery-operated projector.

Family Interlude

On 10 April 1891 JSP lost his last surviving grandparent, Mary Stanley, aged eighty-eight, and he naturally went back to Woodstock for the funeral.[14] She had lived with her son Timothy and his family ever since the death in 1879 of her bachelor brother, whom she had looked after.[15] She had broken her hip at the age of seventy-eight and had gotten around on crutches for the last ten years of her life, propelled by her physical strength, stubbornness, and determination.[16] JSP would display the same traits many years later when circumstances forced him to use crutches for a while.

Back in Toronto, JSP took a fancy to a young woman with a low, sweet voice, just a day shy of being two years his junior.[17] Rebecca Hope Hemley, or Reba as Jack called her, had been born on 16 November 1867, in suburban London, England, but had immigrated to Canada with her parents when she was four.[18] The family settled in Harriston, about a hundred kilometres north of Woodstock. The village was on the verge of a growth spurt due to the arrival of the railway, a promising time for Reba's father to find employment making mouldings.[19] Reba and Jack likely met through JSP's uncle, Joseph Stanley, a carpenter who, with his family, also lived in Harriston.[20] There is some evidence that Reba came to Toronto, perhaps to train as a nurse or to earn a living.[21] Since she and her family were Anglicans like the Plasketts, they may have attended the same church in the city, but when she and Jack were married on 19 July 1892, the venue was naturally her home church of St George's, Harriston.[22] One of her sisters was bridesmaid, and a friend of Jack's since school days, George Cheyne, was best man. George had entered high school at the same time as Robert Ross and became a machinist, probably at Whitelaw's, along with Jack and Robert.[23] By the 1890s, Harriston was in decline; the railway made imports less expensive, thus doing more harm than good to local industry.[24] So in 1895, presumably in search of better prospects, Reba's parents and her three younger siblings moved west to Washington state.[25]

By the time Reba and Jack were married, he had already purchased the couple's first home – a house at 36 Boswell Avenue, a few blocks north of the university, just off Avenue Road.[26] It was a small house with a huge hickory tree in the back yard.[27] The purchase price, $2500, was roughly three times JSP's annual salary, quite a bargain by today's standards.[28] Buying a house might seem like a big step for a technician and his wife, newly married, but their neighbours were similarly situated tradesmen and clerks, some about the same age as the Plasketts. A year after their marriage, Reba gave birth to their first son, on 5 July 1893. The attending physician was Dr Augusta Stowe-Gullen – one of those remarkable people whom we met in Chapter 1. The baby

2.2 Boswell Avenue, Toronto, in 1904. The Plasketts' house was on the north (left) side. Though not visible in this photograph, it was similar in design to those which can be seen. (City of Toronto Archives, S0376-fl0004-it0079)

was christened Harry Hemley after Jack's younger brother Harry and Reba's family name.

The baptism took place at the Church of the Redeemer, located at the corner of Bloor Street and Avenue Road, just a few blocks south of the Plasketts' home.[29] The Plasketts were pew-holders there, along with several Middletons and Millmans, some of whom JSP knew from Woodstock days. Indeed, when Jack and Reba needed a babysitter for Harry they sometimes called on one of the teenagers at the church, Edith Middleton, who was related through marriage to JSP's uncle, Timothy Stanley.[30] Who could imagine that many years hence, Edith's own son (Peter Millman) would be a student of Harry Plaskett's?

JSP's only sister, Josephine, graduated from Woodstock Collegiate in 1896 and moved to Toronto. She got a job as office assistant to Dr Thomas Millman (Edith's brother-in-law) at the International Order of

Foresters, an insurance society headed by Oronhyatekha, whom we also met in chapter 1, and who was by then "one of the strongest and greatest builders of fraternalism in America."[31] Annie Plaskett, with all the children grown and away from home, joined her daughter in Toronto. At first they lived in a flat on Robert Street, then shared a house with Annie's son Tom and his wife, Mabel, for a couple of years at 63 Albany Avenue. Later, when JSP and his family moved to Ottawa, Annie and Josephine assumed responsibility for the house on Boswell Avenue.

The Plasketts fitted comfortably into life in "Toronto the Good." Torontonians were proud of their city's churches, its moral rectitude, and its staunch British imperialism. The poor, of course, lived precariously, as the Plasketts were reminded one evening. Returning home they were shocked to find a bundle on their neighbour's adjoining porch. Wrapped in a blanket was an emaciated baby – the sixth to be abandoned in the city in the past two weeks. Jack and Reba turned the wee girl over to the police, who put her in the Infants' Home and Infirmary.[32]

By the 1890s middle-class Canadians were succumbing to the call of the beautiful lakes north of the city. Those whose employers actually gave them paid holidays went by train to resorts and cottages in Muskoka and on Georgian Bay and stayed for as much of the summer as they could afford. JSP's aunt and uncle (the owner of the bookstore on Yonge Street) owned Bohemia Island in Lake Rosseau.[33] JSP's brother Tom recalled, "I ... spent a week at ... my cousin's summer home. This was my first vacation in years, and I was very thrilled with the swimming, boating, and fishing, and the beauty and the sunshine. To me, Muskoka was paradise."[34] It is not hard to imagine that Jack had similar feelings. Another favourite destination in Muskoka was Fairy Lake, convenient to the town of Huntsville, connected to Toronto by the Grand Trunk Railway. Steamships set out from the town dock for resorts and other destinations. JSP and Tom shared in the purchase of two islands in Fairy Lake called the Twin Sisters, though later, maybe when JSP moved to Ottawa, Tom bought the eastern one from him for $100.[35]

Technician, Student, and Graduate

In the 1890s the faculty in the Department of Physics at the University of Toronto comprised James Loudon (professor), his nephew William J. Loudon (demonstrator), J.C. McLennan (assistant), and C.A. Chant (lecturer). The importance they attached to laboratory work can be gauged by a 300-page manual describing the experiments that undergraduates were expected to perform.[36] Though the authors, the younger Loudon and McLennan, made no mention in their book of Plaskett or his role, he presumably was responsible for setting up equipment for these experiments. Some concerning image formation by lenses and concave mirrors included an investigation of the aberrations that affect the sharpness of the image. Through such experiments, JSP himself learned the principles of optics that would prove essential to him in the future.

JSP also assisted the faculty by arranging demonstration experiments. When, in 1895, Chant gave one of his most memorable lectures, he credited Plaskett in large measure for its success.[37] Entitled "Electrical Science," it involved a remarkably early demonstration of radio waves and their properties. (This was the very year that Guglielmo Marconi astonished the public by transmitting radio signals over a distance of 1.5 km.) In general, Chant recalled, Plaskett "was not a listless helper but was alert to understand what was under discussion. He was also on familiar terms with the staff and students in the Mathematics and Physics Course, and in a few years made up his mind to take the course himself."[38] Encouraged by Reba, JSP first obtained his matriculation from high school and then enrolled as an undergraduate, continuing his work all the while.[39]

JSP, twenty-nine years old, married, and the father of a two-year-old, started his formal university studies in the fall of 1895. For the most part he was unable to attend lectures because of his duties, but of course he would have had to write exams.[40] In first year, besides four math courses and four languages (Latin, English, German, and French), elementary physics and chemistry were normally required but Plaskett's student record suggests that he was excused from the latter two.[41] In subsequent years his physics studies included mechanics, elasticity, hydrodynamics, acoustics, physical optics, thermodynamics, electricity, and magnetism. Math included statics and dynamics, differential and integral calculus, as well as the theory of least-squares.

The prescribed textbooks, especially in physics, were predominantly British.[42] The physicist who wrote four of the texts, Richard Glazebrook of the Cavendish Laboratory at the University of Cambridge, based his books on experiments "which can be performed by the learners themselves," quite in keeping with the pedagogical philosophy at U of T.[43] By looking at the books, we can see the topics to which Plaskett

2.3(a)

2.3 (above and opposite) JSP, about 1897, at the table in his workshop at the University of Toronto. (Photo by A.H. Abbott, published as Plate 13, *JRASC* 35 (1941): 412.) Note that the table and chair are identical to those shown in the accompanying photo of the physics laboratory, University College, 1897. C.A. Chant described it as "a single large room immediately west of the main tower ... lighted by four electric chandeliers suspended from the ceiling." (UTA, A1965-0004 [1.91])

2.3(b)

was exposed, especially those areas that would be vital in his future career. Even in the introductory text, Ganot's *Elementary Treatise on Physics, Experimental and Applied*, Plaskett would have read a lot about the nature of light.[44] He would have learned that light waves are very short, approximately 500 nanometres (nm) (or, to put it another way, 20 000 wavelengths per cm), and would have appreciated why a beam of white light passing through a prism is dispersed into a spectrum of rainbow colours from red (with the longest wavelengths of about 700 nm, or 700 billionths of a metre), through yellow and blue to violet (with the shortest wavelengths of about 400 nm). Even if he had not already seen for himself, JSP would have read that the spectrum of sunlight was split by many dark lines, rather than being a continuous band of colour. The English chemist William Wollaston may have been the first to record this characteristic in 1802, but it was the German optician Joseph Fraunhofer who, in 1814, famously labelled the most prominent dark lines with letters A through K and calculated their wavelengths.[45] At about the same time, he was apparently the first to observe the spectra of some bright stars, noting their different lines. (The first photograph

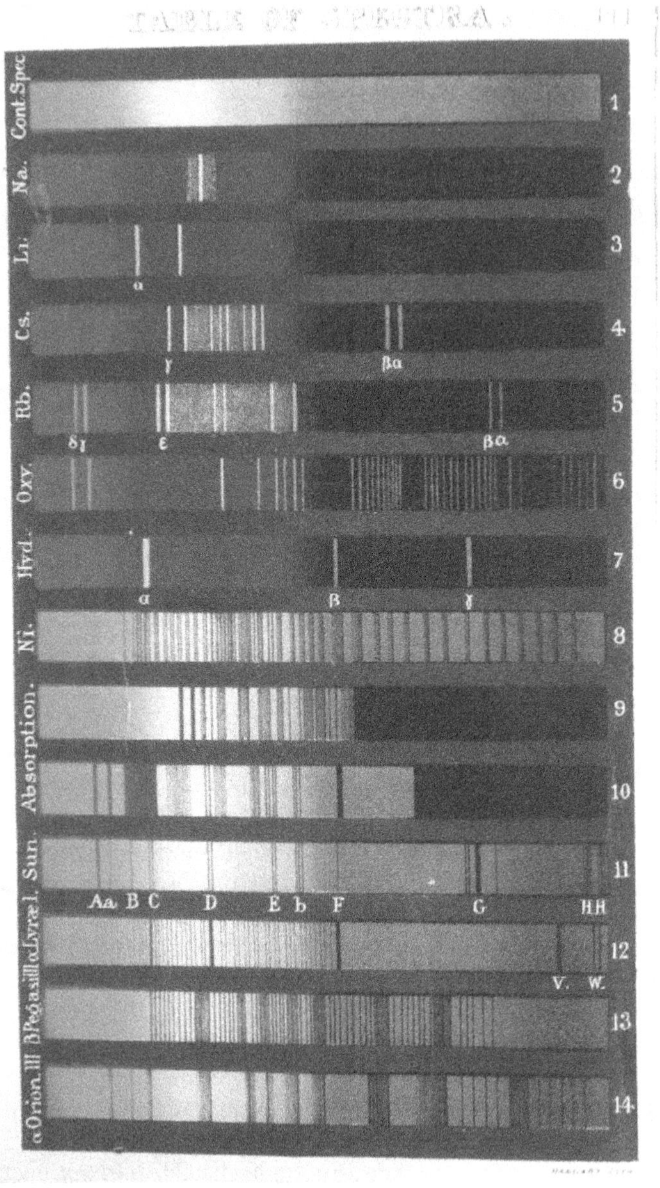

2.4 The frontispiece (originally in colour with red on the left and violet on the right) from *Elementary Treatise on Physics* (1893 ed.) showing (1) a continuous spectrum as produced by a hot solid, (2–7) emission line spectra of hot gases, (8–10) absorption spectra produced when continuous radiation passes through a cooler gas, (11) the solar spectrum with Fraunhofer's lines labelled, and (12–14) spectra of three bright stars.

to show these lines, and more, was a daguerreotype of the solar spectrum made by Edmond Becquerel in 1842, and the first satisfactory photograph of a stellar spectrum was taken by Henry Draper thirty years later.[46]) The frontispiece of Ganot's textbook was an impressive colour plate showing continuous, emission, and absorption line spectra along with examples of spectra from three different stars.

In his courses JSP would have studied the theory of optics, including the way that reflection gratings, formed by ruling thousands of grooves per cm on a polished metallic surface, can disperse light just as well as prisms.[47] He would have learned that scientists, even fifty years earlier, had begun to identify the dark absorption lines in the solar spectrum with bright emission line spectra created by flames and sparks in the laboratory, thus revealing the chemical composition of the Sun. By the early 1860s the German physicist Gustav Kirchhoff recognized the presence of sodium, iron, magnesium, chromium, and nickel in the Sun. Kirchhoff articulated the basic principles of spectrum analysis, leading him to conclude that the Sun was an incandescent solid or liquid (where a continuous spectrum originates) surrounded by a cooler atmosphere (where absorption by specific gases occurs at the same wavelengths that would show up as emission lines when the gas was heated). By measuring the amount of solar radiation arriving on Earth, it was clear that the Sun was very hot, but by how many thousands of degrees was unknown. In any case, it was hot enough to allow the theoreticians to prove that the Sun was gaseous. Almost immediately others began to examine the spectra of stars, finding that some had lines like those in the spectrum of the Sun while others were quite different.[48] This led to various taxonomies of stellar spectra – the system that still survives originated at Harvard in the 1880s and used capital letters, later rearranged in the sequence OBAFGKM. (These capital-letter designations of spectral types should not be confused with the letters A through K used as identifiers of prominent individual "Fraunhofer" lines in solar and stellar spectra.)

Another theoretical development covered in JSP's textbooks had a profound effect on astronomy.[49] In 1842, Christian Doppler spoke to a meeting of the Royal Bohemian Society in Prague "on the coloured light of the double stars and certain other stars of the heavens." Though many of his ideas proved later to be wrong, the kernel of truth in Doppler's principle was that waves of light would be compressed or stretched if the source and the observer were moving towards or away from one another. When the distance between them was increasing, the wavelengths appeared longer or redder; when the distance was decreasing, the light appeared bluer. Doppler thought this shift would affect the colours of stars. Actually the change in wavelength was much too small to be noticed by the eye, but he was right in principle, so the shift has since come to be known as the Doppler effect. In his lecture, he made an astonishingly prescient comment, that "it is almost to be accepted with certainty that this will in the not too distant future offer astronomers a welcome means to determine the movements and distances of such stars which, because of their unmeasurable distances from us …, until this moment hardly presented the hope of such measurements and determinations."[50] By mid-century, the proposal had been made to use the shift in Fraunhofer lines to measure radial velocity – namely, how fast the star is moving along our line of sight, towards or away from us.[51] One of the great British amateur astronomers of the Victorian era, William Huggins, showed in 1868 that the technique was a practical possibility. [52] His pioneering results for Sirius, the brightest star in the night sky, were very discordant but his attempts at measurement opened up a new field.[53] The Doppler effect became one of the most useful phenomena in astrophysics and would be the basis for most of the work carried out at the major Canadian observatories in the twentieth century.

The University of Toronto offered no separate course in astronomy during JSP's time there, but faculty in the Department of Mathematics (likely Alfred Baker) did teach some elements of the subject. The calendar for 1899 listed twenty-eight recommended textbooks for fourth-year mathematics, six concerning astronomy. Only one of them, by the American astronomers Simon Newcomb and Edward Holden, covered any astrophysics, and that was in the last fifty pages of a 500-page text.[54] Probably the math professors would not have had much time for those topics, especially as physics of any kind was not their department. So, with astronomy accounting for, perhaps, a quarter of the fourth-year mathematics course, Baker probably restricted himself to "classical" astronomy: calculating positions of the Sun, Moon, stars, and planets, and perhaps including some instruction on how astronomical instruments could be used to observe these positions.

A government observatory had been on campus since 1840, dedicated to the collection of data on terrestrial magnetism, seismology, and meteorology.[55] The imperial government in London had established it, but in 1853 had handed over financial responsibility to the province of Canada, though the university actually administered it and appointed a professor of meteorology to oversee it.[56] By the 1870s the observatory was in the hands of the federal government, fulfilling a practical role by providing weather forecasts as well as accurate time.[57] A 15 cm refractor telescope was installed for the 1882 transit of Venus and was later used mainly for sunspot observations. Members of the local astronomical society and the public were occasionally given the privilege of an evening visit, but by and large the Toronto observatory stuck to its duties and had little interaction with either the public or the university. The university calendars of the 1890s did specify that, "By courtesy of the authorities of the Observatory, students have access to and use of the astronomical instruments." Unfortunately there seems to be no documentation that any students availed themselves of the opportunity.

When Plaskett was in his final year, J.C. McLennan went overseas to pursue graduate studies at the University of Cambridge, leaving JSP in charge of several lab courses previously part of McLennan's responsibility. McLennan did his research at the Cavendish laboratory in Cambridge on the electrical conductivity of gases traversed by cathode-rays. His supervisor was the renowned Joseph John Thomson, who had only recently discovered that these rays were made up of tiny charged particles called electrons. Thomson sent McLennan's results to the Royal Society of London, which published his paper in their *Philosophical Transactions*.[58] On the basis of this work, the University of Toronto granted McLennan its first PhD in physics upon his return. Stimulated by his overseas experience, he was more set than ever on doing original research at Toronto and Plaskett had a role in it. In one project he assisted McLennan in September 1902 at the university and at the foot of Niagara Falls in measuring the effect of water spray on the ionization of the air.[59] Working for McLennan was not easy; JSP referred to him as "the essence of selfishness."[60] When Tom Plaskett replaced his brother as chief mechanician of the department in 1903, he spoke of the continual grind of pleasing McLennan, "a temperamental autocrat"; even McLennan's biographer referred to his "untiring persistence that took small account of the limits of human endurance."[61] Though JSP may have picked up McLennan's determination and love of research, he certainly did not adopt his autocratic style in dealing with his colleagues in later years.

Through student experiments, demonstrations, courses, and perhaps some actual astronomical observing, Plaskett became acquainted with many of the concepts that he would eventually use, though, as far we know, at this point he had no ambition to become an astronomer.

Throughout his studies and in his graduation year of 1899, he attained first-class honours, outdone only by Norman Wilson in his second and third years and by John Hogg in fourth year.[62] Wilson and Hogg became professors of mathematics and physics at Canadian universities.

Earning a degree opened a new world to JSP as he began to carry out his own investigations, write papers, and lecture about his work. His first published paper originated in a talk he gave in December 1899 to the university's Mathematical and Physical Society on mathematics in the mechanical trades. It was printed in the *Canadian Engineer* in January 1900 and also in England in the February and March issues of the *Practical Engineer*.[63] JSP spoke from first-hand knowledge when he said that the average apprentice in the mechanical trades had only a short and inadequate public school training in arithmetic. He argued that machinists would be able to work more efficiently and with more intelligence if they had a thorough understanding of descriptive geometry and the elements of plane trigonometry. He gave as examples machining metal screws on a lathe, making mechanical drawings to design equipment, and dividing circles into equal parts.

This interest in pedagogy, his increased responsibilities in the laboratory, as well as the opportunity he was given to set the 1901 examination in mechanics all suggested that he would have made a first-rate addition to the teaching staff. As mechanician he was earning $1000 annually. Chant's salary as a lecturer was $1800, so appointment to the faculty would have been a great financial boost for Plaskett. Most of his colleagues hoped and expected that he would join them, but McLennan was not very sympathetic to the idea and JSP himself thought there was no chance of advancement.[64] In 1903 James Loudon expressed his willingness to recommend him or another outstanding graduate, R. Meldrum Stewart (BA 1902), for a teaching position at the University of Missouri but, as it turned out, both men moved to Ottawa and began their careers at the observatory in that city.[65]

Photography and Public Speaking

JSP's inclination towards doing his own research was fostered by his interest and talent in photography. In 1899 he joined the Toronto Camera Club and was soon on the executive.[66] At the annual meeting in November 1902 he won out over five other nominees and was elected first vice-president. This early involvement in the administrative work of a society and the opportunity it soon gave him to lecture would provide valuable experience for the years ahead. The club members enjoyed excursions and social evenings of euchre as well as regular meetings and exhibitions, and they had the use of a well-equipped darkroom in their facilities at Yonge and Gerrard Streets.[67] Several members participated in the popular "American Lantern Slide Exchange"; JSP had about a dozen of his photos included each year from 1900 to 1902. The bucolic theme of much of his work was in line with the club's focus on photography as an art form.[68] A few of his pictures were taken while on holiday in Muskoka and at least one received an award of merit in 1900.[69] At the university he took photos of campus buildings for calendars and the like, and he even included a couple of them in an exhibition many years later.[70] Unfortunately, none of Plaskett's original artistic compositions seems to have survived, just a few poor-quality half-tone reproductions, mainly accompanying an article he wrote on landscape photography in 1902.[71] This article, however, does shed light on some of JSP's views on his hobby. Photography leads its practitioners, as he put it, "into closer touch with nature, it encourages them to visit places that they would never otherwise have seen, and it trains them to perceive beauties, before unobserved, in the simplest scenes." But, he asserted, technical skill was even more important than artistic taste.

Although George Eastman had invented rolled photographic film and the Kodak cameras to use them in 1888, serious photographers continued to use glass plates coated with light-sensitive emulsion. Orthochromatic plates, popular with JSP, were more sensitive to the red and yellow part of the spectrum than "ordinary" photographic plates, which reacted more quickly to the blue. Consequently the ortho photos enhanced the contrast between the blue sky and white clouds, making for more dramatic landscapes.

JSP's photographic proclivities went far beyond taking pleasing pictures. In his lectures to the camera club he spoke about stereoscopic photography, the optical properties and practical uses of photographic lenses, and his experiments in producing colour images. In fact his colour photographs are the earliest known in Canada.[72] Plaskett used the three-colour process, originally proposed by James Clerk Maxwell in 1855.[73] This cumbersome method involved taking three black-and-white photos through red, green, and blue filters and then projecting the negatives through complementary filters to create a combined colour image on a screen. This procedure seemed to be the only reliable way to produce a realistic result until the autochrome process was introduced to the public in 1907. As early as 1901 Plaskett carried out an extensive examination of colour techniques, experimenting with a spectroscope and with filters to see how light of various wavelengths affected different photographic plates.[74] As he explained in a lecture in March 1902, all colours except deep red and extreme violet are compound sensations based on the fundamental three: red, green, and blue-violet.[75] He showed the three fundamental sensation curves as determined by William Abney in 1899, which were centred on wavelengths of 670, 512, and 460 nm. To illustrate the three-colour principle, JSP combined a spectroscope with the projector and made a special brass frame with three slits whose position and width could be adjusted, allowing varied proportions of two or three spectrum colours to be used. This three-colour process was probably the technique that JSP used to produce slides to illustrate a public lecture on Renaissance artist Raphael by Professor William Fraser.[76] The views so impressed the audience that several of them had to be shown again after the lecture, and Fraser repeated his presentation on six other occasions at various venues.

The tricky part in preparing slides using the three-colour process was adjusting the filters so that the photographic plate would mimic the human response as closely as possible. One way to do so was to fiddle around with different combinations until the result was pleasing, but Plaskett realized that the scientific approach involved measuring the photometric densities of negatives obtained with different filters and photographic plates. In this way he hoped to be able to construct an ideal set of filters. The extent to which JSP was influenced by Abney's research is uncertain, but it is worth noting that Abney's interest in scientific photography paralleled a deep interest in astronomy.

Besides the camera club, catering mainly to its members, there were other societies in Toronto giving well-qualified speakers like Plaskett an opportunity to address broader audiences on various scientific topics. Since 1849 the Canadian Institute had been a meeting place for the city's intellectuals, especially those with scientific interests. Between 1899 and 1902 the usual attendance at its twenty-two weekly meetings each year was around fifty, though there were eight occasions when attendance topped a hundred – Plaskett's lecture on colour photography was one of them.[77] Evidently Plaskett had acquired a reputation as a good public speaker though having a hot topic certainly helped. His communication skills, essential to his future success, were already evident.

He was not the only one to speak to the institute on colour photography around this time. David J. Howell, a stalwart of the camera club and the astronomical society, spoke on this topic on 3 February 1900, as did Albert H. Abbott of the university's psychology laboratory on 29 January 1898 and 1 February 1902.[78] But it was JSP's second presentation (on 15 March 1902), "Photography in Natural Colours," that attracted the large crowd mentioned above. It was illustrated with twenty colour slides showing reproductions of the spectrum, colour charts, flowers and fruit, and views around the university. His lecture was published in the institute's *Transactions*, complete with a unique colour plate.[79] The institute did not pay its speakers, even those from outside its membership, as Plaskett apparently was. He did, however, get reimbursed for considerable expenses – a total of $20.90 for his two presentations.[80]

Another Toronto group that sponsored lectures was the university's Mathematical and Physical Society. Rather like a seminar group, it was composed mainly of senior honours students, with faculty providing some leadership. A committee selected a few of the members' best papers each year and published them in small annuals from 1890 to 1894.[81] Three of C.A. Chant's presentations appeared there. He served as the group's president in 1892–3 and again in 1896–7. As we have seen, JSP had addressed the society in 1899 on mathematics; on 5 December 1901 he spoke again to them, this time on colour photography. On this occasion the students were supplemented by members of the Astronomical Society of Toronto. It was a full agenda: in addition to JSP's talk, the audience heard Chant speak on diffraction and diffraction gratings and enjoyed music by the Banjo, Mandolin, and Guitar Club as well as vocal selections by five young men.[82]

As reported in the *Globe* newspaper, JSP also gave a lecture on orthochromatic photography to the astronomical society on 13 January 1903.[83] This group, which was to become very important in the lives of Chant and Plaskett, originated as the Toronto Astronomical Club in 1868. After a number of years of precarious existence it became incorporated as the Astronomical and Physical Society of Toronto in 1890, changing its name to the Astronomical Society of Toronto in 1900. Finally, in 1903, it became the Royal Astronomical Society of Canada, a name that reflected its national aspirations along with the prestige of a royal endorsement. It was a landmark year in the history of the Astronomical Society and a turning point for Plaskett. He was about to begin a new career in astronomy in the nation's capital.

Developments in Ottawa

The government of the Province of Canada had been involved in astronomy at least since the 1850s and the new federal government inherited this responsibility soon after Confederation. The Department of Marine and Fisheries assumed responsibility for the former colonial observatory in Quebec in 1868 and, three years later, took over control of the old Toronto observatory from the university.[84] This latter development marked the beginning of a national meteorological service, which, in 1883, began to supervise the Quebec observatory as well. Observers used telescopes at both locations to record sunspots because of their potential effect on the weather (a role meteorologists continued nominally until 1930).[85] The Department of Marine and Fisheries also supported smaller observatories in Halifax and Saint John, where time balls were operated to provide accurate time to set ship's chronometers, so vital for navigation at sea.

Another ministry, the Department of the Interior, established in 1873, was responsible for immigration, settlement, and Crown lands. It needed correct surveys tied to geographic coordinates and, as accurate time keeping was a prerequisite for longitude determinations, this department also had a legitimate interest in this aspect of astronomy. Eventually its active role in astronomy triumphed over that of its rival department. It contracted out nearly all of the survey fieldwork to qualified surveyors. One who started out this way but later rose to prominence inside the department was Otto J. Klotz.[86] Between 1879 and 1892, when he finally settled in Ottawa, Klotz was involved in assignments of national importance: the viability of a port on Hudson Bay, surveying lands through the Rockies along the route of the Canadian Pacific Railway, and work in connection with the Canada-Alaska boundary. But in all such work there was no Canadian standard point of reference to which local surveys could be anchored. All surveyors, including Surveyor General Edward Deville, recognized there was a real need for a fixed observatory where, at the very least, instruments could be tested and standardized and surveys could be correlated. The first step in this direction was a temporary observatory in the garden of William Frederick King, at that time chief inspector of surveys. By 1888 the government erected a somewhat more permanent structure on Cliff Street in Ottawa (near the present Supreme Court), though it looked more like a wooden cabin than a national observatory. It was equipped with a 7.6 cm astronomical transit and a sidereal clock (essential instruments for determining accurate time) as well as a 15 cm reflector telescope on an equatorial mount, whose main purpose seems to have been occasional public viewing. In June 1889, King became the first chief astronomer of a distinct Astronomy Division within the Department of the Interior;[87] three years later he also assumed the responsibility of international boundary commissioner. He would soon become the central figure in establishing a proper building with equipment fit for a truly national observatory, but such developments would not come to pass without political will.

In the federal election of 1896, Wilfrid Laurier led the Liberal Party to victory. As the son and grandson of land surveyors, he may have had some innate empathy for an observatory; in any case the new prime minister soon made it part of his vision. He had vowed to turn Ottawa from its lumbertown likeness into the "Washington of the North," complete with impressive public buildings befitting a national capital.[88] During his government's fifteen years in office, much was accomplished: the Victoria Memorial Museum, the Royal Mint, the Connaught Building, and the Dominion Observatory were all erected. A key person in the success of the observatory project was Laurier's minister of the interior, Clifford Sifton, born in 1861 near London, fifty kilometres west of Plaskett's birthplace, in southwestern Ontario. Sifton gained early recognition as a brilliant lawyer and politician in Manitoba and was well suited to carry out the mandate of his department – to oversee the settlement of the western prairies. Clearly, a commitment to surveying was part of the ministry's overall policy and tied in well with its wider responsibilities.

Planning for the Ottawa observatory began in earnest in 1898 when King, at Sifton's request, prepared a memorandum.[89] Besides equipment connected with surveying and time keeping, an important and perhaps surprising feature of his proposal was the provision of "an equatorially-mounted telescope, driven by clockwork, and fitted with attachments for micrometric measurements, and for spectroscopic, photometric and photographic work" so that "something could be done in the direction of modern investigations in Physical Astronomy." Evidently King was aware that photography and spectroscopy were revolutionizing astronomy. He also pointed out the value of astrophysical research in stimulating other branches of science and the unforeseen practical benefits that other industrialized nations had found to follow from the advancement of pure science. And he knew the importance of arousing general interest in the "new astronomy" by giving the public opportunities to view the wonders of the skies through an impressive telescope.[90]

Klotz and King had been rivals since their days working on government surveys in the 1880s. Klotz's diary gives a unique insight into the personal dynamics between the two Ottawa men. They were good friends at first but by the late 1890s friction was very evident. King had received positions on the International Boundary Commission that Klotz believed ought to have been his. King had relied on him for cost estimates and a list of other observatories but made no mention of his assistance in the memorandum to Sifton.[91] Most of all, Klotz did not agree with King's vision of a large telescope in an impressive building: "Neither King nor I have ever done any work with such an instrument, it is not in our line, but geodesy is." He resented what he saw as King's ambition to "have something imposing, something to strike the public eye – and then to be able to say 'I am the Director of the Dominion Observatory.'"[92] The rift between Klotz and King was unfortunate, but Klotz did manage to get the site, originally intended to be on Parliament Hill, moved to the Experimental Farm, about four kilometres west, on the edge of the city in those days.[93] It was a much more suitable location, as King himself recognized.[94] The conflict between these two men can be seen as a personal manifestation of the clash between the traditional, old astronomy and the new, a dispute that characterized the development of astrophysics, especially in the United States, as the old century faded and the young one flourished.[95] The new Dominion Observatory, by housing both camps under one roof, would be more akin to government observatories in Greenwich and Potsdam, for instance, than to its American counterpart – the U.S. Naval Observatory in Washington, DC.

One might think that the politicians in Parliament would balk at expenditures for something with no immediate practical payoff, especially one that a senior member of staff, Klotz, opposed. At least publicly, they espoused the observatory's practical potential – though sometimes for reasons that now seem ludicrous. Sydney Fisher, the member whose riding included the Experimental Farm, thought Canadian farmers could benefit from inquiries into "the influence of the sun, moon, and other ... occult forces from above."[96] Sir William Mulock thought the observatory might further his plans to "annex the telephones of Canada to the Post Office Department." Even some members of the opposition favoured the observatory. Thomas Sproule, Conservative member for Grey East said, "I think it would be a great mistake to put up a small [telescope], or one that would not be acceptable at the capital of the country."[97] In the end the proposed aperture of the telescope was increased from 25 to 38 cm. The observatory and its equipment ended up costing more than four times the original estimate of $30 000.[98] Because a telescope's light-gathering power depends on the area of its primary lens, a 38 cm instrument collects over twice as much light as a 25 cm one. In astrophysics, the telescope is in a sense an auxiliary instrument, one whose purpose is to collect as much light as possible from the source being studied and feed it to a spectroscope or other instruments for analysis.

Annual government reports give an idea of the project's progress. For the fiscal year ending 30 June 1901, Deputy Minister of the Interior James A. Smart wrote, "Preparation is being made for the erection of an astronomical observatory at Ottawa, and the installation of a large telescope, provision for this having been made by parliament at the last session."[99] The Department of Public Works prepared the plans for what it called the "Royal Observatory."[100] By the next year the contract had been let, construction had begun, and by June 1903 the 38 cm equatorial refractor had been received and was awaiting installation in the observatory building under construction at the Central Experimental Farm.

Throughout these years of planning and preparation, King was occupied with boundary survey work in British Columbia and on the 45th parallel (separating Quebec and New York State) as well as overseeing telegraphic longitude determinations across the continent. As boundary commissioner he also had to attend diplomatic meetings in London, England. In spite of all these duties he successfully made all the necessary arrangements for the new observatory. With construction on the new building under way by 1902, he recognized the need for extra staff, especially someone with experience in scientific instruments to oversee the installation of new equipment. He knew just where to turn.

King had graduated from the University of Toronto in 1875 and was a good friend of William Loudon, who, since 1881, had been on the Physics faculty.[101] So it was natural that "Billy" King, as Loudon called him, would discuss with him the possibility of prospective graduates working at the observatory. R.M. Stewart was the first to be hired, immediately after graduation in 1902, to get the time service up and running. Plaskett soon followed. Loudon, the important link in attracting JSP to the university thirteen years earlier, recommended him as just the man King needed to maintain the surveying equipment and oversee the installation of the new telescope and its accessories in the observatory. JSP first wrote to King in October 1902 expressing his interest in becoming King's assistant and providing evidence of his qualifications and experience.[102]

2.5 The Dominion Observatory in Ottawa. This photograph, dated 17 July 1903, less than three weeks after Plaskett began work in Ottawa, shows the building just beginning to rise above its foundations. (Photo by JSP; Canada Science and Technology Museum)

He enclosed a copy of his paper on orthochromatic photography along with favourable comments it had elicited from England and suggested to King that this type of photographic plate would be advantageous in capturing a wide range of the spectrum. It took over three months and a recommendation from King's political boss, Clifford Sifton, before King hired Plaskett, apparently not quite sure what this new man would be doing.[103] JSP had to ask him twice to outline the nature and scope of his duties so that he could prepare himself. King advised him to read general works on astronomy and astrophysics, which JSP reported he was doing. King surely could not have foreseen what far-ranging projects his new technician would so capably handle in the coming years.[104]

JSP took up his new duties in Ottawa on 30 June 1903, with an annual salary of $1100, a ten per cent increase over his Toronto pay.[105] He was officially working in the Boundary Surveys Division, and part of his work in the first couple of years involved inspecting boundary monuments, though increasingly his duties revolved around the new observatory.[106] Likely it was he who took a remarkable series of photographs documenting the building's progress from excavation to dome.[107] It had required seven years from King's initial memorandum to Sifton until the observatory was occupied in 1905. Only then did the systematic pursuit of astrophysics make its successful Canadian debut.[108]

Toronto–Ottawa Links

In the early years of the new century, astronomy in Toronto was in turmoil. The citizens, who were proud that their city was the site of the government observatory that served as the headquarters of national weather forecasting, were caught up in speculation that a new observatory in Ottawa would lead to its closure. The rumours were fuelled by the crowding out of the observatory by new buildings on campus.[109] (In the end, the observatory was rebuilt 100 metres to the north, and the Meteorological Service got handsome new quarters even further north, at 315 Bloor Street, now the university's Munk School of Global Affairs.)

And what were those new university buildings that squeezed out the old observatory? Ever since the great fire of 1890 had destroyed the old convocation hall, graduations had been held in the gymnasium northeast of University College. Now, with generous financial support of alumni, plans for a new hall were drawn up by the well-known architects Darling and Pearson. On 10 June 1904, the cornerstone was laid as part of the graduation ceremonies. By coincidence it was a special day for astronomers – three received honorary LL.D. degrees: King, Klotz, and Newcomb.[110] Simon Newcomb, born in Nova Scotia in 1835 and world-renowned for his work at the U.S. Nautical Almanac Office, had been offered the degree in 1900 but only now was able to come to accept it.[111] King was the director-in-waiting of the Dominion Observatory, and Otto Klotz, his assistant, was in fact being honoured as the founder of the university's Alumni Association. No matter, they were now known as Dr King and Dr Klotz, titles befitting their scientific status. Klotz and his wife, Marie, had taken the overnight train from Ottawa and stayed at the new King Edward Hotel.[112] He described the convocation in his diary and seemed especially pleased to be referred to as "the

2.6(a) The old Toronto observatory with the new Convocation Hall rising in the background. (UTA, A1965-0004 [10.4])

astronomical Magellan" for his work in laying the trans-Pacific telegraph cable.

Whether Chant, as a member of the Physics Department, had anything to do with nominating the astronomers for honorary degrees is uncertain, but he had known King for some years. At the convocation, King was presented by Professor Alfred Baker, who remembered him as "one of the best mathematical men who had passed through the University." He also praised King for his role in attempting to defend Canada's position in the Alaska boundary dispute.[113] (Through no fault of his, King was largely unsuccessful, but that is another story.)

Chant had been thinking for some time of starting a new course in astronomy and physics but it was late in 1904 before the university senate passed a statute establishing it.[114] It was to be a specialized option available to fourth-year students in the Mathematics and Physics course from which he and Plaskett had graduated. To inform the

University of Toronto.

ANNUAL COMMENCEMENT.

JUNE 10th, 1904.

I. Admission to Degrees.

LL.D. *(Honoris Causâ).*

Simon Newcomb, Ph.D., LL.D.,
Professor of Mathematics U.S. Naval Observatory.—Retired.
Emeritas Professor of Mathematics and Astronomy, Johns Hopkins University.
Presented by The President.

William Rainey Harper, Ph.D., D.D., LL.D.,
President of the University of Chicago.
Presented by the Premier of Ontario.

Charles Sedgwick Minot, LL.D., D.Sc.,
Professor of Histology and Human Embryology, Harvard Medical School.
Presented by The Vice-President.

Thomas Clark Street Macklem, M.A., D.D., LL.D.,
Provost of Trinity College.
Presented by The Vice-Chancellor.

John Lorne McDougall, M.A., C.M.G.,
Auditor General of the Dominion.
Presented by The President of Victoria College.

William Saunders, LL.D., F.R.S.C., F.L.S.,
Director of Experimental Farms for the Dominion.
Presented by James Mills, LL.D.

Otto Julius Klotz, C.E.,
Astronomer, Department of the Interior.
Presented J. C. Glashan, LL.D., by Dr. R. A. Reeve
Pres. Alumni Assoc.

William Frederick King, B.A.,
Chief Astronomer of the Dominion.
Presented by the Professor of Mathematics.

2

2.6(b) University of Toronto commencement program, pasted in the diary of Otto Klotz. (LAC, MG30, series B-13) undergraduates about the new developments, Chant wrote an article for the student newspaper, the *Varsity*.

Politically, Chant's timing was almost perfect. A new Conservative provincial premier, James Whitney, took over from his Liberal predecessor on 8 February 1905. He quickly acted on his long-held view that the university should be "put on a sound, stable, and permanent footing." He introduced a bill in May removing the existing university debt and generously provided for future funding. The bill included financing to complete both Convocation Hall and a new physics building next door, where as it turned out, astronomy would also be taught for the next sixty years. Whitney's biographer, Charles Humphries, and university historian Martin Friedland concur that Whitney "may well have been the greatest friend the institution … ever had."[115]

Once classes ended in the spring of 1905, Chant set out on a tour of observatories in the eastern United States so that he would be able to make a proper assessment of what material to include in his lectures and what equipment should be purchased to give the astronomy students the necessary practical experience. He found that "a 9 or 10-in. telescope in a pleasing little building was a favourite donation to a college in the United States. It was much superior to anything we had in Toronto."[116] Even such a modest plan eluded Chant for decades.

Plaskett, as we shall see, developed much more ambitious ideas and successfully advanced them by appealing to the federal government rather than looking to private benefactors.

3
Ottawa Advancement, 1903–1907

Ottawa was a city with a population of only 100 000, including the suburbs, when the Plasketts moved there in June 1903. It was dominated then, as it still is, by Parliament Hill, the seat of the federal government. Not far away, near where the national war memorial is now, was the office where JSP initially worked before the observatory was completed.[1] About two kilometres farther east was 402 Daly Avenue, the Plasketts' first Ottawa home.

Their son, Harry, was about to turn ten when they moved. He attended the Model School and later Ottawa Collegiate Institute, where, by his own account, he was a mediocre student.[2] Nevertheless, even during his teenage years and before he went off to the University of Toronto in 1912, he took an enthusiastic interest in his father's work.

The Plasketts' second son, Stuart Stanley, was born in Ottawa on 10 November 1904 and was baptized in January by his newly ordained uncle Frank Plaskett at All Saints' Sandy Hill, the Anglican church attended by the Plasketts, about six blocks away from their home.[3] The building was (and still is) across the street from Laurier House, the prime minister's residence at the time, and was the place of worship of many prominent Ottawa families, including those of Sir Henry Bate, its chief founder and benefactor; Deputy Minister of the Interior William W. Cory (the top civil servant to whom W.F. King reported); and Robert Borden, leader of the Conservative Party after 1901 and later prime minister.[4] It is not certain that JSP hobnobbed with these men, as the church's minute book shows that he was active at vestry meetings and as a sidesman only in 1904–6, which predates mention of Cory and Borden in the church records.[5] The Plasketts changed churches sometime after 1906, when they moved closer to the observatory.[6] They lived at

3.1(a). Reba Plaskett and her younger son, Stuart, in July 1905. The archival documentation states "Mrs. I. S. Plaskett, July 1905" but it seems clear that a handwritten "J" was mistaken for "I." Stuart would have been eight months old, in apparent agreement with the baby in the picture. (LAC, Topley Studio Collection, R639-O-5-E)

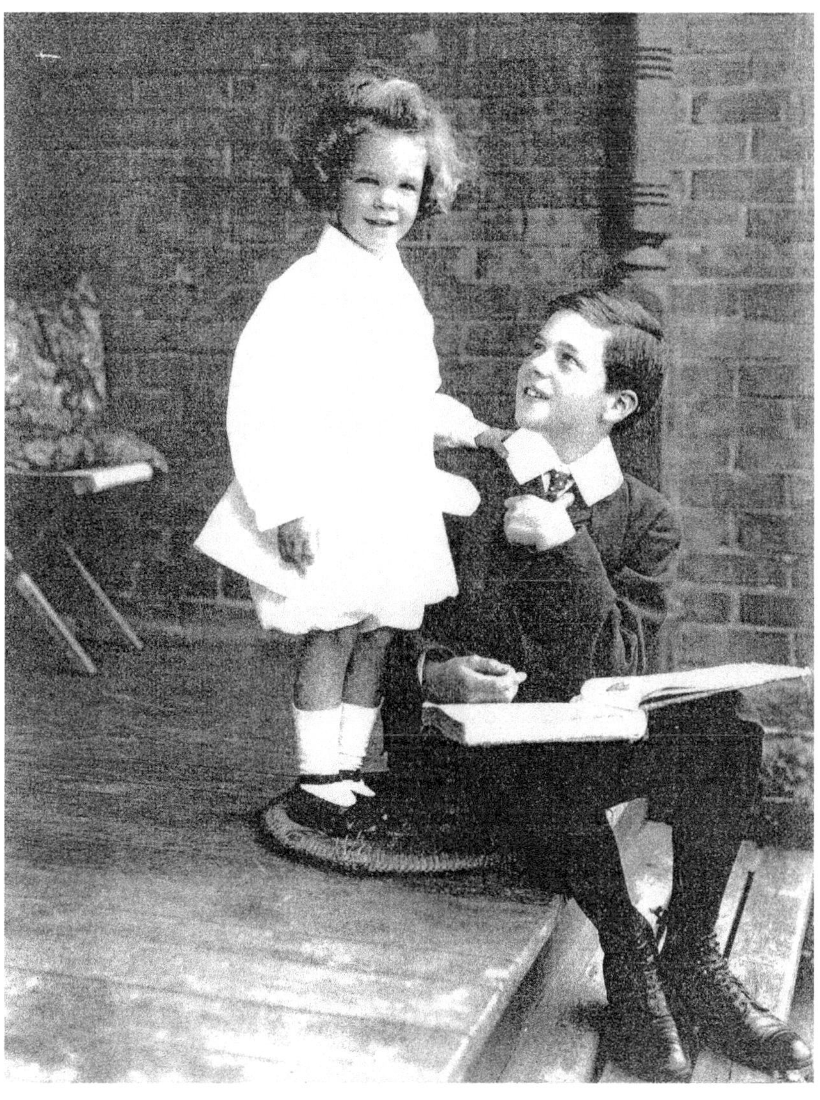

3.1(b). Harry and Stuart Plaskett, probably in 1906 Although the people in the photograph are not identified, it is in a folder in the Annie Jump Cannon papers labelled "Plaskett." Annie Cannon and the Plasketts became very good friends. (HUA, HUGFP 125.12p).

3.2 318 Fairmont Avenue, Ottawa, as it is today. This spacious house just two blocks from the observatory was the Plasketts' home from 1910 to 1917. City directories show that the Plasketts' immediate neighbours were also connected with the observatory. There is a possibility that the government provided this housing so that the staff, particularly the astronomers who had night work, would be only a short distance from home. (Photograph by author)

three different addresses before settling in at 318 Fairmont Avenue, just a block from the observatory, in 1910.[7]

The Dominion Observatory in Ottawa: Staff, Building, and Equipment

According to a detailed report of Chief Astronomer King, the staff of the Astronomical Branch moved from their cramped quarters at 26 Wellington Street into the observatory building in April 1905.[8] King's permanent staff consisted of thirteen people in all: a secretary/accountant and a correspondence clerk, two surveyors in the Boundary Surveys Division, and nine people in the Observatory Division, comprising a chief computer, a record keeper, four observers, a photographer, and two astronomers (Otto Klotz and Plaskett).[9]

Klotz had expected to be named chief astronomer when King was named director but his boss retained both titles.[10] On the salary scale, Klotz had little cause for complaint – less than 10 per cent separated his pay from King's.[11] Yet by far the longest part of the observatory's first official report was Klotz's work, dealing with his observations and determinations of longitudes of Pacific islands between Canada and Australia, which had just been connected by trans-Pacific cable.[12]

JSP's part of the report included a fourteen-page description of the observatory building and equipment, complete with floor plans and sixteen photographs.[13] The impressive sandstone structure comprised two wings, with a central tower crowned by a dome housing the main telescope. The circular room on the ground floor of the central section served as the entrance foyer; the one above it had glass-fronted cabinets to store surveying equipment; and the one below, in the basement, was the clock room, where timekeepers ticked away the seconds in a stable environment. Plaskett's room was in the eastern wing of the first floor – that is, one above the ground floor. Next to it was the room of the chief computer, John Macara, and across the hall was the drafting room.[14] A large photographic room occupied the entire east end of the first floor. To get to the telescope from his office, Plaskett would climb two flights of stairs adjacent to the building's central circular tower. The first flight up would bring him to the "mid-way floor," level with the accessible flat roof of the main part of the building. On this level, in the circular tower, was a darkroom for developing photographic plates exposed during an observing session. One more flight up brought him to the equatorial room, where the telescope, nearly 6 m in length, was located.

This circular room was 9 m in diameter with a hemispherical dome, rotatable on its circular track 2.7 m above the floor by pulling on a rope looped around a pulley. Another loop of rope allowed shutters to open, giving a slit 1.5 m wide for the telescope to peer through. The solid pillar supporting the telescope was independent of the building and extended down three floors through the basement to bedrock.

The basement of the observatory housed a workshop, which JSP equipped with lathes, a milling machine, and smaller tools.[15] Before long the heavy machines had to be moved to a separate building because their vibrations interfered with the clocks, seismographs, and gravity pendulums – scientific instruments requiring great stability – which

3.3 Photomontage showing the newly completed Dominion Observatory in Ottawa, Chief Astronomer W.F. King (left), and his assistant Otto Klotz (right). The handsome sandstone building is still a landmark near the entrance to the Central Experimental Farm. (CSTM)

3.4 The 38 cm telescope in the dome of the Dominion Observatory, Ottawa. Note the observing chair and stairs, which can rotate around the dome. (Photo by JSP, LAC, PA-107533)

were also in the basement.[16] Plaskett gave a thorough description of the intercom telephone system and the electrical system, which he probably helped to design, and of course the telescope and its accessories: the eyepieces, photometer, position micrometer, solar camera, stellar camera, and (most importantly for him) the spectroscopes or, more precisely, the spectrographs.[17] (The term *spectroscope* denotes a device for studying the spectrum. Sometimes the term is used generally and sometimes more restrictively for an instrument used visually. A *spectrograph* is used to photograph the spectrum, and the resulting image is called a *spectrogram*. The advantage of photography is not only the permanent record it provides, but its ability to accumulate the faint light on the emulsion.)

The objective lens at the front end of the telescope, typical of refractors, was made up of two elements – one of crown glass and the other of flint glass. This arrangement had the advantage of bringing the various colours of light to nearly the same focal plane. Without such an achromatic lens, objects would appear to be surrounded by a coloured halo. The highly respected Pittsburgh firm Brashear and Company, established in 1881, made the lens with a diameter of 38 cm and a focal length of 5.8 m. The Ottawa telescope was unusual in having the flint glass component in front of the crown, a feature designed for Brashear by Charles S. Hastings to minimize condensation on the lens from the cold night air. The other end of the telescope, or tail, could be equipped with an eyepiece, or a spectrograph to capture stellar spectra, or a micrometer for measuring small angles separating close stars. When the equipment was interchanged, counterweights had to be adjusted to keep the telescope balanced. The firm of Warner & Swasey in Cleveland made the mounting and drive mechanism, a scaled-up version of a 30 cm instrument described in their catalogue.[18] A drive is needed to keep the telescope pointing at the same object in the sky, thus counteracting Earth's rotation; with an equatorial mount the drive needs to act on only one axis. Optical and mechanical excellence must be combined to produce a really useful telescope.

Since stellar spectroscopy is central to astrophysics and to JSP's achievement, we need to consider what it entails.[19] The whole process seems almost miraculous. Imagine a star, barely perceptible to the unaided eye as a dim dot in the sky. Imagine its light, spreading out in all directions through space. After hundreds of years, an inconceivably tiny fraction of this starlight lands on the 38 cm diameter lens of a telescope in Ottawa. The light, focused by the telescope on the slit of

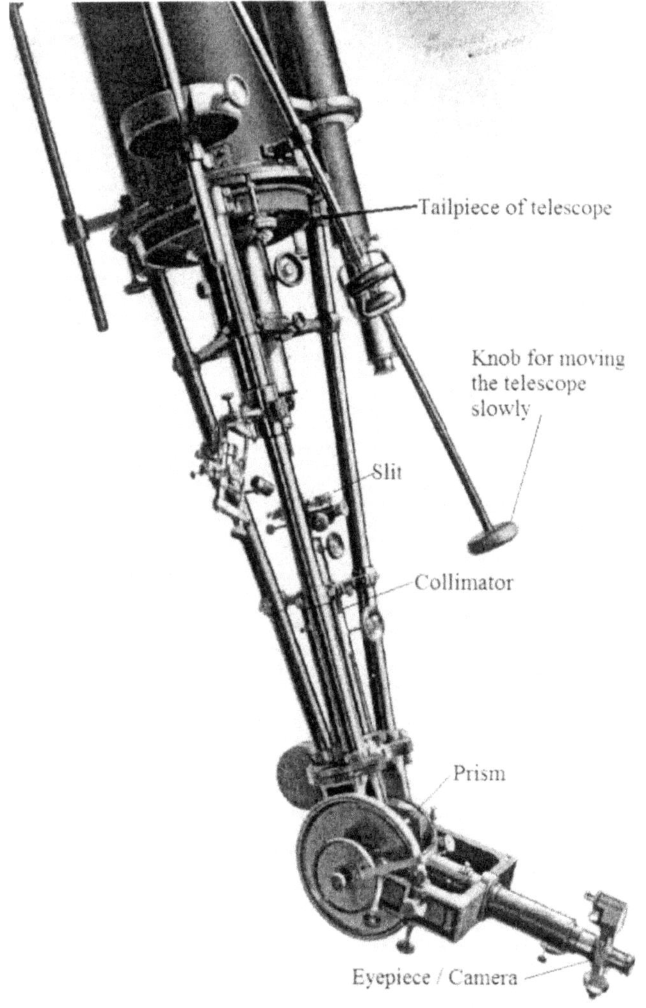

3.5 A Brashear universal spectroscope attached to the tailpiece of a telescope. The starlight emerges from the telescope and is focused on a slit. The slit acts as a light source, illuminating the lens of a collimator immediately below it, and the collimator produces a parallel beam of light. This illuminates a prism (hidden behind the circular component in this diagram), and the light, refracted and dispersed into a spectrum by the prism, is then brought to a focus by a lens in the tube at the bottom, where it can be examined with an eyepiece or photographed with a camera. The spectrum is really just a side-by-side array of images of the slit in different wavelengths or colours. (Illustration adapted from Brashear Company catalogue of 1911)

3.6(a) Toepfer measuring engine used by JSP at the Dominion Observatory to examine spectra and to measure the minute shifts in their lines. It is currently in the Canada Science and Technology Museum, catalogue number 700221. A smaller instrument, a measuring microscope, is also shown. (Reproduced from figure 4 in *Report … for 1911*, 122)

the spectroscope, is then spread out by a prism into a strip about 6 cm long, where it interacts with chemicals on a glass plate, leaving its vital fingerprint in a pattern of light and dark. This light makes its impression not only on the photo chemicals but on the mind of the astronomer, who will wrest every possible detail from it. How did the star produce the energy? Is it different in some respects from our Sun? What is the star made of? What is its mass and temperature? Is it moving? What can such data reveal about the star's distance and age? These were questions that spectroscopy would help answer. On a broader scale, it would confirm the universality of natural laws and even, as some

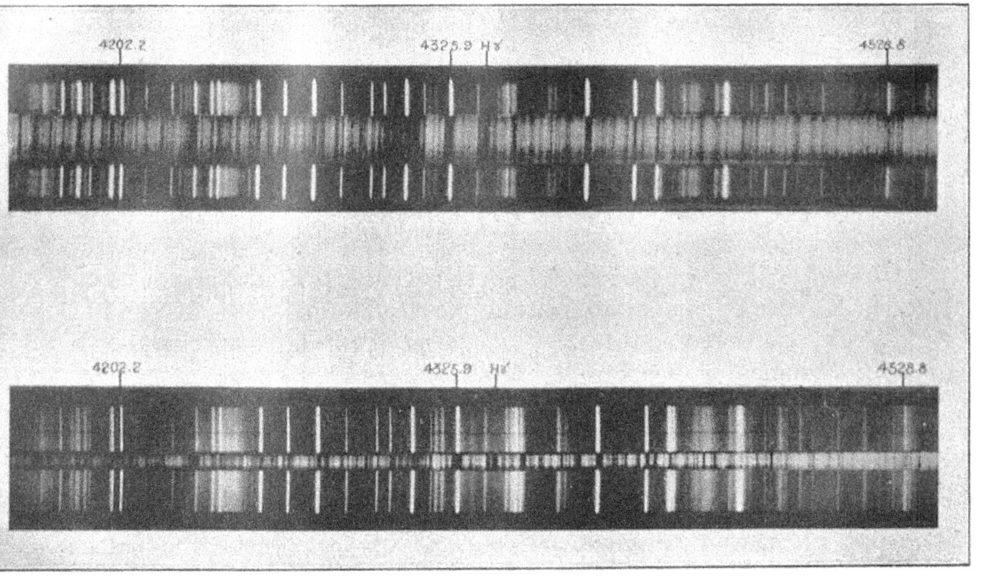

3.6(b) Enlarged images of a 2 cm-long portion of two plates taken by JSP in 1906. In both cases the spectrum being studied is the middle strip. On either side are two identical bright emission line spectra of an iron arc photographed immediately before and after the star's spectrum was captured. The wavelengths of the iron lines are known, and three are indicated (in Angstrom units) on this illustration. Using them for comparison, the wavelengths of the dark absorption lines in the spectrum being studied can be calibrated and identified. Among the myriad dark lines in the spectra, Plaskett singled out a hydrogen line, Hγ, for identification in these images. The top spectrum is that of sunlight reflected from the planet Jupiter and provides a nice illustration of the Doppler shift. Because Jupiter is rotating on its axis with a speed at its equator of almost 13 km/s., the light from the side of the planet that is receding is shifted to longer, redder wavelengths (to the right in this image), while the light from the side that is approaching is shifted to the blue. The result is the tilted appearance of the lines in Jupiter's spectrum. The bottom spectrum records light from the star Arcturus. (Reproduced from figure 6, *Report … for 1906*, 62)

would argue, shed light on the big questions – the relationship among humanity, God and the universe.[20]

An extremely useful feature of astronomical spectroscopy is the Doppler shift in the lines in the stellar spectrum as a result of the relative motion of the star and the observer on Earth. The shift can be detected in comparison with a standard arc spectrum recorded at the start and finish of the exposure adjacent to the star's spectrum on the same plate; from a measurement of this shift, the star's speed along the line of sight (radial velocity) can be found. In this way Plaskett would soon be using the Ottawa equipment to measure the rotation rate of the Sun and to study spectroscopic binaries – stars whose mass, distance, and size could sometimes be inferred.[21] As astronomer Vic Gaizauskas has written, "It was Plaskett who turned King's vague interests in astrophysics into concrete observing programs. Their success caught the attention of the astronomical world."[22]

Though Plaskett and his co-workers moved into the Dominion Observatory (DO) in April 1905, it was mid-November before he was able to actually start observing. He had been fully occupied in erecting, adjusting, and testing the new observatory instruments and preparing for a total eclipse of the Sun, predicted to be visible in parts of Canada, at the end of August.[23]

The Labrador Eclipse

In a total solar eclipse, the Moon moves in front of the Sun and completely blocks its brilliant surface (photosphere) for a few precious moments. For astronomers trying to understand the physics of the Sun, the phenomenon provides a rare opportunity to study the Sun's tenuous corona, or outer atmosphere, normally invisible because of the Sun's dazzling glare.

Plaskett had never witnessed the spectacular sight of a total solar eclipse. There had already been five in North America during his life but he would have had to travel hundreds of kilometres to see any of them. Only two seem to have attracted Canadian participation. The Canadian government sponsored an expedition led by Edward D. Ashe of the Quebec observatory to Iowa in August 1869.[24] And the total eclipse of 28 May 1900 was close enough that two Toronto amateurs, Thomas Lindsay and George Lumsden, joined expeditions of professional American astronomers in the southern States.[25]

It had been determined that the next total solar eclipse in North America would occur on 30 August 1905. The most westerly point from

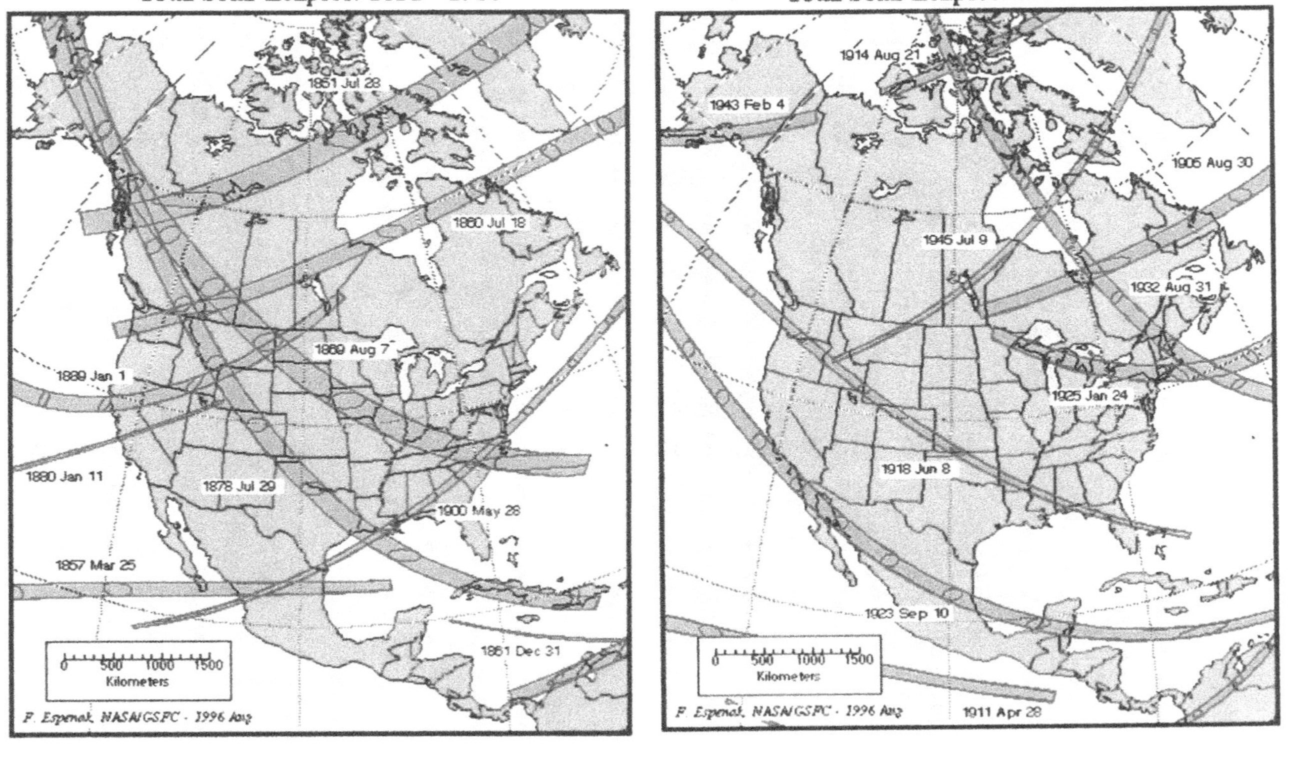

3.7 Maps of total solar eclipses in Canada and the United States from 1851 to 1950. (Courtesy of Fred Espenak, NASA/Goddard Space Flight Centre)

which the complete phenomenon could be viewed was James Bay. From there the path of totality swept across the Quebec interior, crossed the coast of Labrador, made landfall again in Spain and, after crossing northern Africa, ended in Saudi Arabia. If changes in the corona were to be tracked during the eclipse, it was clearly advantageous to make observations from this side of the Atlantic, where the eclipse began, allowing comparison with observations a few hours later from Europe, Africa, and the Middle East. California's Lick Observatory was making arrangements to send an expedition to Labrador, as well as to Spain and Egypt. Planning in Canada began none too soon. In November 1904 the council of the Royal Astronomical Society of Canada (RASC) urged the federal government to organize an expedition and requested that some of the society's members who were qualified observers be invited to go along at no expense to themselves.[26] The government, perhaps feeling some embarrassment because American astronomers already were preparing to come to Canada, agreed with the RASC, approved a grant of $7500, and put Chief Astronomer King in charge of the overall plans.[27] Government financing for eclipse expeditions was the British model; in the United States and elsewhere, individual observatories usually financed and sent out their own parties.[28] King decided that Labrador offered the best prospects of any Canadian site for observation, and he selected a specific spot on the shore of Lake Melville, close to both the mouth of the North West River and a Hudson's Bay Company post.[29] He chartered a small steamship, the *King Edward*, to extend its normal service on the north shore of the St Lawrence in order to transport the party and equipment from Quebec City to their destination and back. Twenty-three people were chosen to make the trip, though a few more went along, partly at their own expense.[30] With professionals and amateurs, men and women, young and old, francophones and anglophones, the makeup was quite egalitarian. Included were staff from the Dominion Observatory, RASC members, and two renowned solar experts from England, Annie Russell Maunder and Edward Walter Maunder.[31] Walter had begun his professional career at the Greenwich Observatory in 1873 as photographic and spectroscopic observer. His wife and collaborator, Annie, was an accomplished astronomical and solar photographer. C.A. Chant was naturally one of the RASC people. His intention was to make visual and photographic observations of polarization of the faint light of the corona. Those not on King's staff were free to devise their own observing program, though some RASC amateurs opted to assist the Ottawa professionals.[32]

With the move into the new observatory in Ottawa imminent in early March 1905, JSP must have had a myriad of details to look after; even so, King sent him to Toronto for a few days to discuss eclipse plans and RASC involvement. He made the most of the opportunity by speaking to both the society and the Canadian Institute about the eclipse expeditions, the preparations, and the objectives.[33] And he surely would have visited family, probably celebrating his mother's sixty-fifth birthday on 12 March.

At the Dominion Observatory JSP was responsible for the astronomical apparatus as well as the surveying instruments when they were not being used in the field.[34] Not surprisingly, then, King charged Plaskett with designing the equipment and preparing for the eclipse observations to be carried out by the observatory staff.[35] At the same time he promoted JSP to the rank of astronomer at an annual salary of $2000, a 66 per cent increase over his previous earnings as mechanician.[36] With only two years of service, he was now equal in rank (though inferior in salary) to Klotz, a situation that contributed to a grudge against Plaskett that Klotz harboured for the rest of his life. King's perceived favouritism towards astrophysics and its chief practitioner, Plaskett, fuelled Klotz's resentment. There are two sides to every argument, and there can be no doubt that Klotz had a great deal of responsibility as well as seniority, but JSP had become accountable for the development and direction of a new field of work at the observatory. To modern eyes it was an astonishing advance for a man in his fortieth year who had only a bachelor's degree and who was just becoming exposed to observational astronomy. In fact Plaskett's lack of any formal postgraduate education was the norm for North American astronomers and remained so until after the Second World War.[37] Astrophysics was a brand new field in Canada, and JSP was the ideal person to carry out the approach originated in Europe and pushed by the visionary George Ellery Hale in the United States – namely, that innovation was essential in a sort of laboratory environment where astronomy and physics merged. JSP was always grateful to King for the opportunity to spread his wings, and he consistently showed how well deserved his chief's confidence was.[38]

His promotion was also a remarkable testament to King's prescient recognition of and trust in his very capable assistant, and an indication of JSP's ability to learn quickly and to produce a detailed plan of action. Plaskett, having never even seen a total eclipse of the Sun and with almost no astronomical experience, had to start from zero. In the course of his background reading, he learned more generally about solar astronomy – what observations were being made, where, by whom, and with what equipment. He began by studying what had been accomplished on previous eclipse expeditions and by trying to ascertain what other astronomers were planning for this one. The corona had often been recorded on previous occasions but always on photographic plates most sensitive to the blue end of the spectrum. Plaskett's earlier experiments with orthochromatic plates suggested there were better options.[39]

Part of JSP's plans included photographing the Sun's corona in monochromatic red, yellow, and green.[40] By doing so, he hoped to gain some idea of the distribution of colour in the corona and to distinguish between sunlight reflected from particles in space and emission from luminous gas. This necessitated a lot of preliminary work in choosing the most suitable plates and filters.

Another specific problem he hoped to study concerned the brightest emission line in the inner solar corona, in the green part of the spectrum at a wavelength of 530.3 nanometres (nm).[41] No one had been able to identify this particular line with any known chemical element in laboratory spectra. Some thought it arose from a new element, tentatively named coronium. Plaskett was keen to find out how the source of this emission was distributed in the solar corona, in the hope of furthering some understanding of its origin. But to photograph the corona in this light he had to find plates sufficiently sensitive to that part of the spectrum and filters allowing only a narrow band of light centred on 530 nm to pass. Unable to find any commercially available plates with the required sensitivity, he had to bathe plates in various dyes to find the best combination – a German plate sensitized with a chemical known as Orthochrom T. Fortunately this setup, along with a green filter, required exposures of only a minute, well within the two and a half minutes of eclipse totality.

JSP also proposed photographing the spectrum of the so-called reversing layer: in this region of the Sun's atmosphere, what are normally dark absorption lines in the Sun's spectrum appear as bright emission lines during a total eclipse because the dazzling light from the Sun's photosphere is blocked by the Moon. As JSP explained in his official twenty-four-page report, careful planning was needed to ensure that the proposed observations could be made during the brief period of totality.[42] First of all, specialized equipment had to be designed and made – another area where Plaskett's previous mechanical experience proved invaluable. Although some standard instruments from the observatory could be used, modifications were essential, as everything

3.8(a) The coelostat and long-focus camera set up for the total solar eclipse in Labrador, August 1905. The mirror, seen in the distance, was driven by clockwork and tracked the Sun, reflecting its image down the long covered path to a camera operated by the person standing on the left. Note the very large plate holders. The man near the coelostat mirror with his hand raised is W.F. King, chief astronomer. (Courtesy RASCA, 1905 eclipse album)

had to be transported to Labrador, carted close to a kilometre from the ship to the actual site, and then mounted on stable concrete piers with operable driving mechanisms to keep the instruments pointing accurately. Plaskett wisely opted for a system called a coelostat, which used a clock-driven mirror to reflect the sunlight into a fixed horizontal telescope and camera.[43] This set-up allowed for a much longer focal length and a larger image of the Sun, and was a more stable arrangement than an equatorially mounted telescope. Besides, it was much simpler to operate at a convenient, fixed location where plates could be loaded in rapid succession during the short period of darkness. Plaskett proposed, and King concurred, that the additional cost of the coelostat, to be built by Brashear for $2665, was justified, because it would subsequently be used in Ottawa for studies essential for understanding the link between solar changes and meteorological and climatic changes

3.8(b) Near the mirror of the coelostat, astronomers rehearse for the eclipse. The prismatic camera (left) and the grating camera (right) are attended by two men with their backs to the camera. The two people facing the camera are King and Plaskett. Notice the lanterns, which would have been used during the darkness of the eclipse, and the metronome for beating seconds. (Courtesy RASCA, 1905 eclipse album)

on Earth.[44] Previously, Canadian astronomers and meteorologists had merely recorded sunspots but Plaskett would be instituting solar studies of a much broader scope.

In addition to the scientific apparatus that he would transport to Labrador, Plaskett needed to take a photographic darkroom and other structures to protect the equipment from the elements. All of it was set up in Ottawa on the grounds of the observatory and tested in every possible way, ensuring that any shortcomings were eliminated before it was all dismantled, crated, and shipped. For the instrumental needs alone, Plaskett reckoned that there were eighty-eight boxes, bundles, and crates, weighing about five tons. In addition, of course, there was all the camping equipment and provisions needed to keep the party dry, well fed, and content.

King and Plaskett set out in late July and were already at the Château Frontenac, their hotel in Quebec City, when Chant got there on 1 August.[45] With a couple of days on their hands, they enjoyed sightseeing together in the historic city and on a day trip to the magnificent Montmorency Falls. Joining them was JSP's brother, Frank. He had graduated from Bishop's College, Lennoxville, Quebec, in 1903, and had been ordained in Montreal as an Anglican priest on 21 December 1904. His first parish was the remote 800 km stretch of coast along the Gulf of St Lawrence from Sept-Îles to the Strait of Belle Isle, so he had some familiarity with the area the astronomers would be sailing past. Indeed, JSP had asked his brother for advice on Labrador weather and supply lines during the planning stages of the expedition.

Most of the astronomers and their equipment set sail on their 1700 km journey from Quebec City to Labrador on 4 August. The rest, with less-demanding set-ups, would leave a couple of weeks later. Although packed into smelly, narrow quarters and enduring bad food and seasickness, they enjoyed some redeeming sights as they steamed up the Labrador coast.[46] They could see snow still lingering in the valleys and were awestruck by icebergs on all sides of their little ship. On shore they met local Innu, Hudson's Bay Company employees, and missionaries, and saw codfish by the hundreds split and laid out to dry on flakes. From the many snapshots taken of the memorable trip, about a hundred were organized into souvenir albums – one for each person on the expedition. JSP contributed thirty-three of the pictures.[47]

The travellers arrived safely at their destination, North West River, a week after their departure, giving them about three weeks to prepare the site and apparatus for the big day. They were happy to find Professor Louis B. Stewart of Toronto already there; he had gone in advance to scout out the site. After Plaskett and the first contingent arrived, the steamer made a return trip to Quebec City to pick up the other astronomers, who reached the eclipse camp on the 28th.

Frank Plaskett had told his brother about the brilliant northern lights, which rustled like a silk dress.[48] Indeed, when the astronomers got to Labrador they saw some magnificent auroral displays. They also experienced something that Frank may have wisely kept to himself – the "biggest and fiercest mosquitoes in the world," and little blackflies that made the mosquitoes "seem friendly by comparison." At least that was

3.9 (a) The S.S. *King Edward* at one of its ports of call; stacks of dried cod are visible on the dock. (Photo by JSP, courtesy RASCA, 1905 eclipse #20)

the way that the English astronomer Walter Maunder portrayed the pests, adding that nearly everyone was "scarred and pock-marked by their bites almost out of recognition."[49] And, as the astronomers drifted off to sleep, serenaded by the howl of huskies, they felt dozens of mice scampering over them.[50]

Misery loves company, and the astronomers enjoyed each other's friendship. Chant described how, "on cold evenings as we gathered about the campfire and joined in the general entertainment with song or joke, Dr. King threw aside somewhat the cloak of reticence in which he was usually wrapped and gave us some of Bret Harte's poems. He stood the camping test; the nearer we came to him, the more did his sterling character show itself."[51]

During the day everyone was kept busy with tasks – erecting temporary structures, mixing concrete, mounting equipment, and so on.

3.9(b) Members of the Canadian eclipse expedition gathered in their Sunday best for this group photo at North West River, Labrador. In the front row, left to right, are C.A. Chant, JSP, and W.P. Near. In the middle row are Mrs Codd, Winnifred King (daughter of W.F. King), E. Walter Maunder, Annie Maunder, Father J.J. Kavanagh, and A.T. DeLury. Standing behind Winnifred King and Walter Maunder is W.F. King. A more complete group photo was taken aboard the *King Edward* and published with a key in *RASC Selected Papers and Proceedings for 1905* (1906): 3. (Photo by JSP, courtesy RASCA, 1905 eclipse #62)

Once everything was in place and the instruments were adjusted and focused, Plaskett led rehearsals, each person having very well-defined duties to carry out. He wrote:

> Each one practised with his own programme until he could perform it satisfactorily. On the morning of the 29th [the day before the eclipse] a complete rehearsal was held under as nearly as possible the same conditions as would prevail at the eclipse. The signal was given for totality, after a warning 30 seconds previous, and then the time was called every ten seconds until 150 seconds had elapsed, the intermediate intervals being given by the beats of a metronome. So well had every one practised that the first rehearsal went through without a hitch. The following rehearsals served to perfect the movements, until, after fifteen or twenty, the whole programme went with machine-like regularity. Another set of rehearsals was held in the evening by lantern light to accustom the operators to working by artificial light which might be required during totality.[52]

The day before the eclipse was rainy with a falling barometer, indicating more of the same to come, but, as Plaskett said, "We still hoped, even against our better judgment, that it might clear up for the time, but at daylight on the 30th, although the clouds were more broken than on the previous day, there did not seem much chance of observing the eclipse. However, all preparations were made, the plate holders placed in order in their positions on the stands provided for them, the canvas roof rolled back, the gable end dropped, and the ridge and rafter removed."[53]

Sadly the clouds never parted. Undoubtedly the mood was as gloomy as the landscape during totality. Plaskett put the best possible face on it:

> Naturally it was a bitter disappointment at having practically no result for six months' work, except for the experience in preparation and the useful knowledge gained of colour sensitive plates and absorbing screens [filters]. If everything had not been in such first-class shape for the observations, if the perfection of adjustment to focus and working of the camera shutters and plate holders, if the running of the coelostat, or the quality of the specially sensitized plates had not come up to my required standard, probably I would not have felt the disappointment so keenly; but, when the prospects of obtaining some original and useful results were so good, it seemed too bad there was no chance to try.
>
> However, nothing remained to be done but dismantle and pack up all the instruments and appliances. This we entered into with such vigour that

> little remained to be packed after the evening of the 30th. The rest of the packing was done on the morning of the 31st, and by evening everything was loaded on a schooner in readiness to be transferred to the steamer.[54]

Two hundred kilometres to the east, the Lick Observatory eclipse team was stationed at Cartwright, another Hudson's Bay Company Post on the Labrador coast. Their intentions were to photograph the corona and to search for intra-Mercurial planets, but they too faced dense clouds on the day of the eclipse.[55] The Canadians learned of the Americans' frustration when they called at Rigolet, Labrador, to meet Wilfred Grenfell on their homeward journey.[56] The RASC had elected the renowned medical missionary to associate membership in May when plans for the expedition were well under way.

After an uneventful trip up the St Lawrence the *King Edward* delivered everyone safely back in Quebec City on 7 September. Before returning to Toronto, Chant stopped in Ottawa, where Plaskett took him through the new Dominion Observatory and showed him some views through the telescope. The next day Chant was on his way home to Toronto, and within a month he was speaking to the RASC about his experiences.[57]

Chant in Toronto

Back at the University of Toronto, Chant began teaching a new astronomy course in the fall of 1905. He recognized that he could cement links to the DO through the RASC and in the training of astronomy specialists at the university. On this last point he did have precedent on his side, because it seemed almost a tradition that anyone employed in astronomy at the DO had to have studied at U of T. King and Klotz, Plaskett and R. Meldrum Stewart, were the main luminaries but there were lesser lights too – Frederick W. Orion Werry (BA 1897), W. Maxwell Tobey (BA 1900), and F. Archie McDiarmid (BA 1902).[58] Better-qualified graduates would flow once Chant began teaching the new course.

JSP's Work at the Dominion Observatory

After JSP returned to Ottawa from Labrador in September 1905, he had to write up detailed accounts of the eclipse and also of the observatory itself (these reports were quoted at length earlier in this chapter).[59] Then he was finally able to get down to the business of observing stars.

JSP's night schedule at the observatory involved photographing stellar spectra on Monday, Wednesday, and Friday nights – whenever it was clear. (Tuesday and Thursday nights were used by his colleague Maxwell Tobey to do photometry – that is, to measure the brightness of stars.)[60] Like any organized observer, JSP knew what stars he hoped to observe each night, selecting those that were well above the horizon at that time of year and time of night. He would have written down their celestial coordinates (declination and right ascension) from a catalogue and, using a celestial atlas, made a sketch of the field of view centred on the star he would study. Armed with these data, a logbook, and a light-tight box of photographic plates, Plaskett went up to the unheated dome as night fell. He pulled on the rope opening the shutters to the sky. It could be bitterly cold in winter, but heating was out of the question, as the resulting air currents would have tossed the stellar image around like a cork on an ocean. Opening the glass doors on the side of the telescope pier, he started the telescope's drive, which was powered by falling weights, and then set the zero point on the right ascension circle to the correct sidereal time. He unclamped the telescope and tilted it by hand until the pointer on the declination scale, illuminated by small electric lamps, matched the star's declination. After reclamping the telescope in declination, he released the right ascension clamp and, using the hand wheel, swung the telescope around until the pointer above the wheel showed the correct right ascension. He then pulled on the other rope loop, causing the dome to rotate with a great rumbling until the slit was in position directly above the telescope's objective.[61] After rolling the movable steps so he could climb to the eyepiece, he was then ready to look through one of the two finding telescopes attached to the main instrument to see if the star identified on his sketch was in the crosshairs. If not, he would turn the knobs of the slow-motion controls until it was centred. Observing could then begin. In his usual work, Plaskett recorded stellar spectra on little glass plates measuring 5 × 7.5 cm, no bigger than the palm of his hand. During the exposure, sometimes lasting an hour or more, JSP kept watch through the eyepiece to ensure that the star's image fell centrally on the spectrograph's slit and trailed along the length of the slit to give some width to the photographed spectrum. Guiding the telescope for such an exposure was a tiresome chore. No wonder Plaskett was eager to improve the design of the spectrograph so images could be obtained more quickly. Developing the plates in the darkroom under the dome was usually done by an assistant right after the exposure was completed so that, if there was an obvious mistake, another spectrogram could be obtained.

3.10 JSP holding the spectrograph frame of the 38 cm refractor at the Dominion Observatory, Ottawa (City of Toronto Archives, James Collection, f1244_it2209)

After the telescopic work was done and the plate had been developed and labelled, the next step involved measuring the spectrum. This was a daytime task, which involved placing the spectrogram on the stage of a travelling microscope, setting its crosshairs on each line in the stellar spectrum (whose shifted wavelengths were to be found) and on the lines in the comparison spectrum (whose standard wavelengths were known) (see figure 3.6b). Measuring displacements of less than one-thousandth of a millimetre was the meticulous part of the work; after that, straightforward calculations led to a radial velocity (in km/s) for each line measured in the stellar spectrum, and the results for several lines were then averaged to give a better estimate of the radial velocity of the star at the time of the exposure.[62]

The Press and the Public

Opportunities for the public to visit the Ottawa observatory were always part of the plan. Shortly after the staff had moved in, the director, King, got public relations off to a great start by hosting a tour for the parliamentary press gallery, on 29 April 1905. Lubricated by King's scotch and mineral water, a *Toronto Star* correspondent enthused that the facility was "money well spent, when the useful work done by the Observatory staff is considered." He spoke of Plaskett, "an expert in color photography [who] has conducted some valuable experiments in the line of sensitized plates. He has already achieved marvelous effects in the way of chromatic combinations. Spectrum photography is also part of the business of the Observatory."[63] If the reporter was, by association, making the far-fetched suggestion that stellar spectroscopy would improve colour photography, it was nothing compared to his rationale for solar studies at the observatory: "The day is not far distant," he wrote, "when weather forecasts of approximate accuracy will be given for two years ahead, and be of almost absolute accuracy for periods of three months." As well, he noted the observatory's important geodetic work: "Correct latitudes and longitudes are very essential in the art of navigation, and the more accurate they are the smaller become the rates for marine insurance." Finally, lest anyone doubt the source of these wonders, this reporter for the Liberal Toronto newspaper concluded, "If the Observatory at Ottawa ever hits upon a little stranger in the spacious firmament it might do well to assert Canada's importance in the scientific world by naming it Laurier," after the prime minister whose government footed the bill.

Whether Laurier himself ever visited the observatory is unknown, but Lady Laurier and friends came calling in the spring of 1906,[64] and international astronomers came to discuss research interests from time to time. Every Saturday evening, members of the general public had the opportunity for a guided tour and a look through the telescope if the weather cooperated.[65] Over four thousand people visited during the first year alone, and JSP would have been on duty on some of those Saturdays. Probably a lot of the visitors were young people, as they were on one recorded occasion when they "deluged Mr. Plaskett with astronomical queries, all well put, and very much to the point."[66]

Visitors delight in viewing the shadowed craters on the Moon, the spectacular rings around Saturn, or the sparkling stars in a cluster, but astronomers are seldom interested in seeing anything through the telescope. Their only purpose in peering into the eyepiece is to ensure the telescope is pointed exactly at the star under investigation. Public star nights provided an opportunity to dispel common myths about the work that astronomers do: they do not cast horoscopes (that's what astrologers do!) and they do not spend countless hours gazing through a telescope in the hope of discovering some new object (Laurier or otherwise) that happens to sail into view. In fact a lot of their effort is devoted to mundane matters that, bit by bit, bear on the grand quest for understanding how the universe operates.

Instruments

It has been said that "science, like civilization itself, is a matter of tools."[67] Successful experimental scientists and observational astronomers need to be skilled technicians, continually alert to needed improvements in their equipment. If they can also carry out the necessary modifications themselves, so much the better. That ability, which JSP had in spades, was a very important aspect of his success but it was frequently severely challenged.[68] In the first year at the observatory, he became aware that the main telescope drive was irregular; he had to try a number of remedies, including an electric control for the clock, modifications to the governor, and ultimately re-cutting the worm gear, before the problem was rectified. The solar camera, which could be attached to the tailpiece of the telescope to take large-scale photographs of the Sun, presented many challenges too. After several attempts to get it working successfully, JSP found that the camera's enlarging lens had not been designed

3.11(a) Coelostat house, shortly after completion, slid back to reveal the equipment. Sunlight is reflected by the primary coelostat mirror (facing us in this photo) to a secondary mirror whose back can be seen, then to a concave mirror hidden by the house at the right, and finally to a focus, 24 m away, on the slit of a spectrograph in the basement of the main building at the left. The louvered design and white paint of the buildings was intended to minimize uneven heating and air currents that would distort the image. (LAC, PA-107539)

properly. Brashear supplied a replacement lens, but then the shutter, needed for the short exposures of the brilliant Sun, caused problems.

Then there was the coelostat that had gone to Labrador and back. King and Plaskett were eager to adapt it and put it to use in Ottawa. So JSP wrote for advice to G.E. Hale, the founding director of the Mount Wilson Solar Observatory in California, and to Charles Abbott, director of the Smithsonian Astrophysical Observatory in Washington, DC.[69] Following their recommendations, Plaskett conceived an overall plan, the Department of Public Works prepared the detailed drawings of the structure that would house it, and the government gave its approval in 1907, but four years passed before useful research could be done.[70] The original coelostat mirror, about 50 cm in diameter and housed in a roll-off structure 21 m north of the observatory, reflected the sunlight southward to a new secondary plane mirror, and then to a concave mirror (both made by Brashear in 1907) that formed an image of the Sun, 23 cm in diameter, in a lab in the basement of the main building. Plaskett knew the arrangement was less than ideal but he had nowhere else to install it.[71]

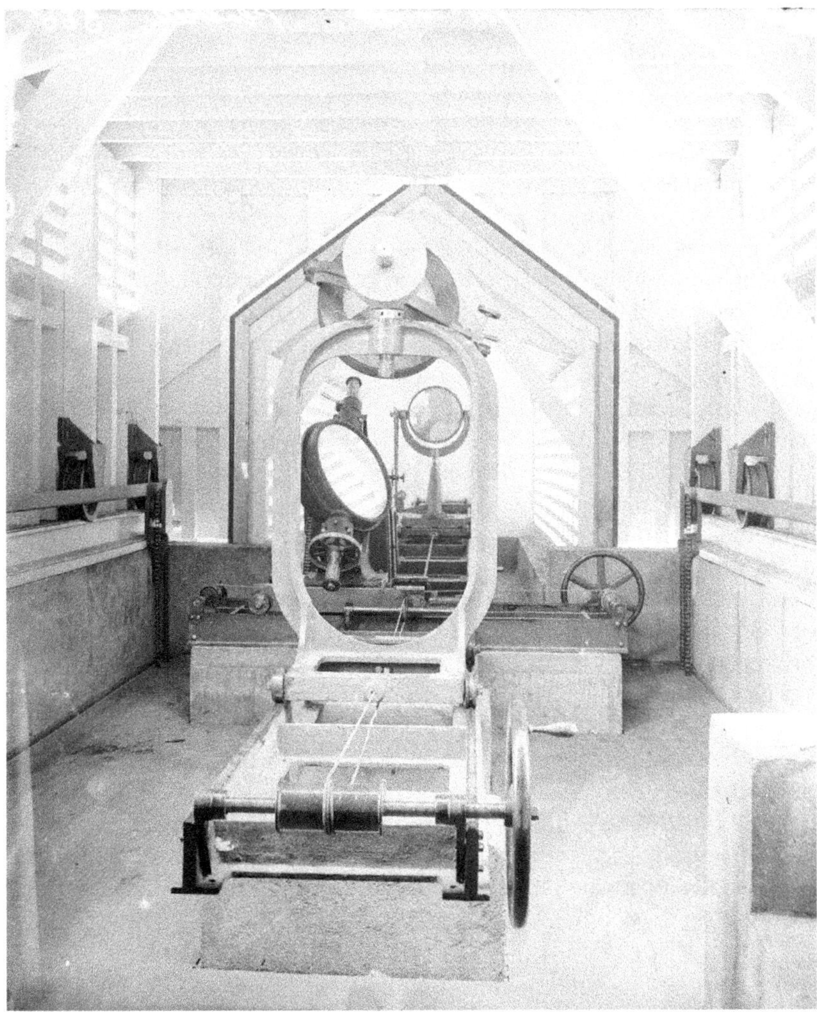

3.11(b) Coelostat mechanism at the Dominion Observatory, c. 1910. In this view, looking north away from the main building, the concave mirror can be seen in the distance. It would send the Sun's image southward under the secondary mirror, whose back is seen in the foreground here. The secondary mirror could be moved back or forth as the Sun's declination changed throughout the year. The primary (coelostat) mirror, driven to compensate for Earth's rotation, could be slid to the right for morning observations or to the left (as shown) for afternoon observations. (Photo by JSP, LAC, PA-107532)

Vic Gaizauskas, who struggled with the equipment in the 1950s and 1960s, wrote that its location in the shadow of the main observatory

> severely reduced observing time except for a few weeks in mid-summer and rendered the telescope useless in mid-winter. The optical path from the image-forming mirror dipped slightly from the horizontal to enter a subterranean laboratory with a horizontal grating spectrograph. Turbulent ground air always degraded the quality of the solar image formed at the entrance slit of the spectrograph ... A few minutes of cloud cover sufficed to cool the 25 metre focal length mirror so that its focus changed. The horizontal spectrograph could not be isolated from the outside air.[72]

Of all the astrophysical instruments, the stellar spectrographs attached to the tailpiece of the telescope proved to be the most productive during Plaskett's time. By measuring the photographs taken with this equipment, he and his co-workers found radial velocities of stars in the line of sight – new data that were urgently needed to complement the much more abundant values of proper motion, transverse to the line of sight, that older observatories had been accumulating throughout the previous century. (In fact, decades were required for the stars' minuscule proper motions to amount to measurable values.)

The DO's original instrument for obtaining radial velocities was a universal spectrograph made by the Brashear Company, similar to those they had built for the Allegheny Observatory in Pittsburgh and the Lick Observatory in California.[73] It was called "universal" because it could be adjusted to view different parts of the spectrum and could be configured to operate with one prism or with three. At first Plaskett usually used the latter arrangement, in which the starlight gathered by the telescope passed consecutively through the three prisms, causing the spectrum to be more spread out than if a single prism were used thus allowing finer details to be resolved. However, this higher dispersion also meant that the light was dimmer, requiring longer exposures – sometimes two hours – if the stellar spectrum were to register clearly on a photographic plate. Plaskett soon realized his results were inconsistent. To isolate the source of the problem, he carried out a series of very thorough tests under varying conditions.[74] In the course of these investigations, Plaskett made hundreds of exposures of spectra, some of planets as standards of comparison, as their velocity relative to Earth was easily calculated. In spite of his best efforts, problems still persisted.

In the early days, Plaskett must have felt quite isolated in Ottawa. He had no mentors in Canada to talk to about his difficulties and no one even to guide him on a suitable program in stellar spectroscopy. So in the spring of 1906, he sought the advice of two acknowledged experts, Edwin B. Frost, the director of the University of Chicago's Yerkes Observatory (with the world's largest refractor), and William Wallace Campbell, director of the Lick Observatory and a recognized leader in spectrographic design and research.[75] These men, like Abbott and Hale, with whom JSP had corresponded earlier, were all about his age, a factor that may have made his correspondence easier than it would have been for a younger neophyte who might have felt intimidated by their senior positions. Frost provided some helpful technical advice and encouraged JSP by saying that excellent work could still be done with a telescope of modest size, though he recommended stars with less sharp lines, where high resolution would be no advantage. This advice may explain how it happened that so much of Plaskett's research would focus on the hot stars of spectral types O and B. For such stars he could use a lower dispersion, single-prism spectroscope, allowing him to reach fainter stars. He hoped to build a better spectrograph but for the time being he would try to obtain the best possible spectra using the existing equipment. In his letters to Campbell, Plaskett outlined in great detail the steps he had taken to deal with technical difficulties and even sent him some test spectrograms. Not surprisingly, Campbell could not provide specific remedies from the other side of the continent, and this fact convinced JSP and King of the value in travelling to some of the major observatories to see their facilities first hand and to discuss problems face to face.

The odds of doing so improved in 1906. That summer marked the graduation of W. Edmund Harper, Chant's first student in the specialized honours astronomy course and the first recipient of the RASC gold medal. On the very day of his graduation ceremony, he was hired by the Dominion Observatory and became JSP's assistant. He was given the task of measuring the backlog of plates that had already accumulated.

Plaskett and Chant Visit U.S. Observatories

Having Harper on the DO staff gave JSP the freedom to visit a number of observatories in the United States in the summer of 1906, including Lick, where he could have direct discussions with W.W. Campbell.[76] Plaskett set out by train for Vancouver in mid-August and then headed south through Seattle and Portland to San Francisco, a city rebuilding after the horrendous earthquake in April. To reach the Lick Observatory, he travelled about 40 km by stage from San Jose up a long zigzag road to the summit of Mount Hamilton at an elevation of 1.5 km. The clarity of the mountain air, making a very high proportion of nights available for observing, and the excellent "seeing," evident in the very small jitter in the star images, impressed Plaskett.[77] His next call, the solar observatory on Mount Wilson northeast of Pasadena, was at an even higher altitude of nearly 1.8 km. There JSP met the director, G.E. Hale, who made the case for the desirability of Canada joining the International Union for Co-operation in Solar Research (Solar Union, for short). It was a perfectly natural suggestion, since Plaskett was hoping to make solar research an important part of his work at the DO and Hale was keen to develop international cooperation. However, the union's constitution required that representatives be appointed by each nation's scientific academy or society, and years would pass before the logistics could be worked out as to which Canadian body would formally do this.[78] At Mount Wilson, Plaskett also talked to master optician George W. Ritchey and inspected his latest masterpiece just nearing completion, the large reflector telescope, 1.5 m in diameter. That wonderful new instrument, conceived by Hale and funded by the Carnegie Institution, in its beautiful mountain site undoubtedly led Plaskett to think of the future. Historians have pointed out that

> many of the innovations which Ritchey brought to the 60-inch are now standard, but in 1904 they were largely groundbreaking: optical testing at the mirror's focal plane; a compensating mirror-support system; a rigid trussed tube; interchangeable secondary mirrors to provide Newtonian, Cassegrain, and coudé foci; a permanent, highly stable spectrograph at the coudé focus; electric slow motions; backlash-free gear trains; and meticulous attention to the effects of temperature.
>
> The huge castings for Ritchey's mounting were fabricated at the Union Iron Works, a San Francisco builder of ships and large machinery.[79]

Next on JSP's itinerary was the Lowell Observatory in Flagstaff, Arizona, where Vesto Slipher was doing landmark spectrographic work on planets. For his part, JSP provided helpful advice on sensitizing plates to the red end of the spectrum.[80] He then took the Santa Fe Railroad, arriving in Chicago on 14 September and spending four days at Yerkes Observatory near Williams Bay, Wisconsin, about 150 km northwest of the Windy City. He stayed with astronomer Edward E. Barnard and his wife, and discussed spectroscopy with Director Edwin Frost and Philip Fox, just returned from working with spectroscopist Johannes Hartmann in Potsdam, Germany. By 20 September JSP was in Pittsburgh, renewing acquaintances with the maker of the Ottawa telescope optics, John Brashear, and his son-in-law and business partner, James McDowell. Brashear also introduced JSP to Frank Schlesinger and Ralph Curtiss at the Allegheny Observatory, an institution that JSP at the time thought to be rather unproductive owing to shortcomings in its equipment. Nonetheless, Schlesinger, the director, would become a mentor and friend in the years ahead. On Plaskett went to Washington, DC, to see the U.S. Bureau of Standards, the Coast Guard Survey office, and the Naval Observatory (USNO). The USNO was the official timekeeper for the nation but Plaskett found its apparatus less complete than what he was used to in Ottawa and its system for controlling temperature rather primitive. After some trouble, he located the Flower Observatory in the Philadelphia area, where he met the father-and-son team of Charles and Eric Doolittle. Their refractor was quite similar to the one at the DO. Plaskett ended his tour on a high note with a couple of days at the Harvard College Observatory in Cambridge, Massachusetts, where he had discussions with the hospitable director, Edward C. Pickering, and his staff. It was Pickering and Antonia Maury, one of his so-called harem of assistants, who were among the first to discover spectroscopic binaries seventeen years earlier.[81] JSP was back in Ottawa by the end of September. It was a whirlwind tour but he found it tremendously stimulating. He returned with the sense that the fine climate and high elevation of the western observatories did wonders for what could be achieved, but at the same time he felt that the Dominion Observatory was doing pretty well relative to some others in the eastern United States in similar situations.

JSP had enjoyed meeting several of the most influential American astronomers and found them all willing in every way to give him help, advice, and the benefit of their experience. He was impressed with their "geniality, kindliness, and good fellowship."[82] The opportunity for discussion that Plaskett had on this trip gave him confidence in himself and his endeavours and helped him see the best way to address various complications. Especially important, he was able, during his visit to Lick, to talk with Director Campbell about the most suitable spectroscopic projects to undertake. And it seems Campbell also benefited from Plaskett's photographic expertise with different emulsions. After Plaskett returned to Ottawa he wrote thanking the director and his wife

3.12(a)

3.12(a), above, and (b), opposite, Two Lick Observatory photographs dated 1907, by G.A. Vogt. One, figure 3.12(a), shows people arriving at the observatory by stagecoach, the other, figure 3.12(b), shows members of the Lick staff, including Robert Aitken (second from left in back row), Director W.W. Campbell (first on the left of the middle row), and visitor C.A. Chant (middle of front row). These images, and hundreds of others of Lick Observatory, can be accessed at digitalcollections.ucsc.edu/, where there is a complete key for 3.12(b). (Courtesy UCSC, ua0036_glp_0218 and ua0036_glp_0270)

3.12(b)

for such a pleasant and profitable time and expressing the hope that someday he would have the pleasure of reciprocating. Campbell, for his part, recognized Plaskett's "great determination and ability" and expressed confidence that he would "succeed with the difficult problem of determining radial velocities of stars" – true compliments from one not known for his diplomacy.[83]

Chant probably felt envious of Plaskett's wonderful experience and thought of the personal advantages if he too could spend some time at the Lick Observatory. He was delighted to be appointed as a special assistant there, without pay, for the summer of 1907, an arrangement that others before him had enjoyed.[84] Plaskett supported Chant. "I am sure ... you will find him an exceedingly nice fellow and one who will use well the advantages he possesses of spending the summer at the Lick," he wrote to Campbell. "Tell Chant I am looking for a letter from him soon."[85] Chant obliged: by 3 August Plaskett heard "frequently of the good time he is having and I envy him his chance."[86] Campbell noted that "Chant has made many friends on Mount Hamilton"; indeed Chant felt close to Dr and Mrs Campbell and their boys and to the Lick Observatory for the rest of his life.[87] It and its administrative parent, the University of California, would become the destination for advanced training of two of Chant's students who later worked very closely with Plaskett – Reynold Young (starting in 1913) and Joseph Pearce (from 1924 on). While Chant was in California, he continued to Mount Wilson, where he met Hale, who again brought up the idea of Canada joining the Solar Union.

After his return to Toronto on 22 September 1907, Chant was promoted to associate professor of astrophysics. The future looked bright for him and for the university. The inauguration ceremonies that autumn of a new president, Robert Falconer, were impressive and his comments very promising.[88] He advocated an increase in postgraduate courses and research and had a vision of the University of Toronto as a national institution attracting graduates from all parts of the country. The first decade of the new century was one of rapid growth, with enrolment reaching over 4000 students – by far the largest of any Canadian university. Chant also was eager to have his students try their hand at research. He acquired an expensive new Toepfer measuring microscope and had his students measure spectrograms supplied by Plaskett and Campbell. He reported that they found the work very interesting and got very fair results.[89]

PART TWO

Budding Scientist

4
The Sun and the Stars, 1906–1911

With the ill-fated eclipse expedition and the worthwhile tour of American observatories behind him, Plaskett was ready to begin his research in earnest. Most of his efforts over the next few years can be categorized as improvements in instruments, solar studies, and stellar spectroscopy.[1] Though he managed to juggle all these activities simultaneously, we can understand them more easily by discussing them separately. At the end of the chapter we will see how he was able to present his findings using various associations and societies, through their publications, meetings, and conferences.

Instrumental improvements

At the beginning of his career in Ottawa, Plaskett took great pains to investigate the equipment he was using. His prior years of practical experience as a technician meant that he was well suited to tinker with apparatus, taking it apart, and making adjustments and modifications. By testing it thoroughly to see if improvements were possible, he became a partner with the manufacturer, not simply a critical consumer. By insisting on knowing the limitations of his equipment, he had a clear understanding of the sources of error in any results he derived. His conclusions had the potential to be a real boon to other astronomers who had comparable equipment but who may have lacked the knack or patience to carry out similar trials themselves.

The spectrograph, attached to the telescope, was central to all Plaskett's research. As he explained in a lecture at the Royal Astronomical Society of Canada (RASC), a spectrograph has four essential parts: slit, collimator, prism (or grating), and camera.[2] The main telescope focuses the star image on an adjustable slit whose width is usually between 0.025 and 0.051 mm; the light passing through the slit is made into a parallel beam by a collimating lens; the parallel beam falls on a prism and is dispersed into a spectrum, which the camera photographs. As a result of radial velocity – let's say 20 km/s for a star – the spectral lines shift, in the case of Plaskett's equipment, by a tiny but measureable amount (0.01 to 0.03 mm depending on the instrumental setup). Previous investigators, the pioneers in astronomical spectroscopy, recognized that such a minuscule displacement could be caused by a number of other factors, including flexure or bending of the spectroscope as the orientation changed during a long exposure, a slight variation in temperature, a minute error in focusing the camera, and poor guiding by the observer trying to keep the telescope pointing exactly on the star.[3] JSP spent years studying, adjusting, and modifying every component, even replacing parts and designing new spectrographs for the Dominion Observatory (DO) telescope. Of course he recognized that even a perfect spectroscope with perfect lenses could not bring the spectrum to a focus in a flat plane such as a photographic plate.

By June 1906 Plaskett solved the serious problem of flexure in the universal Brashear spectrograph by adding stiffening trusses and firmly clamping the prism cells to the case. He built a temperature control and an insulated case for the spectrograph to counteract differential thermal expansion of various parts of the instrument. Plaskett put a positive spin on all the difficulties he encountered when he said, "They have certainly been an education on spectrographic peculiarities and causes of error, which could not otherwise have been obtained."[4] These challenges were only the beginning.

The next problem JSP faced arose because the telescope was designed to be used both visually and photographically. Because all colours are not brought to a focus at the same distance from the telescope's objective lens, an image that appears to the eye to be in focus in yellow light is out of focus for photographic emulsions that are more sensitive to blue light, for example. To get around this problem, common to all refractors used for both purposes, a correcting lens is usually inserted in the telescope inside the focus. Plaskett studied the problem extensively and ascertained that his correcting lens, made by Brashear and Company, had a serious problem: the focus for the edge rays was 2.5 mm longer than for the central rays.[5] Charles Hastings, the optical designer used by Brashear, thought the problem could be solved with a larger lens, aperture 96 mm, which he designed. Brashear produced and delivered it in August 1907 but it was no better. Plaskett decided to go to Pittsburgh personally, taking the new correcting lens with him, so that he could be involved in the testing first-hand while Brashear's optician, James McDowell, tried reshaping it. This time everything went well. After just a few minutes, McDowell skillfully "figured" and polished the lens by just the right amount. JSP was full of admiration and acclaimed the Brashear Company for its ability to produce near-perfect optical surfaces and for their generosity in striving to correct any wrong.[6] He was equally pleased when he returned to Ottawa and re-installed the lens. He found that the spectrum of fifth magnitude stars (i.e., somewhat brighter than those at the limit of naked-eye visibility) could be photographed with a two-hour exposure.[7] This, he claimed, showed "the relative efficiency of the Ottawa installation to be equal if not superior to that of any other."[8] Surely other observatories, like Yerkes and Lick, would do well to take similar steps to improve their own systems, he thought.[9] His career as an important astronomer was well and truly launched and he clearly was on a familiar footing with his "good friends at Mt. Hamilton."[10]

Nevertheless Plaskett still had difficulties with the universal spectrograph. So, with extra help available from technician A.S. Mackey, who joined the staff in 1906, he decided to abandon the old instrument and design a new one that would operate in two modes – single or triple prism.[11] Soon after it was completed in May 1907, he once again encountered problems, mainly because the temperature-controlled case had to be temporarily removed when the mode was changed. Uniform temperature throughout the spectrograph was essential for accurate radial velocities. JSP determined that a temperature variation of one Celsius degree in the optical parts could shift the spectral lines equivalent to a radial velocity of 20 km/s. Therefore he decided to avoid the problem by using this new instrument only in three-prism mode and building a separate spectrograph to be used with a single prism for low-dispersion spectra.

The new single-prism spectrograph was put into service in March 1909, designed by JSP and built, except for the optical parts, in the observatory workshop. He proudly declared: "It has fulfilled all expectations both as regards shortening of exposure time and in respect to its stability and freedom from flexure ... There is no question but that it is the most stable single-prism spectrograph ever constructed."[12] JSP was building not only his equipment but also his reputation in optics. Perley Nutting, in his early attempts to found the Optical Society of America in 1910, singled out Plaskett and W.F. King as the two Canadians who might wish to be involved in such an organization.[13] A century later, JSP's success was still recognized as a model of achievement for the time.[14]

The two spectrograph solution seemed to be ideal.[15] Each instrument was extremely rigid and the interchange between the two took only a couple of minutes. The temperature problem was overcome because each spectrograph had its own temperature-controlled case. Mackey did all of the mechanical work on both spectrographs, and JSP generously attributed the ease of operation and the accuracy of the results to Mackey's skill and craftsmanship. Though he did not say so, JSP knew from first-hand experience how to provide the necessary tools, guidance, and design to get a first-rate job done. Comparing the two spectrographs, the single-prism instrument naturally gave a spectrum with lower dispersion but, since the light was less spread out, the time required for the photographic exposure was less. Plaskett discovered, to his surprise, that the statistical scatter (or probable error, to use the technical term) in the measured radial velocities at low dispersion was not much more than at higher dispersion.[16]

A further aspect of JSP's instrumental experiments was his investigation of the effect that the width of spectrograph slit had on the accuracy of measured radial velocities. It was an important concern because a slit that was too narrow blocked a significant fraction of the starlight falling on it, but a slit that was too wide produced a spectrogram in which the lines were too broad to be measured accurately. He looked into this interesting question over a period of four years (1906–10), using close to 200 plates taken with the various iterations of the DO spectrographs.[17] He came to an unexpected conclusion that the optimal slit width was 0.051 mm, about twice what was customary, meaning that spectra of fainter stars could now be captured and that spectra of brighter stars could be photographed with shorter exposures. As a result, the equipment at the DO, and by extension at other observatories, could be operated more efficiently. As Plaskett put it in 1909, "Much of my time during the past year has been devoted to investigations bearing on improvements in instruments and methods ... It is, in my opinion, time well spent, if through such investigations and experiments we are in a position to do a larger quantity of more accurate work."[18]

But Plaskett was still not finished. He tried ten different camera lenses for the spectrograph.[19] He found that by far the best for the triple prism arrangement was a four-element lens specially made by the English firm Ross Limited. However, it had a rather narrow field and therefore he used this setup only for stars with many lines in their spectra, stars for which a relatively narrow range of wavelengths sufficed. The best camera for the single-prism spectrograph had a Brashear lens figured by McDowell. It had a wide field of view, making it ideal for the type of stars with only a few lines in their spectra. Another problem, however, reared its head when Plaskett noticed that the violet end of the spectrum, where one of the most useful lines in the spectrum (the K line) lay, was much weaker than expected. After eliminating a number of possible causes, he determined that the correcting lens that had caused him so much grief earlier was not aligned properly. Though the image it produced appeared to be perfectly central in green light, to which the eye was most sensitive, it was not so for violet light, which the eye could barely detect but which was an essential part of the spectrum. Plaskett realized that he needed to make an adjustment of less than a millimetre in the alignment of the correcting lens to ensure that the slit was illuminated in all wavelengths.[20] All this information was very useful to at least one other astronomer: Vesto Slipher of Lowell Observatory ordered a lens from Brashear "similar to one Mr. Plaskett describes."[21]

JSP's diligence in analysing and improving the spectrographs characterized much of his work at the Dominion Observatory until 1911, by which time he realized that his equipment was too limited for useful new endeavours. Nonetheless, he had learned a great deal about spectrograph design – useful lessons for the rest of his career.

Solar Studies

Plaskett's solar labours began with the Labrador eclipse of 1905. Perhaps the leaden clouds of that sad day were an ill omen of what was to come over the next few years. The equipment problems that were to plague solar research at the DO were even more severe than those he experienced with the stellar spectrograph.

Plaskett had initiated solar investigations at the DO, but Ralph E. DeLury was given responsibility for it, with JSP's supervision and collaboration, when he joined the staff in 1907.[22] DeLury had earned a PhD (the only DO staff member to have done so) at the University of Toronto and had spent a postdoctoral year at Princeton University on scholarship.[23] Though his background was in chemistry, his research capabilities appeared to qualify him well for his new solar work. At first he only used a camera attached to the main telescope to take daily photographs of the Sun for recording the size and position of sunspots; the coelostat was not installed at the DO until 1910. In the meanwhile DeLury put his chemical expertise to use by setting up a laboratory for silvering mirrors in the basement of the observatory and looked into wavelength standards in the iron-vanadium arc used for comparison spectra.[24] As the coelostat installation neared readiness, DeLury tested the plane grating that would be the heart of the instrument and found it was unsatisfactory.[25] Ruling thousands of grooves per cm with perfect regularity was an extremely difficult task, so it probably was no great surprise that a new grating ruled by J.Y. Lee, assistant to Albert A. Michelson, Nobel laureate of the University of Chicago, still had some faults.[26] Nonetheless it turned out to be superior to another one sent by Michelson in December 1910. DeLury spent a great deal of time and effort testing these gratings and doing experiments to see what types of photographic plates and developers were best for different regions of the solar spectrum. JSP, however, found his assistant not very productive and decided to devote more of his own time to working with him.[27] Finally in 1911, with a superior grating supplied by John Anderson of Johns Hopkins University, they were able to put the coelostat telescope into regular use to study the rotation of the Sun by measuring the difference in Doppler shift of the spectral lines at the opposite (approaching and receding) limbs of the Sun.[28] The two spectral images were brought together by a device that JSP had designed: it employed reflecting prisms that put the spectrum from the Sun's east limb between two images of the spectrum from the west limb, so that the relative separation of the lines could be more easily found. But with a rotation speed of less than 2 km/s at the solar equator, even after doubling to 4 km/s by the above arrangement, the effect was tiny. On the other hand, the image of the Sun was 23 cm in diameter, sufficiently large to see how the rotation rate varied with solar latitude and possibly with different spectral lines. Spectroscopic measurements had advantages over rates found from visual sunspots – they could be made at almost any time and at any solar latitude, whereas sunspots were often scarce and were usually confined to the mid-latitudes on the Sun. The collaboration between JSP and DeLury did produce some scientific publications but, as we shall see in the next chapter, problems of one sort or another always persisted.

Stellar Spectroscopy

By late 1906 Plaskett was beginning to get reasonably good spectrograms of stars. C.A. Chant suggested that the prominent variable star Mira would be an interesting candidate to examine.[29] As this star brightens dramatically every 332 days, its spectrum changes. Mira was due to reach maximum at the end of 1906 and would be bright enough to have its spectrum captured with the Ottawa equipment. JSP liked Chant's idea, and during December and January he and his assistant, W. Edmund Harper, obtained eleven spectrograms using the three-prism setup with the modified universal spectroscope. JSP published his results in the RASC *Journal*, which Chant edited, and sent a copy of his paper to W.W. Campbell, the director at the Lick Observatory.[30] Though Plaskett freely admitted there were still snags (both ends of the spectrograms were out of focus), Campbell congratulated him on the promising start he had made in spectroscopy.[31] Campbell also had planned to study the spectrum of Mira but had been plagued by bad weather all fall and winter. On the few plates he had obtained, his radial velocity and JSP's (from Harper's measurements) agreed almost exactly with each other (and with the presently accepted value).

Mira would prove to be an exception to JSP's main interest in stellar spectroscopy. Mostly, the stars that he observed during his Ottawa years were binaries whose spectral lines shift back and forth periodically, revealing variable radial velocities. This periodic change arises because a spectroscopic binary is really a pair of stars in orbit around each other, though they cannot be resolved as separate images by a telescope. Sometimes the spectra of both stars register simultaneously and (since one star recedes as the other approaches) the spectral lines are Doppler-shifted in opposite directions and appear double. Usually though, one of the stars in the pair is much brighter than the other, so only the brighter spectrum is recorded on the photographic plate. Even so, by determining the shift in its spectral lines and converting these measurements into radial velocities, a lot of information can be deduced about the shape and orientation of its orbit.

Finding the period (the length of time for one star to orbit the other) is easily understood. It can be ascertained by comparing the times at which the radial velocity has the same value and is changing in the same sense, either increasing or decreasing. The difference between these two times might be one or more periods, and a bit of experimenting is required to find how many periods elapsed. But ultimately all the observations are brought back to a reduced time, arranged as if they had all been made during one revolution. The result is called the velocity curve. There is no need here to go into the way in which the other orbital characteristics (called "elements") are found from the velocity curve except to say that an unknowable aspect of the orbit is its tilt, or inclination, to the line of sight. Of course, if the plane of the binary system's orbit is at right angles to the line of sight, there will be no orbital radial velocity, but the further the angle of inclination is from 90° the more pronounced the radial velocities will be. The primary importance of all binary star systems is the information they provide on the mass of the stars compared to the Sun. With a large sample of completed binary orbits, astronomers could begin to answer questions such as the following: What is the range of stellar masses? What types of stars are more (or less) massive than the Sun?

JSP recognized this as a fruitful field of research as, by 1906, about 150 spectroscopic binaries were known, though orbits had been calculated for less than twenty.[32] So for now there was no need for him to find new stars of this kind. He thought he would have plenty to do in acquiring sufficient plates of those that were previously identified and in measuring and analysing their spectra. Ideally, several lines were measured in each spectrum and a mean radial velocity for the particular star and date was calculated by averaging the results. (Naturally, the velocity of Earth as it rotated on its axis and revolved around the Sun had to be subtracted out, a process called "reducing the radial velocity to the Sun.") JSP found that the measurement and reduction of each plate required upwards of a day to complete. The observatory purchased another device in 1908, a spectro-comparator, invented by Johannes Hartmann and made by the famed German optical firm Zeiss.[33] After replacing some faulty prisms, JSP and the DO staff used it to find radial velocities by comparing stellar spectra with a standard solar spectrum. As expected, it was useful only for solar-type stars, but in those cases the comparator greatly reduced the measuring time without increasing statistical errors.

After several plates were taken and the radial velocities calculated, and if it seemed that the results changed cyclically, a velocity curve could be found and an actual orbit could be calculated. There were several methods of obtaining the orbit from the velocity curve and W.F. King devised a geometrical method that saved time by simplifying the process.[34] The first spectroscopic binary to be tackled successfully at the

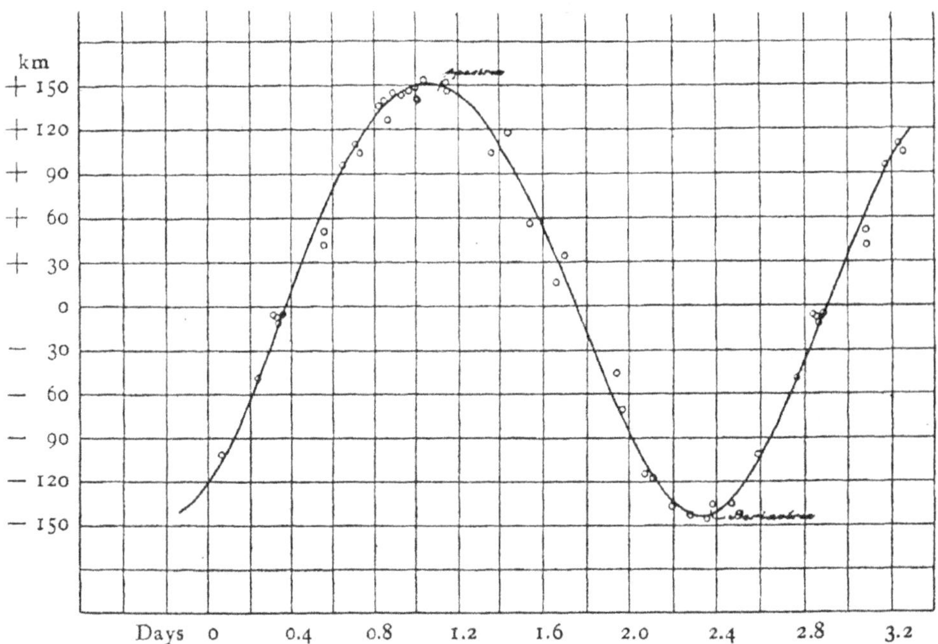

4.1 A velocity curve for one of JSP's earliest spectroscopic binaries, Psi Orionis. The star goes through a large range of velocities in a period of about 2.5 days, receding by as much as 150 km/s and, about a day later, approaching at 150 km/s. The little circles show the radial velocities calculated from the measured shifts on many spectrograms. The curve, fitted to the points, leads to the elements of the orbit. (Copied from *ApJ* 28 (1908): 269)

DO, starting in July 1906, was a bright star in the constellation Draco, Alpha Draconis. It was a star that the Chicago astronomer Edwin Frost had suggested as an example within reach of modest equipment.[35] Plaskett and Harper observed it, measured the plates, found the radial velocities, and calculated the orbit by two different methods. Though the two astronomers were obviously partners in the endeavour, JSP generously let Harper publish the results as the sole author in the RASC *Journal*.[36] The paper was completed in July 1907 and was Harper's first. There were to be many other papers in the coming years by Harper and others in which JSP had a hand, though he was not always named as a co-author.

Another early example of a binary that they studied was Iota Orionis. Embedded in the nebulosity of Orion's sword, it was a star of spectral type O with only a couple of measurable lines in its spectrum, and even those were diffuse. JSP found its radial velocity varied greatly, from −50 to +100 km/s, much more than its discoverers had reported.[37] An unusually long span of time was needed to secure a sufficient number of plates because of consistently cloudy weather just at the time, every twenty-nine days, when the star's velocity changed rapidly. But finally, after securing 107 plates, JSP considered the results were satisfactory enough to publish.[38] Frank Schlesinger, his adviser at the Allegheny Observatory, read the paper and wrote to Plaskett, strongly urging him to use a statistical technique, the method of least-squares, that produced an objective list of orbital elements complete with error estimates. Plaskett was familiar with the basic principles from his university studies, and Schlesinger was just then writing up the process as it applied to spectroscopic binary orbits.[39] King himself, in a paper crediting Plaskett's assistance, said, "The method of least squares may be employed to correct the first values of the elements, and to give the most probable values."[40] The method had the advantage of varying all the elements simultaneously to find the orbit that best fit the observations and was also useful in deciding if there was evidence of a secondary oscillation, perhaps caused by a third star in the system.[41] JSP appreciated the advice, telling Schlesinger, "It is by such help that beginners are aided in making their work better."[42] He then proceeded to use what he had just learned to publish a revised orbit and, after he and Harper found another nearby binary with very similar elements (HR 1952), a third paper comparing the two.[43] Though the similarity was surely just a coincidence, it did lead them to conclude that both systems showed evidence of side effects whose origin, perhaps a resisting medium or tidal interactions, they could only guess at. The possibility of these features are still of great interest, and astronomers of the present day continue to study these two fascinating systems with all the modern means at their disposal.[44]

An unexpected sidelight in the early study of spectroscopic binaries is the research of Joseph Miller Barr, an amateur astronomer in St Catharines, Ontario.[45] He had been fascinated with binary stars for several years when he noticed something peculiar in the way the tabulated orbital elements of spectroscopic binaries were distributed. The values of ω, called the longitude of periastron, were not randomly scattered between 0° and 360° as one might have expected but seemed to be concentrated around 90° – a peculiarity still known as the Barr effect.[46] It was as if the orbits had a tendency to align themselves with the direction towards the observer on Earth – an absurdity that cried out for an alternate explanation. Barr wrote about his finding to Chant, editor of the RASC *Journal*, and offered a possible hypothesis to explain what he had found. Chant sent the paper to Plaskett and King, asking for their comments, and then, to Plaskett's annoyance, published their responses along with Barr's paper. Barr was not pleased with the reaction he got and was probably right when he suggested that Plaskett was too hasty in dismissing his idea, but Plaskett, like many busy professional astronomers, was probably wary of amateurs devising theories or interpreting results in novel ways.[47] By the twentieth century such suspicion was almost always well justified – the tradition of gentleman scientists in Victorian Britain making important advances had passed. In Plaskett's day, as now, there were always a few eccentric cranks who think that they, with little relevant experience or training, have come up with the perfect theory to explain something that has puzzled the best efforts of professionals for years or decades. Barr, however, had carefully studied the literature and had assembled data gathered by professionals (including JSP) and pointed out an interesting peculiarity for the first time. Where he went astray, in Plaskett's opinion, was in his explanation for the anomaly. Barr postulated that if the binary's brighter star (the one giving rise to the spectrum) did not emit its radiation uniformly in all directions, and was rapidly rotating, the spectral lines would be unsymmetrically broadened. The *effective* result, he said, would be a periodic shift of these lines as measured on the spectrograms. It was an ingenious explanation, not that far off the mark as it turned out. In a study of the Barr effect as recently as 2009, astronomer Helmut Abt showed there was strong statistical evidence for the Barr effect in certain types of spectroscopic binaries, an effect he found could be plausibly interpreted in terms of gas streaming from one star of the pair onto the other.[48] But in the early twentieth century, no one had evidence to support such a hypothesis. With a much smaller sample of stars and without the sophisticated methodology and equipment to look for physical causes, Plaskett could see the effect only as a statistical anomaly. In fact, though no one realized it, some of the stars included in Barr's investigation were not binaries but were single pulsating stars (a type known as Cepheid variables).[49] It is somewhat ironic that it was these very stars that Barr singled out as needing a new interpretation, but he only went so far as to hypothesize that their orbits were circular rather than elliptical and that their light and velocity curves were distorted by tidal effects and uneven brightness of the stellar surfaces.

By 1911 research on spectroscopic binaries at the DO was beginning to decline. In his report for the year ending 31 March 1911, JSP summarized the extent of the achievement in stellar spectra during the year as well as the two previous years.[50] Though five new spectroscopic binary orbits were computed during the year prior to his report, the highest number of plates came from two years prior, when 1011 spectra had been obtained on 158 nights. Then the numbers dropped so that only 782 spectra were recorded on 140 nights. Bad weather accounted for some of the decrease but, as Plaskett noted, the average number of spectrograms per night also declined because fainter stars were being studied requiring longer exposures. Besides, there was a diminishing supply of unstudied binaries that were bright enough for the Ottawa instruments and yet had sufficient range of velocities to make orbit calculations worthwhile. Another trend is noticeable in the decreasing number of papers by JSP himself. The astrophysical part of "Notes from the Dominion Observatory," which he had faithfully written for the RASC *Journal* from its inception in 1907, began by 1911 to be written by others on the DO staff. JSP was becoming an administrator, supervising his growing staff and laying grander plans for the future.

By concentrating so closely on Plaskett's activities, we may leave the mistaken impression that operations at the DO were pretty much a one-man show. Far from it. For the year ending 31 March 1911, Director King reported dealing with 5500 items of correspondence and 864 financial transactions, presumably with the help of a secretary and an accountant.[51] A part-time librarian and book binder cared for a library of 4000 books and pamphlets, and a photographer developed hundreds of negatives and produced thousands of prints and blueprints. Two technicians, Mackey and Lucas, were kept very busy with repairs to surveying instruments and improvements to the spectrographs, the solar equipment, and pivots and other parts for the meridian circle. Otto Klotz, by then assistant director, headed the geophysical division dealing with seismology, terrestrial magnetism, and gravity; R. Meldrum Stewart was in charge of the meridian telescope and time service; J. Macara was responsible for latitude and longitude work; and Plaskett headed up the astrophysics division of five astronomers: Harper, DeLury, John B. Cannon, T. Harold Parker, and Robert M. Motherwell – all University of Toronto graduates.

Maxwell Tobey, having tried photometry up to 1908, was no longer mentioned. Apparently neither the original wedge photometer nor the polarizing photometer designed by JSP behaved well, so it seems the program was terminated and Tobey transferred to another department. Motherwell routinely predicted and timed lunar occultations, measured the positions and separations of visual binary stars with a micrometer, and took some direct photographs, most notably of three comets – Daniel in 1907, Morehouse in 1908, and Halley in 1910 – though Plaskett also captured Comet Daniel on at least one occasion, 20 July 1907.[52]

Societies

JSP belonged to a couple of Ottawa societies – the Field-Naturalists' and the Camera Club – which had little to do with his regular employment. His active membership, like that of other civil servants in these and similar organizations, strengthened the rather fragile intellectual life of the small capital city.[53] Yet, he probably joined, not out of a sense of civic duty, but primarily because he enjoyed the camaraderie of friends with similar interests.

On the other hand, the professional societies to which he belonged were essential to his career. At conferences arranged by such organizations, scientists not only air their own research and hear what others are doing, but they are able to debate with their colleagues the strengths and weaknesses of each other's methods and results. Younger scientists may be more adept with new technology and methods, while their elders have experience and contacts to offer. Such meetings are the forge on which advances are hammered out and collaborations are formed. Of course researchers do discuss their problems with co-workers at their own institutions but they usually need to cast a wider net to find those with closely related interests.

As we shall see, the societies having the greatest impact on Plaskett's career were outside the country. Nonetheless there were three societies in Canada attractive to Plaskett: the Canadian Institute, the Royal Society of Canada (RSC), and the Royal Astronomical Society of Canada (RASC). The first two were generalist in nature; the last was specialist, though even it included "allied sciences" in its astronomical mandate.

The Canadian Institute (which became the Royal Canadian Institute in 1914) was broad in scope. Its membership was based in Toronto, so it was important to Plaskett mainly during the period when he lived in the city. As already noted, he spoke to institute audiences in 1901, 1902, and 1905.

The RSC was founded by the Governor General Lorne in 1882.[54] Membership – or fellowship, as it is called – in this select group was, and is, an honour for a limited number of distinguished Canadian scholars in the arts, humanities, and sciences. Candidates had to be nominated by at least three existing fellows. The society was subdivided into a number of sections. Section III was devoted to the physical sciences, including astronomy. King, as the director of the DO, was named a fellow in 1908.[55] He contributed annual reports and presented papers by his staff, including JSP. At the RSC meeting in May 1909, seven papers originated with Plaskett and his assistants. Three of the papers, including two by JSP, were published in full in the society's *Transactions*, paving the way for Plaskett's own admission to the society the following year. Once he (along with Klotz) became a fellow (an FRSC) in March 1910, Plaskett was able to attend their annual gatherings.[56] This he did every May for many years and he delivered at least one paper at most of them. Though fellowship in the RSC was prestigious, it often seemed to be of limited professional benefit. JSP had few colleagues in the RSC with whom to discuss his work. There were fewer than forty fellows in section III, and about half would typically attend any given meeting. The membership was composed of physicists, chemists, surveyors, meteorologists, and a handful of astronomers, and these few were mainly his associates at the DO.

The only Canadian organization dedicated to astronomy was the Royal Astronomical Society of Canada.[57] During Plaskett's lifetime, an association of professional astronomers would have been out of the question – there were so few in Canada that even national meetings would have been too small to be useful. Chant was the one who envisioned how a critical mass could be achieved by bringing amateurs and professionals together and it was he who developed the RASC into a national organization with international recognition. From the time he joined in 1906, Plaskett contributed a great deal to the RASC. Over the years the membership mix would prove to be a happy synergy as public interest in the observatory would be a positive factor in the eyes of government officials.

Developments in the RASC strengthened ties between the astronomical communities in Toronto and Ottawa. There was a general feeling in Ottawa, once the observatory was in operation, that there should be an astronomical society in the capital city to promote an understanding of the science and to provide a venue for interested people to meet and discuss astronomy.[58] A meeting was called to consider the possibility on 20 December 1906 in the recently opened Carnegie Library downtown. Though an independent organization was debated, the advantage of becoming affiliated with the RASC was clear. The RASC was about to launch its bi-monthly *Journal* and was intending to begin issuing an annual astronomical handbook, publications which all members would receive. Besides, as Klotz pointed out, a united society would have much greater influence in advancing astronomical interests than a number of separate bodies. The assembled group unanimously approved a motion by Plaskett "that an Astronomical Society be formed in Ottawa as a Section of the Royal Astronomical Society of Canada." An executive committee, including King and Klotz as president and vice-president and Plaskett as secretary, acted very quickly. On 31 December they had a constitution and bylaws ready as well as a program of bi-weekly lectures drawn up. Half of the meetings would be held at 3 p.m. in the observatory and were of a technical nature; the other half were more popular and were held in the evening at the Carnegie Library. There, on 17 January 1907, King gave the inaugural address, "Astronomy as a Science," to so many people that they could not all find seats.[59] Over one hundred signed up to be members of the society. Plaskett made two presentations in the first season: one, "The Star Image in Spectrographic Work," at a 3 p.m. session on 31 January 1907 and the other, "The Optics of the Telescope," at 8 p.m. on 14 March.[60] He also addressed the RASC in Toronto on the same subjects.[61] Not only were there interested audiences in the lecture halls, but talks like these were usually published, or at the very least summarized, in the fledgling RASC *Journal*. It is doubtful the *Journal* could have survived, let alone flourished, without contributions of this sort. And, without these presentations and articles, the public would not have felt pride in the important results in Ottawa that were beginning to put Canada on the map as a nation doing serious astrophysical research.

As the first secretary of the new Ottawa Centre, JSP went with King to Toronto in February 1908 to discuss how the RASC should be organized.[62] They were adamant that Toronto should not have special status but rather that all centres, including theirs and a new one in Peterborough, should be considered as equals. Plaskett served the Ottawa Centre as vice-president in 1909–10 and chair in 1911–12. In 1914 he began a two-year term as president of the national RASC, the first time someone located outside Toronto had held that office. Besides serving on the

OFFICERS AT OTTAWA.

✦

CHAIRMAN
Otto Klotz, LL.D., F.R.A.S.

VICE-CHAIRMAN
J. S. Plaskett, B.A.

SECRETARY
Carl Engler, B.A.

TREASURER
C. C. Smith, B.A.

COUNCIL
R. M. Stewart, M.A.
F. A. McDiarmid, B.A.
J. J. McArthur, D.L.S.

✦

NOTES.

Discussion is invited on each paper and on other timely subjects.

Membership is open to everyone interested in Astronomy, Astrophysics, or allied branches of science.

The annual subscription is $2.00, which includes the publications of the Society.

The Royal Astronomical Society of Canada

MEETINGS AT OTTAWA

PROGRAMME

SPRING TERM
1910

Evening Lectures in Y.M.C.A. Hall

Afternoon Lectures in the Observatory

PROGRAMME OF PAPERS.

✦

JANUARY 13th, 3 P.M.

A Comparison of Astronomic and Geodetic Survey Data, and the deduced truths therefrom.
W. Maxwell Tobey, B.A.

JANUARY 27th, 8 P.M.

Our Earth in the Universe. (Illustrated.)
Otto Klotz, LL. D.

FEBRUARY 10th, 3 P.M.

How the Camera is Applied for Topography.
J. J. McArthur, D.L.S.

FEBRUARY 24th, 8 P.M.

Optics of the Telescope and Spectroscope. (Illustrated).
J. S. Plaskett, B.A.

MARCH 10th, 3 P.M.

Flexure in the Axis of a Meridian Circle.
R. M. Stewart, M.A.

Latitudes.
—F. A. McDiarmid, B.A.

MARCH 24th, 8 P.M.

Determination of the Constitution and Radial Motion of the Stars. (Illustrated).
J. S. Plaskett, B.A.

APRIL 7th, 3 P.M.

Measurement of Geodetic Base-Lines with Invar Tapes and Wires.
C. A. Bigger, C.E.
P. A. Carson, D.L.S.

APRIL 21st, 8 P.M.

Stellar Evolution and Theories of World Building. (Illustrated).
J. S. Plaskett, B.A.

MAY 3rd, 3 P.M.

Accuracy of Observations with a Small Transit.
C. C. Smith, B.A.

Diffraction Grating of the Solar Spectroscope
R. E. DeLury, Ph.D.

MAY 19th, 8 P.M.

Comets and Halley's Comet. (Illustrated)
Otto Klotz, LL.D.
R. M. Motherwell, M.A.

✦

The papers presented at the afternoon meetings will be of a technical character.

Members are invited to send questions to the Secretary, and the same will receive attention at the following evening meeting.

4.2 RASC Ottawa Centre program for the spring term of 1910. Plaskett was the speaker at three of the ten meetings. (LAC, MG30, B13, vol. 5) local and national executives, he spoke frequently and in 1910 gave a miniseries of lectures to the Ottawa Centre.[63] Klotz, never missing an opportunity to be critical of Plaskett, thought that his talks were too technical for a popular audience and that JSP showed too many slides of spectra.[64]

Publications

JSP's organizational affiliations had some bearing on the important decision of where to publish, but his first obligation was to his employer. As part of the federal government's Department of the Interior, the Dominion Observatory issued reports every year. Until 1914 they were called *Reports of the Chief Astronomer*. These annual records give a very thorough account of the achievements of the various branches of the observatory, with Plaskett writing the section on astrophysics. Because these reports were also made available as separate publications, without content from other divisions of the Department of the Interior, they could be distributed to suitable observatories and institutions. This method of disseminating research was hardly ideal, since the reports came out only once a year, often many months, and sometimes over a year, after the end of the period they covered. Moreover the scientist, in the role of civil servant, had to explain a lot of technical terms that would have been clear to astronomers. And, for internal consumption, the emphasis tended to be on how resources were being used, what was being accomplished, and what the hopes and needs for the future would be.

Scientific societies whose raison d'être was to promote their members' research through meetings and publications provided an alternative to the official reports. The Canadian societies each had pros and cons as vehicles for publishing. The Canadian Institute issued its *Transactions* and the RSC its *Proceedings and Transactions* annually but, because of their eclectic scope and infrequent appearance, these publications had trouble reaching any specialists.[65] On the other hand the Canadian Institute did have a large exchange list, ensuring wide geographical distribution, and the RSC had the additional advantage of being recognized as the senior national body of eminent scholars. The RASC's *Journal* was a timely and important vehicle for publicizing research and, over the years, JSP published dozens of papers in it. Chant, the founding editor, was eager to have professional and amateur astronomers contribute to the new RASC *Journal*, noting that "there are few technical articles, which, if clearly written, have not a real value to the amateur, while the work of the latter is always of interest to his professional brother."[66] To encourage professional papers, Chant enlisted King and Plaskett, along with R.F. Stupart (director of the Meteorological Service of Canada) as associate editors of the *Journal* and put their names on the cover.[67] Each made a substantial contribution to the first issue – Plaskett's was his paper on Mira Ceti. The *Journal* was sent to sister societies and observatories around the world and as a result became well known outside Canada. For instance, the widely read, highly respected, and influential scientific weekly *Nature* noted the *Journal*'s debut. Frequently *Nature*'s founding editor, Sir Norman Lockyer, used articles from the *Journal* for items in the weekly's regular feature "Our Astronomical Column." So Canadian astronomers felt confident that their work was visible to their international colleagues. Still, the great majority of members of the RASC were armchair amateurs and so the *Journal* aimed at an informed but non-specialist readership and was not really suitable for publishing detailed investigations.

Therefore, the very small professional Canadian astronomical fraternity often turned to international societies and journals. Overseas meetings, except on rare occasions, were unaffordable luxuries requiring ocean travel at considerable expense in terms of time and money. Naturally, then, Canadian astronomers looked south for affiliations. The *Astrophysical Journal* (*ApJ*), then owned and published by the University of Chicago Press, was valued for its high standards and wide and timely circulation. Its strong focus on spectroscopy attracted JSP during the formative years of his career. When Plaskett first published in the *ApJ*, Edwin Frost, director of the University of Chicago's Yerkes Observatory, was the journal's managing editor. He had succeeded George Ellery Hale in this capacity in 1905 after the latter had left Yerkes to establish the Mount Wilson Solar Observatory in California. Both were willing advisers to JSP in his early career.

Meetings

The first broadly based organization for professional astronomers in the United States, the Astronomical and Astrophysical Society of America, traces its origins to a conference organized by the young George Hale in connection with the inauguration of the Yerkes Observatory at Williams Bay, Wisconsin, in 1897.[68] Plaskett and Chant joined in 1906.[69] Chant went to the meeting at the end of that year at Columbia University in New York City and realized that an account of it would make for interesting reading in the RASC *Journal*, which was due to make its debut in 1907.[70] It was the start of a tradition lasting for fifty years.

Whenever Canadians attended an astronomy meeting, Chant would encourage them to do a write-up for the *Journal*. Full reports could be found elsewhere but that was not the point – Chant saw the *Journal* as a vehicle to inform members and subscribers of the latest developments in astronomy, particularly as it reflected on Canadian interests. His colleagues sometimes balked at his requests but they nearly always succumbed to his charm and influence.[71]

King and Harper were admitted to membership in the American society at the next meeting, on 25–7 August 1908, but Plaskett was the only Canadian to attend. He enjoyed the informal setting at Hotel Victory, Put-in-Bay, on Lake Erie, although the facilities were less than ideal. JSP presented two of the fifteen short communications, with the aid of a very inadequate projector (or lantern, as it was called in those days), and one of the thirteen longer papers. His topics were the coelostat at the DO, camera objectives for spectrographs, and the effect of increasing slit-width upon the accuracy of radial velocity measurement.[72] This was his first time at a gathering of professional astronomers, but he found their presentations to be very clear and easy to grasp. He wrote a colourful description for the RASC *Journal*, giving the uninitiated a good idea of the tenor of the meeting:

> Previously the meetings had usually been held in a large city during the Christmas vacation, but some members thought it would be desirable to try holding the meeting in the summer time and at a place where all members could live under one roof ... The experiment undoubtedly proved successful and I am sure all the 25 or 30 members present will agree ... [The] exceeding leisureliness of the service in the dining room, where the members were grouped together at three or four adjoining tables, compelled them to spend, besides the time in the meetings, at least four hours each day together, thus cultivating that spirit of friendliness and good fellowship which was so marked a feature of the meetings ... It was even rumored that, not satisfied with the scenery, some of the older members were absenting themselves and indulging in flirtations with some of the fair inhabitants, – but this must be scandal.

This last aside may have been an attempt at humour, for elsewhere in his account Plaskett spoke of the members' faithful attendance at sessions. He further commented that he would have liked more discussion but "most of the members seemed too modest to say anything. The President, Professor Pickering [director of the Harvard College Observatory], made an ideal presiding officer and that there was not more discussion was certainly not his fault, as every encouragement was given and a good example set by him."[73]

The members had to conduct some business, including nomination and election of officers. During these sessions, Plaskett tells us, there were discussions of the society's cumbersome name. When the group had formally organized in 1899, astrophysics was new and considered distinct from traditional positional astronomy. Within a decade, however, it was well established and was in fact the main activity in most major observatories. Those active in the field tended to call themselves astronomers so there seemed to be support for W.W. Campbell's proposal to alter the name of the organization to the American Astronomical Society, which ultimately occurred in 1914. (For the sake of simplicity, I will refer to the organization as the AAS from here on.)

Canada, and Plaskett in particular, began to play an increasingly important role in the AAS. The arrangements for the 1909 meeting – in August at Yerkes Observatory near Chicago – were chosen to allow members of the British Association for the Advancement of Science to attend the AAS before moving on to their meetings in Winnipeg. The Canadians, however, seemed to choose one or the other. Plaskett went to Yerkes and presented two papers of his own on instrumentation and one by Motherwell.[74] King went to Winnipeg and presented Plaskett and Harper's joint paper on the two curiously similar spectroscopic binaries, described above.[75]

The Extraordinary Meetings of 1910

The 1910 meeting of the AAS was a very special one. About sixty of its members met at the Harvard College Observatory in Cambridge, Massachusetts, from 17 to 19 August.[76] This meeting was followed two weeks later by the fourth triennial convention of the Solar Union at the Mount Wilson Solar Observatory in southern California.[77] As Hale, the director of Mount Wilson urged, this timing allowed leading astronomers from Europe as well as America to attend both meetings. At Harvard, Plaskett presented his own paper "The Probable Errors of Radial Velocity Measurements" and read (in the authors' absence) papers by two California astronomers – W.W. Campbell, director of the Lick Observatory, and Walter S. Adams, second in command at Mount Wilson but increasingly in charge as Hale's health declined. Asking JSP to read these contributions suggests that these men already considered him a close associate. A year later he would write a review of Adams' book on the rotation period of the Sun by spectroscopic methods for the leading American journal *Science*.[78]

At the Harvard meeting, JSP was placed on three committees – one on solar rotation, another on spectral classification, and on a new committee struck to consider cooperation in radial velocity determinations. He felt especially honoured to join this latter group, whose members were all directors of major observatories: Campbell (Lick), Frost (Yerkes), Schlesinger (Allegheny), Karl Schwarzschild (Potsdam), and Hugh F. Newall (Cambridge, England).[79] Plaskett's appointment spoke eloquently of the rising importance of the DO and his role there. His status among his associates was undoubtedly a factor when they accepted the official invitation, which he conveyed to the AAS, to come to Ottawa in August 1911.

Immediately after the meeting at Harvard, about thirty of the astronomers, including JSP, made the long train journey across the continent to marvel at Mount Wilson's impressive new 1.5 m reflector and its mammoth solar telescopes. En route, the group made some stops – at Niagara Falls, the University of Chicago (where JSP discussed the DO's defective grating with Michelson), Lowell Observatory at Flagstaff, Arizona, and at the Grand Canyon. At least from Chicago on, the heat was almost unbearable. Pickering noted, "the thermometer went down when you put the bulb in your mouth!"[80] At the Grand Canyon the German astronomer Karl Schwarzschild, determined to walk to the bottom, suffered some good-natured razzing from the others when he fainted from the high temperature and exertion.[81] Father Aloysius Cortie, musician and astronomer from Stonyhurst College in England, kept his companions diverted by singing such songs as "The Wild Man of Borneo Has Now Come to Town," with its cumulative lyrics, "The wind through the whisker of the flea of the hair of the tail of the dog of the daughter of the wife of the Wild Man."[82] On a more serious note, the long train ride in two special cars afforded plenty of opportunities to talk over common concerns.

Other astronomers joined the party at various points along the way. Chant came from Toronto and joined the rest of the group in Pasadena. For this trip he had persuaded his wife to leave the children behind "for once" and go with him.[83] Their itinerary was pretty much the reverse of Chant's earlier journey, this time travelling by ship from Owen Sound on Lake Huron to Fort William on Lake Superior, then west on the Canadian Pacific Railway before heading south to California. Of course he couldn't resist calling in at Lick Observatory, where his former student and RASC gold medallist, Reynold K. Young, was pursuing postgraduate studies.

A total of 83 members went to the five-day conference of the Solar Union at Mount Wilson; they included 46 from the United States, 8 from Great Britain, 24 from continental Europe, and 3 from Russia.[84] Plaskett and Chant attended as representatives of the RASC, which at this time was considered an adhering body of the union.[85] It was an unprecedented opportunity for JSP to meet so many astronomers from abroad whom he had known only by reputation.

On the first full day, 29 August, staff of the Mount Wilson Observatory showed the guests around the offices, laboratories, and machine shop at the headquarters in Pasadena. JSP was enthralled with the spectrographic lab, where various means were employed to obtain comparison spectra and where research was being done on the Zeeman effect (the splitting of spectral lines in a magnetic field). He was especially delighted to inspect the just-completed spectrograph for the 1.5 m telescope. The astronomers also looked with amazement and disappointment at the flawed 4 tonne, 2.5 m glass disk from France, intended for the world's largest telescope. The grinding machine, which was to shape the mirror for the telescope, sat idle, awaiting another disk at some undetermined date. That evening Hale and his wife entertained everyone at a garden party at their home.

The next day was spent largely in reaching the summit of Mount Wilson, elevation 1800 m. Most of the astronomers, including Chant and Plaskett, elected to make the ascent in a stagecoach pulled by a team of horses. It was a nerve-wracking ride, as Pickering described it, on a steep, dusty trail so narrow that the wheels of the carriage were "within a foot of the edge (and death) for a large part of the way." The trail was only 15 km long but stops to refresh the horses and passengers were frequent. When they reached the summit late in the afternoon, some were able to find accommodation in cottages at the Mount Wilson hotel while others had to make do with tents. After a rest, a chance to clean up, and dinner, they were rewarded with a wonderful view in the evening. Chant wrote, "the stars [shone] with extraordinary brilliance in a clear sky above and the multitude of electric lights in Pasadena, Los Angeles and the smaller towns in the valley below presented a picture which will not soon be forgotten."[86] He probably did not foresee the immensity of the problem that artificial light would cause for astronomers in the decades ahead as urban light pollution mushroomed relentlessly.

The next three days and nights were divided between lectures, committee reports, and inspection of the telescopes on the summit. The committees of the International Union dealt with a wide range of solar research: standard wavelengths, measurement of solar radiation, spectra of sunspots, eclipse studies, spectroheliographic investigations, and the spectroscopic determination of the rate of solar rotation. JSP was a member of the committees dealing with the two latter areas. The solar rotation committee decided to share its work among observatories at Pulkovo (Russia), Edinburgh, Cambridge (England), Allegheny, Mount Wilson, and Ottawa.[87] Each was allotted a different wavelength range (as well as a common range so their results could be compared), and each was to investigate the radial velocity of the Sun to see how its rotation rate varied with latitude and possibly with wavelength. The astronomers knew that the equatorial regions of the Sun rotated fastest but they hoped that, by studying the dependence on solar latitude more closely, they would get a better understanding of the Sun's physical nature.

Hale, a leader of great drive and influence, had conceived and designed the Mount Wilson Solar Observatory and persuaded the Carnegie Institution to finance it in 1904. Its equipment was unequalled anywhere in the world. There were three solar telescopes with very long focal lengths – the horizontal Snow telescope (similar to the Ottawa instrument) and two other "tower telescopes" oriented vertically, an arrangement intended to minimize the effect of air currents that had plagued the horizontal instrument. At the base of each tower telescope was a deep well, at a virtually constant temperature, where a spectrograph received the solar image from the optics at the top. The shorter of the tower telescopes, at 18 m in height, had been in successful operation for three years and was about to be supplemented by a grander version, 45 m in height. Hale hoped to have an elevator in operation to carry conference attendees to inspect the top but he had not been able to get it ready in time. Instead, a large bucket hoisted those daring enough to ride in it. Chant, with characteristic understatement, said "the outlook [from the top] was somewhat startling and the rapid descent to the ground was equally so."[88] This magnificent telescope, nearly complete, formed a solar image 43 cm in diameter, large enough for even small regions of the solar surface, such as individual sunspots, to be studied spectroscopically.

4.3 The 45 m tall solar telescope that enthralled the visitors to Mount Wilson in 1910. (Hale Observatories, AIP Emilio Segrè Visual Archives)

Even though the Solar Union was nominally dedicated to solar research, its members acknowledged their common interest in stellar spectroscopy as well and unanimously agreed to expand the organization's mandate to embrace that area. A committee of fourteen members, including JSP, was formed to consider the classification of stellar spectra. Among them, Campbell, Frost, Hale, J.C. Kapteyn, Newall, H.N. Russell, and Schlesinger would become especially important to JSP. Pickering was chosen as chair and in spite of the 38°C heat, he sat his members down in a circle to discuss the merits of various schemes. "I shall probably never preside over a committee of such eminent astronomers," he said. To his delight, the system used at his Harvard College

Observatory was adopted.[89] For years, he and a large team of women astronomers headed by Annie Cannon had been classifying thousands of stellar spectrograms. This project was financed by Mary Anna Draper, who had assisted her husband, Henry Draper, an American pioneer in astronomical photography, and then, after his death, had created a foundation in his name to fund research. The Harvard astronomers initially used alphabetic categories, but the classes were later rearranged into the OBAFGKM system when it became clear that the new order signified a real physical progression from the hottest bluish-white O stars to the coolest red M stars.

As impressive as the solar instruments were at Mount Wilson, Plaskett reckoned that "the 60-inch [1.5 m] reflecting telescope was the instrument which created the greatest interest and admiration, and excited the greatest envy among the visiting astronomers."[90] It was, at the time, the largest working telescope in the world.[91] Its designer and optician, George Ritchey, had shown it to JSP as it neared completion four years earlier, and now he proudly showed it off over two nights, giving the astronomers ample opportunity to view star clusters, nebulae, and the planet Saturn. He also attached a low dispersion spectrograph at the prime focus. As Plaskett noted, "The star images were beautifully hard and crisp and the exposures required to get measurable spectra surprisingly short. A spectrum of a fifth magnitude star, requiring upwards of an hour with our refractor, can be photographed with the same linear dispersion in about five minutes, an efficiency that particularly excited my admiration and even envy when I thought how our work could be extended if we had such an instrument."[92]

It was a dream that would propel Plaskett for most of the next decade. He argued that "the determination of the radial velocities of stars fainter than the fifth magnitude is one of the most pressing problems of modern astronomy, as upon the knowledge of such radial velocities depends the solution of many statistical studies of the constitution, motions, and dimensions of the sidereal universe."[93] To reach fainter stars, a larger telescope was needed and Plaskett began promoting the need for such an instrument. He reported to his superior:

> If our Observatory could take part in such work it would place it in the first rank among observatories, and would undoubtedly give Canada a very high standing in the scientific world.
>
> The desired increase in telescopic aperture can be most economically obtained by the use of a reflecting telescope, which can be erected for less

than a quarter the cost of a refractor of the same aperture, and which, for spectroscopic use, is almost equally efficient and indeed possesses some advantages over the refractor, notably in that it is perfectly achromatic and that shorter, photographic, wavelengths of light are not absorbed to the same extent as when passing through glass. As to the size of the aperture desirable I would say, after the performance of the 5-foot reflector on Mount Wilson, that we should not be satisfied with a smaller aperture, but, on the contrary, perhaps aim at something greater, 6-foot say, which would give us the distinction of having the largest in the world and, a far more important consideration, enable us to reach fainter stars and to obtain sufficient exposure on the brighter ones in considerably less time. The question of covering such an instrument by a movable roof, which can be rolled back out of the way when observations are made, instead of by the ordinary dome, is worth considering, for, if a suitable wind shield could be devised, there is a decided advantage so far as the seeing is concerned in working in the open, and in addition, a building with a movable roof would only cost a small fraction of one with a dome.

I would, therefore, strongly urge upon you [Dr. King] the desirability of the installation of a large reflecting telescope principally for radial velocity investigations, though it would be desirable to make it suitable for other lines of work also, especially as this can be done without much additional cost. Such an instrument would place our Observatory in the first rank, so far as equipment goes, among observatories, and would enable our staff, who have already obtained an enviable record for the quantity and quality of the work done with very modest equipment, to excel that record and to place our Observatory in the forefront in the production of valuable scientific work. There is, as I have previously stated, a pressing need for just the kind of work that we would be best prepared to do with such a telescope, and our taking up of this work would add much to our prestige as an Observatory and as a nation. It may not be amiss to point out that, as it would take two or three years to construct such a telescope, all that would be necessary in the meanwhile would be to have the construction authorized, no money would require to be voted for the present.[94]

King fully supported Plaskett's vision and in his report to Deputy Minister Cory on 1 May 1911, he wrote, "The aperture of the [present] telescope is relatively small, compared with the instruments used in many of the observatories engaged in this branch of astronomical work. A larger telescope is much to be desired."[95] King's report was the first official hint that big ideas were beginning to take root.

1 A one-room schoolhouse typical of Oxford County, ca. 1872 (AO, S14861, Acc. 9434)

2 University College, University of Toronto, shortly after the fire of 1890, showing (at the left side) the undamaged west wing and "Round House" (detail from a photograph by Herbert Simpson, Toronto Public Library, E 3-14d)

3 J.C. McLennan at his desk in the University of Toronto's new physics laboratory. On the wall is a photo of the old west wing of University College where the Physics Department had been located. (UTA, McLennan scrapbook, B65-0012/001)

4 Harry Plaskett's class at the Ottawa Model School, 1904. Harry is third from the right in the back row. Clifford Sifton, son of the minister of the interior, is fifth from the left in the middle row. Many of the boys in this photo later served in the First World War, including Harry, young Sifton, and Alan Beddoe (fifth from the right in the front row), a distinguished war artist. (Credit: Beddoe, Alan / Library and Archives Canada / PA-111974)

5 Several foreign astronomers attended the American Astronomical Society meeting at the Harvard College Observatory in 1910. (Courtesy Hayden Memorial Library at Citrus College, Schlesinger Collection)

6 Members of the International Union for Cooperation in Solar Research at Mount Wilson, 1910. Among them are the following who are referred to in the text: Father Aloysius L. Cortie, S.J. (5), Edward C. Pickering (6), Clarence A. Chant (8), Philip Fox (12), Frank W. Dyson (13), Alfred Fowler (15), John S. Plaskett (17), Aymar, comte de la Baume Pluvinel (23), Vesto M. Slipher (26), Frank Schlesinger (27), Henry Norris Russell (28), Johannes F. Hartmann (30), Edward Emerson Barnard (34), Walter S. Adams (39), Charles E. St. John (44), Herbert H. Turner (48), and Karl Schwarzschild (66). A complete identification key may be found at photoarchive.lib.uchicago.edu/db.xqy?one=apf6-04317.xml. (UCL)

7 AAS group photo at the Dominion Observatory, August 1911. There seems to be no contemporary key, but some of the people mentioned in the text can be identified. Starting with the man near the middle with the prominent watch fob and going from left to right: W.F. King, E.C. Pickering, D.B. Marsh, W. Bruce (with large white moustache), and JSP. Frank Schlesinger is in the front row, surrounded by women whom I have not been able to identify. Annie Cannon appears to be dressed in black and centred in the front row. W.E. Harper is the bald man in the upper left, R.E. DeLury is the seated man in the lower left, with R.J. McDiarmid beside him, and C.A. Chant is on the extreme right, holding his camera. Harry Plaskett may be peeking over H.N. Russell's right shoulder at rear of the picture. (UCLUniversity of Chicago Library, Special Collections Research Center; http://photoarchive.lib.uchicago.edu)

8a Group photo in two halves, above and opposite, of the International Union for Solar Cooperation at the Physical Institute in Bonn, 1913. Among those mentioned in the text are the following: on this side, front row, Alfred Fowler is at the extreme left; E.C. Pickering is holding a white hat; next to him is K.F. Küstner (with walking stick) and then Frank Dyson. At the extreme right is Annie Cannon, with Frank Schlesinger directly behind her in the second row. In the back row is Ejnar

8b On this right side, JSP, the man with the dark moustache, is near the middle of the second row. W.W. Campbell is directly in front of him with a white hat resting on his knees; J. Hartmann has his arms folded and a cane between his legs, and A. Eddington is at the extreme right of the front row. F. Stratton is the shortest man in the back row, between the two windows. The complete photo with key is available online at www.repository.cam.ac.uk, (University of Cambridge, Institute of Astronomy Library)

9a Dr William Roche, minister of the interior from 1912 to 1917 (LAC C-020866)

9b William Cory, deputy minister from 1905 to 1930 (William J. Topley/LAC PA-167436)

10 The first executive committee meeting of the Royal Astronomical Society of Canada, Victoria Centre, 1914. (Courtesy National Research Council of Canada, DAO)

11 In this photo of members of the AAS gathered on the steps of Yerkes Observatory in August 1914, JSP is holding his straw hat and is seated in the foreground while his son Harry is almost hiding in the background behind everyone else, perhaps suggestive of their contrasting personalities. Some of the other astronomers who are mentioned in the text include Annie Cannon (fourth woman from the left in the back row), Harlan Stetson, Henry Russell, and Edward Barnard (the three seated men in the fourth row in front of Harry). Seated in the second row is Edwin Hubble (at the left) and Edwin Frost, the director of the Yerkes Observatory, with his foot beside his hat. This photo with a complete identification key may be found at photoarchive.lib.uchicago.edu

12 Victoria's crowded inner harbour, ca. 1912 (Cambridge University Library, Royal Commonwealth Society Photograph Project Y30695A-001)

13 The RASC Victoria Centre picnic in July 1914, at the site selected for the new observatory. Plaskett is prominent in the foreground with, it seems, A.W. McCurdy and W.F. King behind him and higher on the hill. (Courtesy National Research Council of Canada, DAO Historical Photo 001 [10])

14 Although this photograph was often thought to have been taken at the opening of the DAO, in fact it was taken when members of the RASC Victoria Centre and their guests assembled on 21 October 1916, a year and a half before the mirror was installed in the telescope. Knowing some of the prominent people who were present, it is possible to make a likely identification of some figures in the front row (from left to right): Napier Denison (with coat and cane), JSP, an unidentified girl, Stuart (age 11), and A.W. McCurdy (holding his coat). The man in uniform is certainly not Harry Plaskett, who was then stationed at Kingston, preparing to go overseas. (Courtesy National Research Council of Canada, DAO Historical 001 (149))

THE DAILY COLONIST, VICTORIA, B.C., SUNDAY, JUNE 4, 1916

Little Journeys out of Victoria

1. To Little Saanich Mt. and Brentwood Bay in a Studebaker "Six"

The Studebaker Car at the Crest of the Grade to the Top of Little Saanich Mt

Arrival of the Studebaker Car at Brentwood Hotel

A Glimpse of the Road to the Top of Little Saanich Mountain.

WILL BE POPULAR SHORT MOTOR TOUR

Scenic View From Top of Little Saanich Mountain Is Unequalled Anywhere on Vancouver Island.

Attention is attracted to Little Saanich Mountain at the moment on account of the expected arrival in this city some time next week of Dr. J. S. Plaskett, which presages the speedy completion of the great Dominion Government Observatory at its summit to house what is to be the largest reflecting telescope in the world.

The mountain is of special interest to motorists, not alone on account of the structural feature which will soon attract universal attention to it but as [illegible]

[illegible] The train runs over smooth tracks and as we had to contend with deserts, mountains and [illegible] roads, we think the motor did pretty well against the locomotive. Our route from Los Angeles was through Flagstaff, Ariz., Albuquerque, N. M., [illegible] Dodge City and Emporia, Kan., Kansas City, Mo., St. Louis, and [illegible] to Indianapolis, Ind., [illegible] and via Wheeling to Pittsburgh, and then to Philadelphia and through Trenton to Jersey City.

Mr. Baker drove the car himself all the way. In computing the new coast-to-coast record, three hours in time was subtracted because of the difference between Eastern and Pacific time.

ROAD BULLETINS

Compiled by the Island Automobile Association

Island Highway.—To Nanaimo very good. From Nanaimo to Campbell River good except a few miles' stretch between Nanoose and Englishman's River.

[illegible]

15 Even before the observatory was complete, it was a popular destination for a Sunday outing. (*Victoria Daily Colonist*, 4 June 1916)

[illegible] globular star cluster M13 in Hercules, taken by JSP soon after the DAO telescope went into service in 1918. (Courtesy DAO, CANFAR pl13_07_016)

17 JSP at his desk (Courtesy RASCA)

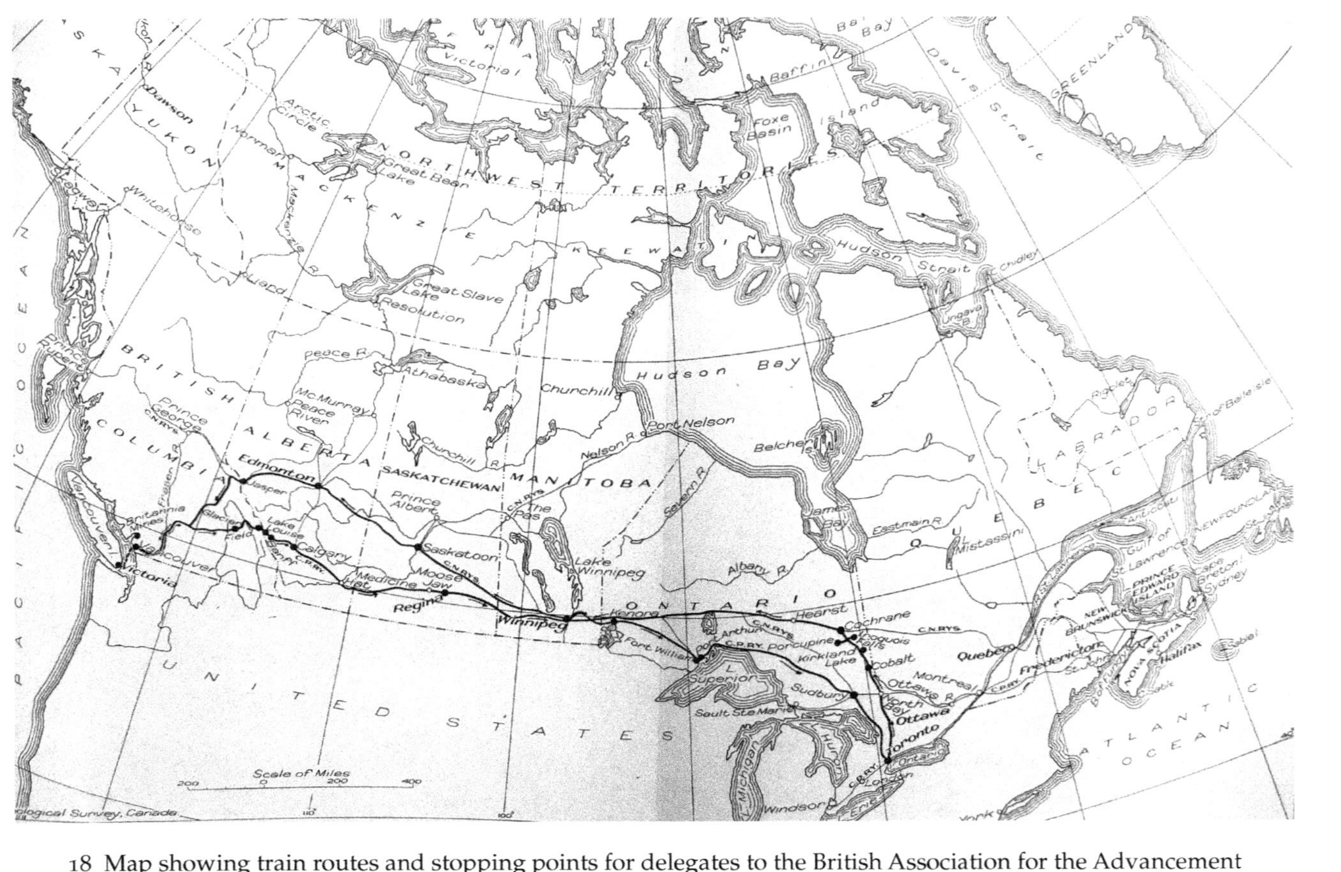

18 Map showing train routes and stopping points for delegates to the British Association for the Advancement of Science meetings, 1924. (Copied from J.C. Fields, Appendix II)

19 (a and b) Some fellows of the Royal Society of Canada on the front steps of the Victoria Museum, Ottawa, 21 May 1925. The above photograph shows: in the back row, JSP at the extreme left, R.K. Young (the shortest man) with J.C. Patterson, a meteorologist on his left; in the middle row, A.T. DeLury is the shortest figure; and in the front row are mathematicians J.C. Fields (leftmost figure), A.E. Baker (looking sideways), and physicist J.C. McLennan, president of RSC (extreme right of front row). The photograph on this page shows two figures from the extreme right of the original photo: F.T. Shutt (l.) and H.M. Tory (r). (UTA, A72-0035/0008++)

20 (a and b) These two recent images by outstanding Canadian astrophotographer Jack Newton show, above, a wide angle view of the Milky Way – an "inside" view of our home Galaxy – and, below on the facing page, an "outside" view of the nearest large spiral galaxy, M31, in Andromeda. Though even a superficial look at these pictures suggests that we may reside in a system similar to M31, the research of Plaskett and Pearce in 1928–35 enhanced our understanding of that fact. Notice that M31 can be seen as a faint smudge near the lower left corner of the wide-angle photo.

21 Inside the Plasketts' home. An Edison phonograph, Chippendale model, is a prominent feature. (Courtesy National Research Council of Canada, DAO hist-89)

22 An International Astronomical Union gathering for a canal cruise in Holland on 7 July 1928. JSP is the figure on the extreme left, supported by crutches and looking down at his camera. (Photograph by M.V. Vereenigde Fotobureau Neuwe Heerengracht 165, Amsterdam, courtesy AIP Emilio Segre Visual Archives, W.F. Meggers Collection)

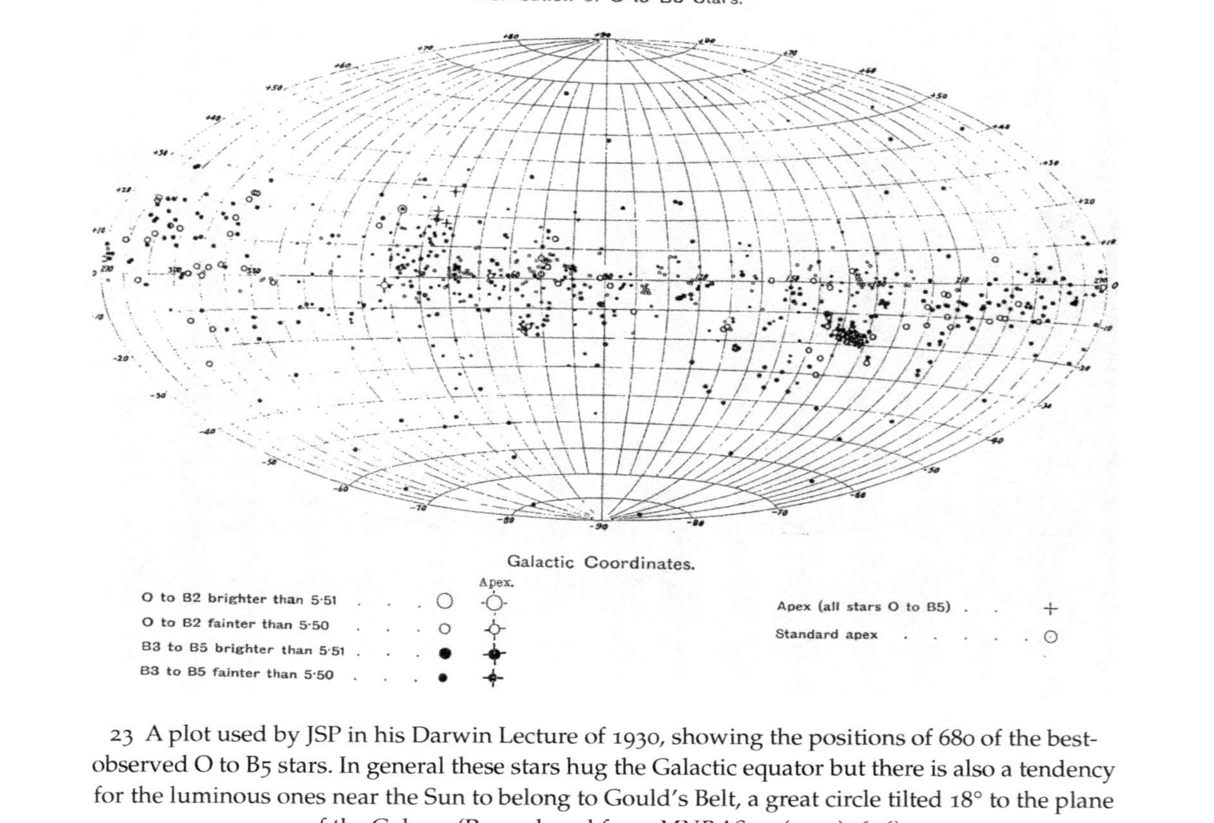

23 A plot used by JSP in his Darwin Lecture of 1930, showing the positions of 680 of the best-observed O to B5 stars. In general these stars hug the Galactic equator but there is also a tendency for the luminous ones near the Sun to belong to Gould's Belt, a great circle tilted 18° to the plane of the Galaxy. (Reproduced from *MNRAS* 90 (1930): 616)

24 The ceremony on 6 August 1932, at Bamfield, BC, marking the thirtieth anniversary of the Trans-Pacific cable. Judge Frederic Howay is speaking and JSP is on his right. The top of HMCS *Vancouver* can be glimpsed at the dock below. (Courtesy of Paul Church)

25 A distinguished group of astronomers on an outing to Plymouth, Massachusetts, in September 1932. Left to right are A. Eddington, JSP, W. Adams, J. Oort, H. Russell, H. Shapley, W.K. Miller of the Plymouth Chamber of Commerce, F. Dyson, F. Slocum, and B. Lindblad (AIP Emilio Segre Visual archives, Shapley Collection)

26 A 1935 meeting of the Royal Astronomical Society, Burlington House, London. James Jeans is speaking. Seated at the table are William Smart (secretary), William Greaves (chief assistant), and John Reynolds (president, in the chair, centre of table). Although the exact date is not specified, all the circumstances suggest that it was the meeting of 8 March, which JSP attended. (RAS/Science Photo Library V700/278)

27 Formal portraits of JSP and Harry, in middle-age. (Courtesy RASCA)

5
The Dream of an Upright Man, 1911–1913

In his first few years at the Dominion Observatory, Plaskett had made fine progress, given the limitations of the equipment in place. In a number of useful papers, he had described his efforts to improve the instruments at his disposal and especially to make the spectrographs more efficient. He had begun to produce valuable results through the study of spectrograms of a variety of sources: a well-known variable star (Mira), a few spectroscopic binary stars, and our local star, the Sun. Although he continued to work in those fields, he recognized the diminishing returns and began to yearn for the greater yield attainable only with a much bigger telescope.

Meetings in 1911

JSP and his chief, W.F. King, sought support for a large Canadian telescope in a number of ways. King was incoming president of the Royal Society of Canada in May 1911, when he reported to them on the state of spectroscopic binary research at the DO. "As work progresses, fainter stars have to be observed upon, requiring much larger exposure time. To meet this, increase in telescopic aperture is needed, which however this Observatory is not at present in a position to provide for."[1] It was a remark that planted the idea of an important need in the minds of the fellows.

Three months later the American Astronomical Society (AAS) held its first Canadian meeting, in Ottawa at the Dominion Observatory.[2] JSP and King had managed to get "an appropriation to meet all necessary and some unnecessary expenses in connection with the entertaining of the members."[3] And, as an indication of his growing profile in the AAS,

JSP was elected as one of its councillors, a position he held for the next seven years. Naturally, at this Ottawa meeting, there was a stronger Canadian contingent than usual – nine from a total attendance of forty-eight, but it was still a very small group, indicative of how tiny the active astronomical community was in Canada. The group comprised six people from the DO (including Plaskett), C.A. Chant from Toronto, and two prominent amateurs, William Bruce (Hamilton) and the Reverend Daniel B. Marsh (Springville, near Peterborough). Notable American astronomers were there, of course, including Henry N. Russell from Princeton, and a trio from Harvard – Edward Pickering, Annie Cannon, and Henrietta Leavitt. Both women were severely deaf, making larger gatherings challenging, although one-on-one conversations were possible. Cannon became fond of the Plaskett family, as her many letters to JSP show.[4] Pickering and JSP had gotten to know each other personally at earlier meetings and on the transcontinental train trip the previous summer. Russell had also impressed Plaskett at those times with his efforts to understand how the spectra of stars might change as they age, and JSP was pleased that they could talk over such matters in Ottawa.[5] The Plasketts' son Harry, then about to begin his final year of high school, was quite taken with the stellar gathering and recalled, many years later, that Russell was at their house for meals.[6] Perhaps it was on such a visit that the distinguished guest made paper animals for Harry's young brother, Stuart.[7]

The Plasketts had invited John Brashear to stay with them but he decided not to come. He had lost his beloved wife, Phoebe, the previous year at their Muskoka summer home and was planning a trip to Scotland. So the Plasketts asked the Schlesingers instead. Frank Schlesinger wondered if he needed to bring a tuxedo for any formal reception; JSP advised him that "a frock coat would be the correct thing but failing that a business suit ... will not be considered dreadfully out of form."[8] Schlesinger came alone by train from Pittsburgh and apparently had a good time in Ottawa. Mrs Plaskett said he was "a delightful guest, so thoughtful and careful,"[9] but she was disappointed that his wife did not come too. For his part, Frank enjoyed his sojourn "in the very pleasant Plaskett household ... Mrs. Schlesinger and I are taking it for granted that you will both attend the 1912 meeting [in Pittsburgh], and please tell Laddie that Wobbles (he will know that's our cat) is anxious to know whether Rex is coming!"[10] Laddie was the name everyone used for the Plasketts' younger son, Stuart, and Rex was apparently his pet. All in all, the reunion seemed to have had the warmth of a family gathering.

It was at this AAS meeting that a significant motion, orchestrated by Plaskett and King, was passed unanimously. It stated that the society

> examined in detail the work of the observatory and expresses its very favorable opinion of the character of the investigations carried on in all of its departments. This is particularly the case with the determinations of radial velocity, from which unusually valuable results have been obtained by means of a telescope of comparatively small size. In view of the pressing need for such data, the society hopes that a more powerful telescope may soon be provided, and one in keeping with the standing now attained by the national observatory of Canada.[11]

Evidently the resolution was intended to serve as an independent assessment of the situation by a prestigious body of astronomers that could impress the minister of the interior. It was presented to him but the timing was awful. Canada was in the midst of a federal election campaign. Laurier was defeated at the polls on 21 September 1911, after fifteen years in power, and the Conservatives under Robert Borden won a majority. Plaskett wrote to Schlesinger just days after the election: "Nothing has yet been done on our reflector project and it will likely have to be postponed for a while until things become somewhat settled after the change in government and until we get a new minister who I hope will be an educated man and not a mere politician."[12]

Of course Plaskett did not sit on the sidelines waiting for something to happen. On at least two occasions, 23 November 1911 and 8 February 1912, Brashear came to Ottawa, and undoubtedly the two men discussed the telescope project. Brashear made presentations to the Royal Astronomical Society of Canada (RASC) in both Ottawa and Toronto. With the title The Making of Giant Telescopes and What They Will Do," it was a subject perfectly planned to whet everyone's appetite for the day when Canada would have such an instrument.[13]

Another opportunity to keep the project top of mind came in December 1911, when the AAS met jointly with the American Association for the Advancement of Science (AAAS) in Washington, DC.[14] King, Plaskett, and Chant were the Canadian delegates at the astronomy meetings at the Carnegie Institution and the U.S. Naval Observatory. King and Chant had no active role at the meetings, but JSP reported on his efforts to determine the rate of solar rotation spectroscopically. It is not hard to imagine that he and King also used the occasion to talk up the reflector project among their associates whom they would soon ask for individual letters of support.

Asking favours was not difficult for Plaskett, whose gregarious nature shone at these relatively intimate gatherings. An indication of flourishing friendship was a little gift from Schlesinger, who was also in Washington – a piece of petrified wood that he had picked up when they were in Arizona in the summer of 1910. When JSP got home and gave it to his wife, she was charmed with the present and looked forward to having the Pittsburgh astronomer come again. "I like to have Dr. Schlesinger, he is so thoughtful and congenial in the house," she told her husband.[15] Undoubtedly the two families felt quite close to one another.

Developments in 1912

Besides the large telescope proposal, Plaskett naturally had his regular research and other interests on his mind. Hints of his wider understanding and philosophy are revealed in a couple of speeches he gave in 1911 and 1912. As chair of the RASC Ottawa Centre, he addressed the group on the progress of astronomy during the past year, or at least those aspects of interest to him. The society was pleased to publish his speech in its *Journal*; the Smithsonian Institution in Washington considered it of sufficient importance to reprint it in their *Reports* of 1912, and the editors of *Scientific American* issued an abridged form in their *Supplement*.[16] Plaskett opined that there had never been a time in the history of astronomical science when advances had been so rapid or more likely to produce generalizations. He attributed this in part to the recent and unprecedented interdependence of astronomy and other branches of science – physics, chemistry, geophysics, geology, and even biology. He noted that astronomy's long-term projects required faithfulness, unselfishness, and untiring perseverance, especially because data collected in one era might become useful only for later generations. He described how improvements in instruments, spectroscopy, and photometry all served the overarching goal of understanding the extent, form, and motions of the sidereal universe. An example was the discovery of the Sun's course (and Earth with it) towards an apex close to the bright star Vega. The conclusion was based on hundreds of observations that showed that stars in the hemisphere of the sky centred on Vega seemed to diverge from the point to which the solar system was heading, while stars in the opposite part of the sky appeared to converge towards the antapex. Evidence of other so-called star streams had been found, suggesting the complex structure of the cosmos. Plaskett summed up this way: "It is quite certain then that the visible Sidereal Universe is of almost inconceivable dimensions and of a structure so complex that, although we are gradually obtaining a knowledge of some of the motions and some idea of its form and arrangements in part, we are yet far from any clear notion of its constitution."[17]

Another speech reiterating some of the same thoughts provides insight into JSP's broader views. He joined the Ottawa Field-Naturalists Club in 1912 and delivered a talk to them on 27 February that year.[18] It is unclear if he had a prior interest in natural history or whether he joined the group at the urging of his friend Frank Shutt, the Dominion chemist and an active member of this club as well as the photographic society.[19] Perhaps he had his arm twisted by Otto Klotz, who had also been a member of the Ottawa Field-Naturalists for a number of years and an associate editor of its monthly publication. However it came about, JSP's paper, based on his talk, is rather different from any of his scientific output.[20] It shows his sympathy with the Victorian view, shared at least by many older members of his audience, that science revealed the activity and plans of God in nature.[21] It also demonstrates that he had a philosophical turn of mind and was beginning to ponder the broader implications of astrophysical research. JSP entitled his paper "The Evolution of the Worlds," thinking that his topic might have at least some commonality with the biological interests of his listeners. He explained how evolution of stars or trees can be perceived without experiencing it by asking his listeners to imagine they were in an oak forest: "Though we cannot expect to see the growth of any one oak from the acorn and seedling through small and large to a fully developed tree, and then through the process of decay to a crumbling log, yet we would have no difficulty, owing to the examples in all stages of growth around us, in correctly tracing and arranging the development."[22]

In some respects, he anticipated later thinking very closely. For instance, he said that the great nebula in Andromeda and others like it "are possibly Galaxies or Universes like our own situated at almost inconceivable distances," an idea that would become generally accepted over a decade later.[23] Other aspects of his talk, such as Russell's understanding of stellar evolution, would be superseded.

The paper also revealed JSP's religious beliefs when he wrote:

> Although we, in our feeble way, can trace the process of development from the original primal material in its simplest forms to the very complex manifestations that we see all around us, both on the earth and in

> the heavens, and can see that this development in both [astronomy and natural science] has followed … the operations of laws, which, simple in themselves, are yet so perfect and complete and far reaching as to excite our admiration and awe, yet we have in the very beginning to start with the Creator. Surely there is not one of us but feels that such a plan of creation as is here implied requires a higher, wider, and nobler conception of the Almighty Ruler of the Universe than the one which imagines it to have been made, as it were, in a moment.[24]

His ideas were reminiscent of those expressed by the influential English polymath William Whewell nearly eighty years earlier, that "events are brought about, not by insulated interpositions of Divine power, exerted in each particular case, but by the establishment of general laws."[25]

Clearly JSP did not take the Genesis story literally, yet he steadfastly believed that a supreme power "so ordered and arranged the development of Creation as to make it the result of the action of natural laws." He went further, expressing his wonder of "the Love, which created the human mind and gave to it the power … to reach out to the inconceivable depths of space and reduce the apparent confusion of stars to orderly systems, to deduce the laws which govern these systems and thus unify to a certain degree all the wonderful phenomena of suns, planets and comets, stars, nebulae and clusters into one whole."[26]

Perhaps, in quiet moments, Plaskett called on God to grant him courage, as he sometimes faced indifference and even animosity to his proposals during the coming years. We do know that he and King were "good church-goers" to use Klotz's somewhat sarcastic turn of phrase.[27]

Though JSP's regular work – instrumental improvements, solar and stellar studies – continued, he got an unexpected opportunity for some new spectroscopic exploration in March 1912, when a Norwegian astronomer, Sigurd Enebo, discovered a star in the constellation Gemini at a spot where none was known before.[28] It was called Nova Geminorum because it seemed to be new but in fact it had been there all along as a faint star, too dim to be noticed. At its brightest it was clearly visible to the naked eye (at magnitude 3.5) though hardly spectacular, so only those who had a good knowledge of the sky and happened to be looking in the right place at the right time would have noticed it. Plaskett got the news via telegram from the Harvard College Observatory, which acted as an international clearing house and disseminator of communiqués announcing unexpected astronomical events.

Novas were rare – only four had been seen during JSP's lifetime and the most recent had made its brilliant appearance in 1901, before he had embarked on his astronomical career. So this was his first opportunity to study such an unusual occurrence.

Nova Geminorum's apparent mediocrity was only a result of its great distance. Had it been a neighbouring star, it would have even been visible in full daylight! Plaskett and his contemporaries were agog at the tremendous energy released and puzzled over how it could be explained. As he put it, "When we consider that this outburst, collision, explosion, or whatever it may have been, has set free sufficient energy to increase the light given out by this body ... thousands of times in a few hours, it seems hopeless from any terrestrial experience to attempt to explain the cause."[29] Even so he felt impelled to try; a study of the nova's spectrum held the best hope of finding clues. Time was of the essence, as novas usually fade back to their earlier obscurity in a matter of months and thus Nova Geminorum would be too faint for the Ottawa spectrograph within a couple of weeks. So, after an initial delay of five days as a result of a mistake in the telegram message, and the dubious Ottawa weather, Plaskett was able to obtain only seven spectrograms. The last ones required exposures of over two hours. The spectra, typical for novae, showed many of the usual stellar absorption lines, though greatly shifted to the blue indicative of a high velocity of approach, accompanied by very bright, broad hydrogen emission lines or bands. Plaskett quickly wrote up a report including photographic prints of six of the seven spectra he had obtained, along with intensity curves drawn by visually estimating the brightness of the bands. These clearly showed how the spectrum changed in the course of ten days.

Several observers around the world also obtained spectrograms of Nova Geminorum. From these, they were forced to make an inescapable conclusion that seemed almost unbelievable. As Plaskett reluctantly said, "[Though] it seems most improbable that the widened emission and the largely displaced absorption lines are due to motion of the gases in the line of sight which would require velocities of the order of a thousand kilometers [per second], there is, however, one phenomenon in accord with such an hypothesis, which is, that the width and displacement are proportional to ... the wavelength."[30] This relationship – a property of the Doppler effect – had been noted previously for the nova of 1901, so JSP was providing corroborating evidence. By 4 April, just six days after his latest observation,

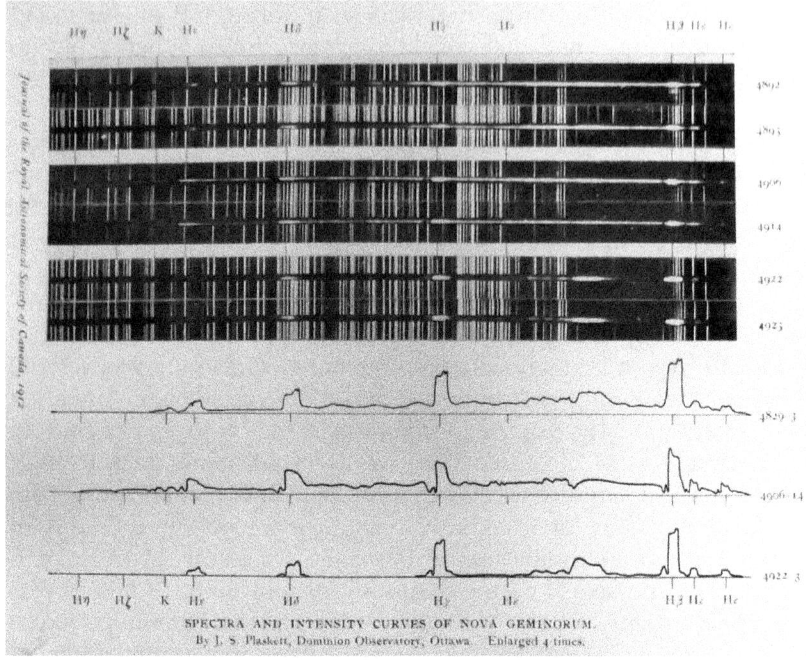

5.1 JSP photographed three pairs of spectra (around 18, 22, and 28 March 1912) of Nova Geminorum and made the hand-drawn intensity curves, shown below, for each pair. As usual, each exposure of the star's spectrum is like a thin slice of meat in a sandwich between two comparison spectra. Very wide hydrogen emission (bright) lines (labelled Hβ, Hγ, Hδ, Hε) are clearly seen in the spectra, especially in the latest pair when the continuous spectrum has begun to fade. Absorption (dark) lines on the left edge of the emission lines arise in the middle pair of spectra. (*JRASC* 6 (1912): Plate III at p.28)

JSP submitted his paper to Chant for publication in the RASC *Journal*. Chant had held up that month's issue to include the timely article, so it was fresh news indeed. Astronomers Walter Adams and Arnold Kohlschütter at the Mount Wilson Solar Observatory were able to cite JSP's work when they found similar results using their much larger 1.5 m telescope.[31] They also used Plaskett's measures of the sharp, dark absorption lines (the H and K lines, to use Fraunhofer nomenclature), which did not share in the high velocity shown by the other lines. These "stationary" lines, seen in many intrinsically bright stars, would be an important feature of much of Plaskett's future work.

That JSP's paper in the RASC *Journal* was cited by the Mount Wilson astronomers and in the AAAS journal, *Science*, was a good example of the advantage of prompt publication.[32] By contrast, he published the identical paper in a new series entitled *Publications of the Dominion Observatory, Ottawa*, but it did not appear until over a year later – and in fact the collected papers making up volume 1 of the *Publications* were not complete until 1916. Clearly, *Publications* served primarily as a record of the achievements at the DO. For transitory events they were of very limited use.

The push for a large telescope got under way in earnest at the RSC meeting in May 1912, when Plaskett, newly named secretary of Section III, drafted a memorial, or statement of facts, to the government.[33] The document noted that the DO, in only a few years of operation, had achieved "an enviable reputation in the scientific world." In particular, the astrophysical division "obtained surprising and valuable results" with respect to radial velocity work and in orbits of binary stars. JSP cited his appointment to three international committees as evidence that distinguished astronomers around the world held him and the DO in high regard. The directors of the leading observatories in England, Germany, and the United States were planning a campaign to find the radial velocities of all stars visible to the naked eye as a means of furthering "knowledge of the structure, constitution and motions of the Universe." But the project urgently required another large telescope, "four or five times the aperture of the small one at present in use [in Ottawa]." The RSC memorial reiterated that a reflecting telescope had many advantages for this type of work and would cost less than a quarter of the amount required for a refractor of the same size.

> For a comparatively moderate outlay therefore, a reflecting telescope can be constructed and installed complete. Canada would then have the largest telescope in existence and would be able to take probably the greatest share in the important work already referred to. There need be no fear of results when the record already achieved by the present staff with their very modest equipment is considered and the addition of one or two observers would enable them to keep the new instrument in active and efficient operation …
>
> The possession of [a large reflecting telescope] would enable our country to take a prominent part in this co-operation between the observatories of which mention has been made, and would convince the world that Canada is energetic and progressive in matters which appertain to the acquirement of scientific knowledge.[34]

As JSP told Schlesinger, he presented "as weighty arguments as we could invent in its favor." He asked him

> to write a friendly letter to Dr. King saying that you have heard of the memorial (as a reason for writing to him) and as it was in line with the resolution of the A & A S last year express the hope that something may be done and then present all the arguments that may occur to you. The importance, value, urgency, timely character of the work may be urged, the prestige accruing to the institution & country that takes it up, especially a new country like Canada, the assurance of valuable results being attained from the record of the present staff, and any others that may occur to you.
>
> Furthermore I would be obliged if you could obtain for me the cost of your large dome. If the cost of the steel work, the covering, and the driving & working mechanism can be given separately without too much trouble, so much the better.[35]

At the same time Plaskett wrote to other leading astronomers in the United States and Europe.[36] Schlesinger complied with JSP's requests and so did many others, including England's Astronomer Royal, Frank Dyson, whom JSP had met in 1910.[37] At this stage in the planning the idea was that the new instrument would be part of the DO in Ottawa. Even Plaskett saw advantages in this, arguing that observatories in a climate like Ottawa's generated more than enough plates for the astronomers to study; he implied that if a new facility were in a better location they would never be able to keep up. But Edwin Frost at Yerkes Observatory was unwilling to support Plaskett's campaign unless the telescope was located at the best possible site, especially one where there was not much diurnal temperature change.[38] (It was desirable to have a small range from daytime highs to nighttime lows, since different parts of a large, thick, glass telescope mirror respond non-uniformly to changes in temperature, causing distortions in its shape and leading to focusing problems.)

J.C. McLennan, the University of Toronto physicist with whom JSP had worked a decade earlier, had similar thoughts to Frost's. As past-president of Section III of the Royal Society of Canada, he presented the memorial to his colleagues.[39] Though Plaskett and King thought the project stood a much better chance of being approved by the government if the telescope were to be located in Ottawa, where the necessary infrastructure was already in place, McLennan urged that the original recommendation for "the installation of a large reflecting telescope ... at the Dominion Observatory" be amended by adding the words "or at some branch thereof to be established should climatic and other conditions warrant." The memorial in its amended form was adopted by 20 June 1912. "Accompanied by strong letters of commendation from the most eminent astronomers of Europe and America," it was presented to the prime minister just before his departure for England, where he was to discuss Canada's contribution to naval defence in the light of the rapid expansion of the German fleet. Prime Minister Borden replied that the memorial would have the careful and earnest consideration of the government, but he obviously had other pressing matters on his mind.

In late August 1912, JSP travelled to Pittsburgh for the annual get-together of the AAS. The highlight was the dedication of the new Allegheny Observatory on 28 August. It was an especially happy outcome for Schlesinger, who had previously threatened to resign over the unsatisfactory state of affairs there.[40] The grand new structure in Pittsburgh's Riverview Park replaced the original 1859 institution and took twelve years to complete. It housed three major telescopes: a 33 cm visual refractor, a 76 cm reflector used for spectroscopic binary research since 1906, and a new 76 cm photographic refractor not yet complete; the latter two were made by Brashear. This great occasion, at which Brashear, Pickering, and Schlesinger spoke, whetted JSP's appetite for similar action north of the border.[41] The time away from Ottawa was a nice break for Reba too, enjoying her first trip with her husband in many years. They stayed in Pittsburgh for a week, partly with John Brashear and partly with the Schlesingers with whom they had "such a good time."[42] The Plasketts took in a baseball game in Pittsburgh and a couple more when they broke their train trip home in Rochester. There they visited relatives and were disappointed to see the old hometown team, Toronto, go down to defeat in both games of a doubleheader.[43] Once he got home, JSP wrote letters of thanks to the Schlesingers and encouraged them to come to Ottawa, expressing the hope that Frank could undertake a speaking tour of RASC centres.

Just prior to Plaskett's trip to Pittsburgh, Minister of the Interior Robert Rogers had requested that King provide cost estimates associated with the large reflector project.[44] Plaskett had anticipated such a request and had many of the figures already at hand. Using Plaskett's figures, King was able to submit preliminary estimates to the government by 19 September 1912. Once again, the timing turned out to be

5.2 Plaskett (holding his folding camera), John Brashear, and Frank Schlesinger – a detail from a larger group photo on the steps of the Allegheny Observatory, Pittsburgh, 28 August 1912. (Allegheny Observatory Records, 1850–1967, UA.5.1, Archives Service Center, University of Pittsburgh, 6422.514.AO)

bad.[45] A cabinet shuffle was imminent; Rogers was about to be replaced by William Roche. The telescope project seemed likely to fall between the cracks but, once Parliament resumed at the beginning of November, JSP "haunted the lobbies practically every evening for two months, attempting to find members on the government side, who were interested in astronomy, to intervene with the Minister ... in favour of a large telescope for Canada."[46] At the same time, McLennan (perhaps at Plaskett's urging) wrote Sir Edmund Osler, member of Parliament for Toronto West, urging him to take an active interest in the matter.[47] He was a brother of Sir William Osler, the famous surgeon, and was one of Canada's wealthiest men. He had provided philanthropic support for hospitals and cultural institutions in Toronto, but apparently no one had considered approaching him about extending such support to an observatory. Nonetheless, as we shall see, involving Osler turned out to be a very good idea.

When the AAS and AAAS met jointly in Cleveland over the Christmas holidays in 1912, Plaskett and R.M. Stewart from the Dominion Observatory and Chant from the University of Toronto all attended.[48] The greatest value in such meetings was, as Chant put it, "the personal contact with working representatives of science, and the mingling together in the corridors, on the streets or in the little dinner parties each evening after the sessions were over."[49] One such meal was at the Duquesne Club, where Plaskett joined Schlesinger, Brashear, Russell, and others.[50] JSP used the opportunity to bring some of his spectrograms so that he could show them to Schlesinger and discuss their features with him before writing up his findings for publication.

The astronomers also enjoyed tours of two factories – the Peerless Automobile Company and the Warner & Swasey works. The latter firm was well known to JSP, as it had built the mechanical parts for the Ottawa telescope, and to astronomers generally for manufacturing the mountings of many large telescopes at major observatories, including Yerkes and Lick.

Of course, there were paper sessions and JSP made two presentations: one on solar rotation, the other on the relative intensity of grating and prismatic spectra related to his research on the efficiency of spectrographs. Both were outgrowths of committee work in which he was heavily involved.

Research during 1912–1914

JSP's research on solar rotation was carried out at first with Ralph DeLury.[51] A paper they published jointly in 1913 was in effect a report on their part of a shared project that had been agreed to by the Solar Union committee (see chapter 4).[52] Astronomers had known since 1850 that the Sun did not rotate as a solid body; sunspot observations showed that the equatorial region rotated faster than the parts closer to the poles. But some wondered if different chemical elements would show different rates of rotation. Plaskett and DeLury found no evidence of this but did get results similar to other researchers for the spectroscopic rotation rate at different latitudes.[53] Nonetheless such measurements were far from an exact science – in the 1913 paper, for example, JSP found the sidereal rotation rate at the solar equator to be 14.37 degrees per day, while DeLury's measurement was 14.04; in other words, JSP found that the Sun rotates once every 25.05 days at its equator, while DeLury's result was 25.64 days.

As a member of the radial-velocity committee formed at Harvard in 1910, JSP had agreed to carry out pioneering experiments comparing the advantages and disadvantages of prisms and gratings in astronomical spectroscopy. A grating he received from Brashear in January 1912 was ruled with 6000 lines per cm at Johns Hopkins University, the premier centre for such work, and was designed to concentrate most of the dispersed light into the first order spectrum.[54] (Normally a grating, unlike a prism, wastes light by dispersing it into several "orders" of overlapping spectra.) The aim of the experiments was to compare the intensity of the spectrum throughout its length with a spectrum of the same dispersion produced by prisms.

JSP eventually published his full results in a paper entitled "Experiments Regarding Efficiency of Spectrographs" in the *Publications of the Dominion Observatory*. It was in three parts (he had previously published the first in the *Astrophysical Journal* of June 1913, and the third in the same journal of July 1914).[55] The first section dealt with the use of a plane grating as the dispersing element in place of the traditional prism or prisms, where Plaskett found the uniform intensity produced by the grating for most of the spectrum was the main advantage, especially for stars of spectral classes O, B, and A having few lines spread over a wide range of wavelength. Uniform intensity was important; otherwise a single spectrogram could be overexposed at one end and underexposed at the other. However, the overall lower intensity of light diminished the advantage of a grating. (This, of course, was no problem in solar work.) In the second part of the paper he discussed the results of experiments he conducted with two different objective prisms of 20 cm aperture (again supplied by Brashear) mounted in front of a 12 cm telescope attached to the observatory's main telescope. He found the accuracy of the radial velocities determined using this setup was less than with standard slit spectrographs, and he understood that the expense and difficulty in obtaining large objective prisms would limit the aperture of the telescope that could be used and put fainter stars out of reach. Finally, in the third part of the paper, he showed that in standard spectrographs, the use of prisms made of lighter flint glass allowed more light to pass, resulting in a significant reduction in exposure time. Coupled with other improvements, he had shown "the importance in stellar spectroscopy, where the light is always meager in quantity, of so selecting the materials and designing the optical parts of a stellar spectrograph that all the losses by reflection and absorption may be minimized … [The] results indicate what a great saving in exposure time and consequent increase in output and range may thus be effected."[56] These papers were clear evidence of Plaskett's expertise in spectrograph design.

The work on spectroscopic binaries, which had done a lot to build the status of the DO, continued but was now carried out mainly by junior members of the staff under JSP's supervision: J.B. Cannon, W.E. Harper, and T.H. Parker. JSP seemed quite discouraged by 1913. Edwin Frost at Yerkes had sent a questionnaire to the main observatories where these investigations were carried out, including the DO.[57] The purpose of the survey was to avoid duplicate effort and to ascertain the present state of research in this area. Plaskett replied that

> the present condition of the orbital determination of spectroscopic binaries is not very satisfactory. The process of selection of binaries for investigation was to choose first of all those with a high range of velocity and with good or moderately good spectra. Consequently the binaries now available for investigation consist practically wholly of stars with poor spectra or with a low range of velocity or with both these drawbacks. I am convinced that all workers in this line of research, especially those like ourselves with small aperture, have felt that the binaries which offer any prospect of yielding satisfactory orbits are very few indeed.
>
> We at Ottawa have made a large number of spectra of several binaries, in one or two cases more than a hundred plates of a star, without being able to obtain a period, and it is very likely that several stars on our present list will prove equally unsatisfactory.[58]

Responses from the other smaller institutions were equally glum, though those with larger telescopes, like Allegheny, Lick, and Yerkes, seemed to have plenty to keep them busy, as did the Royal Observatory at the Cape of Good Hope with its near monopoly on southern hemisphere stars of this type.

The sole star that JSP personally investigated during this time was the spectroscopic binary Theta2 Tauri. Its membership in the Hyades star cluster provided an independent means for finding its distance fairly accurately by the so-called moving cluster method, thus allowing the binary star's actual size and intrinsic brightness (absolute magnitude) to be calibrated. JSP originally spoke about the star to the RSC on 14 May 1912 and published his findings in three places – the RASC *Journal* (1912), the *Transactions* of the RSC (1912 and 1914), and the

Publications of the Dominion Observatory (1915). Once again the *Journal* showed itself as the place to go for a quick turnaround, though the *Publications* provided space for a fuller account. Twenty-five years later, one of Plaskett's successors, R.M. Petrie, carried out a re-analysis of the binary system and added new observations of his own, made, of course, with improved equipment.[59] With the exception of reducing the system's overall velocity by about 10 per cent, Petrie concluded there was no change in Plaskett's elements of this long-period spectroscopic binary on its highly eccentric orbit.

The Success of the Telescope Campaign, 1913

In any effort to get support for a major project, it is important to clarify what benefits it will bring about. What would a large telescope do that smaller ones could not? An eloquent answer was given by JSP's contemporary Robert Aitken of California's Lick Observatory: "First of all, it must be said most emphatically that sensational 'discoveries' are not to be expected … Unexpected discoveries, in the popular sense, may come, of course, but advances in astronomy at the present day are made chiefly by the analysis of great quantities of material accumulated by patient and persistent observation, along lines laid down in carefully matured plans."[60] This point of view was completely in line with Plaskett's attitude. In a document entitled "Some Reasons Why Large Reflecting Telescope Should Be Provided for Our National Observatory," he emphasized that vastly improved equipment was needed to extend the existing efforts at the DO and that it would put Canada in the forefront of astronomical research.[61]

The technical advantages of a bigger aperture are greater light-gathering power (which depends on the area of the main mirror) and an increase in resolving power. This latter characteristic was of less importance to Plaskett. It theoretically allows an astronomer to see two very close stars separately, for example, or to discern fine detail on a planetary surface. However, resolution is more often limited by atmospheric conditions rather than the size of the telescope's aperture, and such disturbances are generally more noticeable in large telescopes than in small ones. (Nowadays this generalization is no longer true for large telescopes equipped with adaptive optics.)

Following the earlier lead of the RSC and the AAS, the RASC, representing 421 members in seven centres from Ottawa to Regina, passed a resolution at its annual meeting held on 14 January 1913, strongly urging the government to fund a reflecting telescope "more powerful than any now in use … and in keeping with our national aims and aspirations."[62] The RASC naturally desired to support King, who was honorary president, and Plaskett, then first vice-president, but approval by its mainly amateur members showed there was extensive public support for the project. This was an important political consideration, as Plaskett and King realized they needed to directly engage the attention of members of Parliament and the cabinet.[63] As a result of all these efforts, especially McLennan's overtures of the previous November, the member of Parliament for Nanaimo, Frank Shepherd, organized a committee chaired by Sir Edmund Osler, and comprising also Agriculture Minister Martin Burrell, from Yale-Cariboo in British Columbia, and Arthur Meighen, a rising star in the Conservative ranks.[64] (Plaskett may have known Meighen since his student days; Meighen graduated in mathematics from the University of Toronto in 1896, and surely would have taken some science courses as well, bringing him into contact with JSP.)

On 12 February the committee met with Minister of the Interior Roche, who shortly afterwards granted approval for a telescope with "a mirror not less than 60 inches [1.5 m] in diameter, and as much larger as possible, consistent with practicability of construction, coupled with consideration of cost." The cost of such a telescope was estimated to be $50,000, exclusive of the observatory building to house it.[65] Less than a month later, Plaskett wrote in excitement to Schlesinger:

> The Minister has consented to the plan of obtaining a large reflector and … we are to go ahead with obtaining information, drawing up specifications and getting prices from several firms, American and English. I have not quite as free a hand as I had hoped in regard to where the instrument is to be made but I will have a good deal to say in regard to the ability of the different people to do satisfactory work and I hope to be able to report conscientiously that it should be given to the Brashear firm for I firmly believe that will be for our best interests.
>
> I am leaving on Sunday night for the west for Lick & Pasadena to get information but will be back in time to welcome you & Mrs. Schlesinger to Ottawa where we can have a good talk about the matter in the light of the information I get …
>
> I did not tell you that it is likely the new telescope will not be located at Ottawa but at the best astronomical situation in Canada possibly in the dry belt in British Columbia. I am to stop over there and look over the ground preliminarily.[66]

King wrote letters to all the astronomers who had supported the project telling them the good news and asking them if they had any features or details that they considered should be incorporated in the design. Campbell, Pickering, and Schlesinger all responded.

Designing the Telescope, Tendering, and Contracts[67]

With government authority to do so and $10 000 budgeted to initiate the project, King immediately made preliminary enquiries to a number of firms in Europe and the United States.[68] Plaskett began considering likely locations for a new observatory. If southern stars were to be visible, the site should not be too far north, and if atmospheric problems were to be minimized, it should be at a fairly high altitude. If the telescope were going to get maximum use, the skies needed to be usually clear, and if the "seeing" were to be good, the range of nighttime temperatures should be small, factors that could be assessed from meteorological records. (When seeing is good, star images appear small and steady; when it is poor, the stars seem to dance around and produce an enlarged blurred image.) Four promising sites emerged in western Canada.[69] Two were in Alberta: Medicine Hat (latitude 50°02′) on the prairies and Banff (51°10′) in the Rockies; and two in British Columbia: Penticton (49° 28′) in the dry interior and Victoria (48° 25′) on Vancouver Island.

Plaskett left Ottawa for the west on 9 March 1913, initially without the financial support of the government, though King soon fixed that problem.[70] He stopped in Medicine Hat, Penticton, and Victoria to have exploratory talks about possible sites. He may have visited family too: his brother Bert was a rancher in southwest Alberta; his sister, Josephine, had married George Fraser in 1907 (JSP's mother lived with them on their fruit farm in the Okanagan region); and his brother Frank was rector of the historic church of St Mary the Virgin in New Westminster (just east of Vancouver).[71] JSP was sorry to leave Reba behind on this trip, but they consoled themselves with plans to go to Europe in the summer.[72]

JSP was also weighing the merits of various telescope designs; he planned to go to major U.S. observatories to see their telescopes in operation and discuss the pros and cons of their construction with the directors.[73] He headed south to California, arriving at Lick, east of San Jose, on 26 March. He stayed there three days and learned about the advantages of their 91 cm Crossley reflector, recently remounted from its old fork style to a cross-axis format. Both arrangements theoretically allowed the telescope to be pointed in any direction but the new set-up proved more satisfactory. Moving down the coast to Pasadena, headquarters of the Mount Wilson Solar Observatory, Plaskett once again saw the 1.5 m telescope that he had admired on two previous visits in 1906 and 1910, and renewed his acquaintance with its maker, George Ritchey, who was at this time working on the revitalized 2.5 m reflector.[74] This brilliant innovator had written to King the previous August, expressing an interest in making the mirrors for the Canadian telescope. "I am developing a new type of reflector, a short Cassegrainian with new curves" (other than the traditional paraboloid and hyperboloid), he told King. "These curves give very much better images and larger fields than the older type, and the telescope is so short that the advantages in economy and convenience resulting from the short tube, small dome, absence of observing platform &c are very great. I expect to demonstrate the new telescope very soon by means of actual photographs of star fields."[75] Plaskett, like everyone else, was wary of trying Ritchey's radical new telescope scheme, but he was delighted to borrow blueprints for the more conventional 1.5 m instrument from Walter Adams, the acting director of Mount Wilson.[76]

Returning to the east, JSP stopped in Ann Arbor, Michigan, to call on Ralph Curtiss, who provided plans of the University of Michigan's new telescope – a 95 cm reflector.[77] Its mirror, made by Brashear, was the largest that company had made. The telescope's simple and proven design ultimately turned out to be the basis on which Plaskett would proceed. Plaskett went on to see Edward Pickering at the Harvard College Observatory. The observatory considered selling him an old 1.5 m mirror that would have needed some refurbishment.[78]

JSP was very grateful for all this valuable advice and the astronomers' willingness to help in every way. He concluded his tour by going to the works of Warner & Swasey in Cleveland and to Brashear and Company in Pittsburgh to discuss mechanical and optical details. In a speech to the upper-crust Cleveland Union Club, he was still saying that the new telescope would probably be mounted in Ottawa.[79]

As he had promised, Plaskett got back home in time for Schlesinger's visit to Ottawa in late April 1913. After several attempts to coordinate schedules, the Pittsburgh astronomer's tour of RASC centres materialized when he tacked it on to a hurried trip to Philadelphia, New York City, and Boston.[80] He spoke on the subject of celestial photography to five Ontario centres in five days. His audience in Peterborough

5.3 The 95 cm reflector at the University of Michigan, completed in 1911; JSP adapted its cross-axis mounting in the design of the DAO telescope. (BHL, Michiganensis Collection, 1912, 204)

was perhaps the largest of any, with four hundred in attendance.[81] Within a month, Plaskett would rejoin Schlesinger in Pittsburgh, on this occasion to receive an honorary doctor of science degree – the first of many honours he would accumulate during his career. As his friend said, it was "intended not only as a well-merited recognition of past achievements but it is hoped that it may indirectly help you, to some small extent at least, to larger opportunities in the future."[82] JSP's old admirer John Brashear, a former chancellor of the university, presented Plaskett to the convocation. He prophetically called him the Keeler of Canada, a reference to the late James Keeler, who in 1898 had moved on from Allegheny to direct the Lick Observatory and to do groundbreaking spectroscopy with its 91 cm reflector, then the largest such instrument in the United States.[83] The Schlesingers once again hosted Plaskett and sent him home with candies for Reba and seeds to plant for Laddie, who was in bed with the measles.[84]

JSP was extremely busy that spring. Besides his twelve-day trip to Cleveland and Pittsburgh in early May and ten days in Toronto in mid-June, he had to prepare for the European meeting of the Solar Union and its solar rotation committee, for which he would act as chair. He had to lay plans for his capable assistant W. Edmund Harper to carry out rigorous testing of potential locations for the new facility, starting in Ottawa in May and progressing to western localities in the summer. Especially challenging, JSP had to take all the information he had gathered on his American tour of observatories and distil it into requirements from which manufacturers could prepare competitive tenders. He wrote:

> In these specifications the purpose was to set forth the general form of the mounting and optical parts, the essential operations to be performed with suggestions as to the means, and the character of the workmanship required, but at the same time to leave the makers of the instrument full scope for the exercise of their ingenuity and experience in working out the details of the mechanism.
>
> In deciding between opposing opinions in regard to the best practice optically or mechanically, especially in the latter, I was possibly better equipped than most astronomers owing to my mechanical training and knowledge and I venture to think that owing to this training the telescope [will be] a better instrument than would have otherwise been the case.[85]

Plaskett's self-assessment was certainly accurate, though he could have added his experience with electrical equipment, spectroscopy, and scientific photography as ideal qualifications for someone assessing the relative merits of various proposals.[86]

The specifications that JSP drew up were clear, well organized, and detailed enough to allow accurate cost estimates to be made. He arranged them under the four headings: Preliminary (4 points), General Specifications (5 points), Optical Parts (14 points), and Mounting (36 points). The estimates were to cover complete reflecting telescopes of apertures 152 cm and 183 cm, and were to show the division of costs between optical parts and mounting, in case contracts were awarded separately. The telescope was intended to be used primarily to obtain spectrograms at the Cassegrain focus but provision was also to be made for direct photography at the prime and Newtonian focal positions at the upper end of the telescope, 7 m from the main mirror. In the Newtonian arrangement, the light returning from the main mirror at the base of the telescope is reflected off a flat secondary mirror placed near the top of the instrument, to a focus at the side of the tube where photographs may be taken. In the Cassegrain arrangement, a convex secondary mirror, in the shape of a hyperboloid, reflects the starlight back down the tube and through a hole in the primary mirror. This latter combination of reflective surfaces has the effect of greatly increasing the focal length of the telescope while keeping a manageable length for the tube (33 m and 7 m, respectively, in the final design).

In the so-called English cross-axis mounting favoured by Plaskett, the declination axis and the polar axis cross at right angles; the telescope, attached to one end of the declination axis is balanced by a counterweight at the other end. For such a massive instrument, however, each end of the polar axis has to be supported on a separate pier. Such an arrangement had been used before, most recently on the Ann Arbor 95 cm instrument whose blueprints JSP had borrowed. Unlike the yoke arrangement that Ritchey planned to use for the 2.5 m telescope at Mount Wilson, the English cross-axis mounting allowed access to any part of the sky, including the region near the Pole Star.[87] So Plaskett opted for the Ann Arbor design but wanted it modified so that the telescope tube would be an open skeleton like the Mount Wilson 1.5 m. This feature not only combined lightness and rigidity but it allowed for easy interchange of mirrors at the upper end. Plaskett introduced a couple of other important innovations. He required that the motions would be electrically operated and controlled, and he insisted that only self-aligning ball bearings (patented in 1907) would be used on both the polar and declination axis.

Eight firms were invited to submit tenders for either the optical or mechanical parts, or for the whole telescope. Five were American: Brashear, Alvan Clark & Sons, O.L. Petitdidier, G.W. Ritchey, and Warner & Swasey. The others were Thomas Cooke & Sons of York, England; Sir Howard Grubb of Dublin, Ireland; and Carl Zeiss of Jena, Germany.

Business Abroad

At the end of June 1913, the Plasketts set sail from Montreal to Glasgow on their first and much-anticipated trip to Europe.[88] The boys stayed home, Harry spending his second summer at the DO, measuring solar rotation plates.[89] The professional aspect of the trip centred on the meetings of the International Union for Solar Cooperation in Bonn, Germany. JSP was a member of three committees and, wishing to be well prepared, he spent six days on shipboard reducing measures of solar rotation made at the DO in 1911 and 1912.[90] While overseas he would discuss telescope designs with Ralph Sampson, Astronomer Royal for Scotland, and he would tour the factories of the telescope makers Cooke in England and Grubb in Ireland to get a first-hand feel for their capabilities. JSP's impression was that "Grubb would make a good mounting though not as well-designed, executed nor finished as Warner & Swasey but I think there might be considerable difficulty with the mirror."[91] Plaskett had learned from Sampson that "Grubb can do and has done good work but that he requires watching," exactly in line with what he had heard in America. In spite of these reservations, JSP found his time with Sir Howard Grubb, his two sons, and their draughtsman to be interesting and profitable. Grubb arranged for Plaskett to meet with Sir David Gill in London. Gill, who having built his reputation at the Cape Observatory in South Africa, was in retirement acting as a consultant in other telescope projects – a role JSP himself would fill in years to come.

The Plasketts spent a week in London. JSP reveled in the opportunity to get advice from Gill ("exceedingly kind and entertaining") and to see Greenwich, where he renewed acquaintances with the Astronomer Royal. There he met Sydney Hough, Gill's successor at the Cape, and Harold Knox-Shaw from the Helwan Observatory in Egypt.[92] With a card of introduction from Gill, he went to see the optical workshop of Adam Hilger and Sons, known for their spectroscopic equipment, and called on Alfred Fowler at the Imperial College of Science in South Kensington. He reported to King in Ottawa that Fowler's spectroscopy lab was "far superior to anything I have seen in America and [equipped with] every appliance for doing the best work. I had the great interest in seeing visually the line 4686 [Angstroms] so universally present in novae, nebulae and early type stars [i.e., those of type O] reproduced in the laboratory, Fowler's latest and most wonderful reproduction of celestial spectra terrestrially."[93] Fowler produced this spectrum by passing a strong electrical discharge through a tube containing helium; yet because the pattern of lines was very close to the spectrum of hydrogen, he thought he was getting another series of hydrogen lines from traces of that gas in the tube. That the lines actually arose from ionized helium seems obvious now, but this was 1913, the very year that Neils Bohr proposed his theory of atomic structure and used it to explain the mysterious series of lines seen in O stars.[94]

Although this sounds like non-stop business, the Plasketts did spend a day in the Lake District, JSP's ancestral home. And, as he told his chief,

> I am trying to see a little of the country on the way from place to place as I find it a most interesting and charming land. Everything is so different, so finished and often so quaint that it is a constant joy even to look out of a railway carriage window, especially when one is interested in photography.[95]

The Plasketts went on from London to Paris, where JSP met Lucien Delloye, manager of the famous old firm of glass makers Saint-Gobain, the only company with experience in producing blanks for very large telescope mirrors. They discussed the possibility of casting a glass disk 1.8 m across with a hole in the middle to accommodate the Cassegrain focus. Drilling a hole after the disk was made would risk shattering the glass, while pouring the molten glass around a core might break the homogeneous flow and introduces strains, so Plaskett suggested a new approach – forcing a core through the molten glass after pouring.[96] Though he must have hesitated to use the renowned supplier of the flawed 2.5 m disk at the Mount Wilson Observatory, there was really no alternative. Delloye assured him that there would not be any serious difficulties and, as it turned out, after one failed attempt, the mirror blank was successfully cast, apparently using JSP's suggestion.[97]

Ninety astronomers attended the Solar Union in Bonn, from 30 July to 5 August.[98] Almost all were from Europe or the United States, with a good representation from Russia, one from Egypt, and Plaskett as the sole representative from Canada. JSP acted as chair of the committee on solar rotation in the absence of the elderly Swedish astronomer Nils Dunér. Perhaps Plaskett was referring to personal experience when he observed that when committees were reconstituted, new names were added and "some names, not enough I think, of those who had never nor were likely to help in the work were left off."[99] Certainly Plaskett and some members of the committee had tried mightily to get good spectroscopic determinations for the solar rotation rate, but they were left with unexplained disagreements of up to 10 per cent (0.25 km/s). As a result of JSP's report, the Solar Union agreed to put future efforts into ferreting out the source of differences between observers, before continuing to measure the solar rotation itself.

Plaskett and most of his American colleagues were not fluent in German and appreciated the trouble the hosts had taken to ensure that the proceedings were in, or translated into, English. Though most of the agenda was focused on solar research, a striking exception was a display of nebular spectra by the Heidelberg astronomer Max Wolf. Some of his photographs required more than twenty-four hours of exposure accumulated over several nights. As for stellar spectroscopy, an important resolution was passed unanimously, endorsing the committee recommendation of 1910 to use the Harvard system of spectral classification, which is part of the familiar OBAFGKM sequence still in use today.

The week in Bonn was rounded out with receptions, banquets, excursions, and general opportunities to socialize. JSP developed an appreciation of the long intervals between banquet courses, allowing for speeches to be interspersed rather than lumping them all together at the end of the evening as was the North American custom. He enjoyed seeing the historic observatory in Bonn and was equally impressed with the very modern Physical Institute. On Saturday afternoon, 2 August, the delegates travelled by electric train to Cologne, where they toured the famous cathedral and the city's art museum. On Sunday they were taken on a 250-kilometre automobile tour through the beautiful Eifel Mountains and the Mosel Valley to Coblenz and back to Bonn. The Schlesingers were in Bonn too, and one can be sure the wives as well as their husbands enjoyed each other's company, especially once the formal part of the meetings was over and the guests had a steamer trip up the Rhine. Plaskett commented that, "If we could only have many such international gatherings as this, more would probably be done to promote international friendliness and

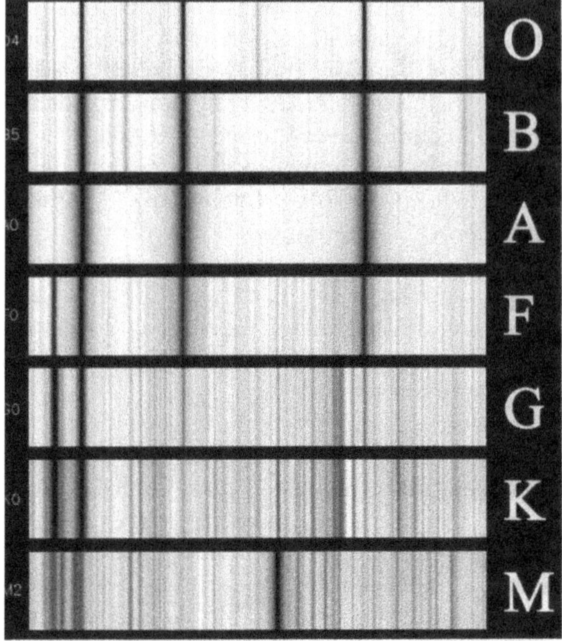

5.4 A spectral sequence for stars from hot, O-types to cool M-types. Note the change in the number and intensity of the atomic lines as the star temperature decreases.

agreement than all the formal peace conferences ever held."[100] This opinion seemed to express the dichotomy between the generous hospitality of the German scientists towards their foreign guests and the growing political tensions between their respective governments. To impress the visitors with Germany's airpower, the kaiser arranged for a flypast of one of his Zeppelins.[101]

After their evening cruise on the Rhine, the Plasketts, the Schlesingers, and sixteen others boarded a special sleeping car for Hamburg, where they attended the fiftieth anniversary of the Astronomische Gesellschaft. Schlesinger recommended Plaskett for membership in this key German astronomical society with international reach. Social functions in Hamburg included a trip to Hagenbeck's Zoo, a banquet in the historic Fährhaus, and a dinner on the *Blücher* in the harbour. The Plasketts and Schlesingers took a train to Berlin for more sightseeing and a

Sunday evening performance of Lohengrin, accompanied by the Pickerings from Harvard and Charles St John, the solar astronomer from Mount Wilson.

By the end of August the Plasketts had returned to England, boarded the *Ascania* in Southampton, and set sail for Quebec, buoyed by their memories of a sensational summer and their dreams of promising prospects for the future.[102]

6
Transition, 1913–1917

When Plaskett returned from Europe early in September 1913, he found that only four tenders had been submitted for the large reflector.[1] The Irish firm of Grubb was the only one to bid on both the mounting and optics. The others were all American: Warner & Swasey bid on the mounting while Brashear and Clark each submitted tenders for the optical parts. Because the cost of a 1.8 m telescope turned out to be only about 30 per cent greater than for a 1.5 m instrument, JSP decided to recommend the larger size. He was delighted that both the Brashear bid of $30 750 for the optical parts and the Warner & Swasey bid of $60 000 for the mounting were the lowest tenders, as these were the companies that had built the Ottawa telescope and JSP knew them well. He had the greatest confidence in their ability to produce an excellent large telescope, and he was pleased that they were near enough to discuss design details face to face.

On 9 October, Minister of the Interior William Roche reported to cabinet on the results of the tenders and recommended that contracts be signed.[2] Plaskett heard there was "a stiff battle" but thanks to strong support for the project by Martin Burrell, the minister of agriculture, success was achieved.[3] A week later cabinet's approval was formalized. JSP expressed his gratitude to everyone who had helped the process over all the hurdles, especially his political chief, the minister of the interior. He ventured to hope "that not one of the smallest of Dr. Roche's claims to the recognition of posterity will be his progressive and public-spirited attitude in regard to the cause of astronomical research in Canada."[4] Once the contracts were signed, the project moved ahead quickly, with Brashear ordering the glass discs from the French firm of Saint-Gobain in November 1913. In hindsight it seems almost certain that if the government had delayed at all, the telescope project would not have been approved, as the boom years rapidly dissipated into a serious recession that winter.

Location, Location

While the U.S. companies were preparing their bids, JSP's assistant W.E. Harper had been in the west from June to October 1913, conducting site tests with an equatorially mounted 11.4 cm Cooke refractor from the Dominion Observatory.[5] On the westward leg of his trip he spent about two weeks each in Medicine Hat, Banff, Penticton, and Victoria; on the homeward leg, he again stopped in each of these locations, in an attempt to reduce the chances that any unusual conditions on his first visit might have given a false impression of any of the sites. In his letters to Ottawa, Harper hinted at the challenge in lugging a heavy telescope and tripod, a camera and photographic plates, and various meteorological instruments to the observing sites. He may well have had a tent and camping equipment too, as he would be observing at night, far from town. In a letter to Chief Astronomer W.F. King from Hotel Incola, Penticton, on 23 July 1913, Harper wrote:

> The next morning after arrival I went up the mountain Dr. Plaskett had suggested [2.4 km due east of town] and set about getting a pack pony to get the stuff up. The packers seem to be all out on the Kettle Valley Railroad, and I could get no one around town to undertake it. That evening I went out to the Indian Reserve and got a Siwash to come in. When we got the boxes part way up the slope he "quit," so I had to set up where I was left. The slope is quite a commanding one though some distance from the top. I will make arrangements to get to the top on my return.[6]

During observing, Harper generally used the highest power eyepiece, 320x, to amplify any unsteadiness in the atmosphere. He observed a selection of double stars whose components had separations ranging from 0.9 to 6.0 arcseconds to find out whether the telescope could be used at its theoretical limit of 1 arcsecond, or if the atmosphere was blurring the images too much to split the components. He tried to observe stars around magnitude 12, near the limit of his telescope, to test the transparency of the atmosphere. All these were very important considerations in assessing the "seeing" at each site. They were characteristics that could be investigated only by an experienced observer on the spot.

A couple of weeks later, Harper reached the west coast and wrote to his boss from Victoria about the crisp clarity of the "copper plate" seeing there. It was the first time he mentioned the location that would ultimately be selected for the Dominion Astrophysical Observatory (DAO).

> It may be possible that this place – i.e. this dry belt on the south eastern part of [Vancouver] island – would show up as favorable as any when the quality of the seeing is considered. It might be advisable to have some check on the quantity [of good nights] as well. I intend before leaving to run over some of the higher elevations ten or fifteen miles north of the city. As you and Dr. Plaskett suggested my going down to Lick [in California] to get a line on the seeing conditions there for the purposes of comparison, I will do so, leaving about the middle of next week.[7]

Although conditions were unusually poor while he was at the Lick Observatory, Harper was struck with the transparency of the atmosphere there on Mount Hamilton, where, with his naked eye, he could easily see eleven stars in the Pleiades cluster, nine being his usual limit in Ottawa.[8] Such clarity was the one feature he had to concede would not likely be commonly found at the Canadian sites he had tested, but of course Mount Hamilton had an elevation of 1300 m.

After comparing his data with similar observations he made in Ottawa, both before and after the trip, Harper believed that Victoria offered the best possibilities overall. It was especially favourable in the steadiness of atmosphere and its low diurnal range of temperatures, allowing a telescope mirror to stabilize quickly as night fell. There were several hills in the vicinity of Victoria that could be considered for the site but, with prevailing winds coming from the southwest, Harper figured the smoke of the city would spoil sites to the north and east. So he chose "a rocky eminence in the midst of a considerable block of timber in the Hudson's Bay Company reserve near Esquimalt."[9] Still, to be assured that the advantages attributed to Victoria were not seasonal, Harper travelled there again in November–December. This time he also considered other sites in the vicinity but did not actually observe from them. He noted that in the Saanich district, north of the city, there were elevations from two to three hundred metres, "free from the city's smoke and lights, and suitable for an observatory."[10] All this and more, Harper put in a report to King in January 1914.[11] He also arranged to have a Mr James Pearce take photographs of the northern sky by simply exposing a camera to the sky for a number of hours each night. Harper concluded that there might actually be fewer clear nights in Victoria than in Ottawa, but the quality of seeing in Victoria was much better and the range of night-time temperature was less.

With Harper's data in hand, Plaskett asked the directors of Lick, Mount Wilson, and Yerkes observatories to give their opinion on the relative merits of the locations.[12] Their replies allowed Plaskett and King to make a convincing case that more work of better quality could be done at Victoria, with the result that Prime Minister Borden agreed to accept their recommendation of the site.[13]

Clearly JSP would need to make several train trips to the west coast as the project developed. On his first trip to Victoria, in February 1914, he met with the premier, Sir Richard McBride, and other members of the Conservative government of British Columbia.[14] McBride had a strong record of stimulating economic development in the province and agreed to contribute $10 000 to purchase land, tentatively identified as fifty acres (twenty hectares) on the top of Little Saanich Mountain, and to build a road to the summit for an estimated $20 000.[15] The word "mountain" might be considered an overstatement for an elevation of only 230 m. However, the site did have the advantage of being far enough (12 km) from the lights and smoke of downtown Victoria yet still accessible by tramline – to the foot of the mountain at any rate.[16] To keep the site confidential and to avoid the risk of bidding up the purchase price of the land, which was already subdivided into city lots, telegrams between Victoria and Ottawa were sent in code.[17]

In his dealings with the provincial government, Plaskett acknowledged the "most effective aid" he received from Arthur McCurdy, "a gentleman interested in scientific pursuits who had considerable influence with the Premier."[18] Indeed, before immigrating to British Columbia, McCurdy had been the private secretary of Alexander Graham Bell for fifteen years and an inventor in his own right. His son piloted the Silver Dart in 1909, the first powered aircraft flight in Canada. Arthur McCurdy, as president of the Natural History Society of British Columbia, had been a driving force in establishing the meteorological and seismological observatory on Gonzales Hill in Victoria in 1913 and he similarly was prepared to do what he could to advance the astronomical observatory.[19] It was he who had set up the meeting with Premier McBride and arranged for Plaskett to give a public lecture on 4 March 1914, which the large audience found very appealing. The lecture led to the establishment of the Victoria Centre of the Royal Astronomical Society of Canada (RASC), officially endorsed by the organization's national council the following month.[20] Plaskett was honorary president of the centre (as well as president of the national society); Napier Denison, who was in charge of the Gonzales Hill Observatory, was president; and McCurdy was vice-president.[21] There were forty-eight charter members. McCurdy showered Plaskett with suggestions about newspaper articles that should be written and meetings that could be arranged to promote the observatory.[22]

Though JSP had not yet completed the land transactions, he returned to Ottawa for a couple of weeks before setting off once again in early April 1914. In light of all the publicity that the Victoria newspapers gave to the new telescope and the speeches by Plaskett on the subject, it seems quite conceivable that the owners of Little Saanich Mountain would have figured out why their property was in demand. Though this suspicion cannot be proven, the best deal that JSP could strike was $14 000 for the twenty hectares. Once the location became public, some saw political motive in selecting a site in the federal riding of Frank Shepherd, the man who served as secretary for the committee of parliamentarians originally championing the cause of a large telescope for Canada. However, at the time of their deliberations, no one knew where the observatory would be situated and no one could legitimately question the extensive and unbiased site-testing that Harper carried out.[23]

When the federal government made the official announcement of the $200 000 facility and its site on 12 May, it was front-page news in the Victoria papers. They proudly boasted that the telescope would be larger than any other in use, that Victoria would become "one of the most important scientific centres in the world." The observatory would be an invaluable asset to the community, for the visitors it would attract and the testament to Victoria's superior climate that it provided.[24] The provincial minister of education characterized the new observatory as "a nucleus of a cultured class," and saw both it and the Gonzales Hill Observatory as institutions in keeping with recent steps to establish a provincial university.[25] An editorial in the city's *Daily Colonist* claimed that the project was "the largest single contribution ever made by any government in the world to the cause of pure science."[26] It was seen simultaneously as a source of immense pride for all Canadians, a reinforcement of recent advances in British Columbia, and a springboard for Victoria's ambitions. Plaskett himself saw the observatory and its proposed park-like setting as "a great attraction, not only for astronomers but also for visitors and residents of the vicinity."[27]

When the announcement was made, Plaskett was still in Victoria, on hand to provide any required details. At McCurdy's suggestion, he prepared a column on the life and accomplishments of Chief Astronomer King and provided facts for a similar article about himself. The resulting article described JSP as "eminently successful" whose "very complete and valuable investigations" were "energetically and enthusiastically" prosecuted, bringing "very widespread recognition of the work to all civilized nations."[28] The newspaper articles got back to Ottawa, where Mrs King thought they reflected Plaskett's "splurge and impudence." Otto Klotz wrote in his diary that "few men get as much rope as Plaskett has got from Dr. King, and Plaskett is on the road to put a noose around his neck."[29] (As we have already seen and will continue to see, Klotz's diaries often betrayed his mean spirit and jealousy, which occasionally bubbled into the open. It may seem unfair to quote extensively from what is essentially a private document, but I am convinced that he expected, and perhaps even hoped, that others would read it and learn the "truth" about his associates. There seems to be no other explanation for his consistent clarifications of the identity of people who were well known to himself.)

Later in May, JSP headed back east.[30] After calling in at Warner & Swasey in Cleveland to check on the telescope's progress, and giving two papers to the Royal Society of Canada (RSC) meeting at McGill University in Montreal, he was back in Ottawa by the end of the month. He told Frank Schlesinger, "I have been away in Victoria in the real estate business, and a devil of a job it was I can assure you … I have only been home about two weeks in the last three months."[31] Before long he was off to Victoria once more, this time accompanied by King, "to definitely locate the building on the 50 acres purchased when I was last out there and also to see about getting road construction started. We will probably be gone about 3 weeks and will call at Cleveland and I hope Pittsburgh on our way back."[32] The visit to Victoria included a picnic for the RASC centre amid the beautiful surroundings of the new observatory site on Little Saanich Mountain.[33]

While the structural work was progressing in Cleveland, the huge glass blank for the main mirror, weighing over two tonnes, and a smaller, half-metre disk for the secondary mirror were successfully cast and annealed at the Saint-Gobain Glass Company in Franière near Charleroi, Belgium.[34] They were shipped from Antwerp on 1 August 1914, just days before the British Empire – Canada included – entered the war in Europe. Around the middle of August, the blanks reached the Brashear factory in Pittsburgh, where JSP, accompanied by his older

6.1 John Brashear in his Pittsburgh shop with the 1.87 m glass for the Victoria telescope before it became a mirror. The tapered central hole, which would allow light to pass through to the Cassegrain focus, is seen at its narrow (rear) opening. (Courtesy National Research Council of Canada, DAO Historical Photo 001 (45))

son, Harry, went to inspect them.[35] He reported enthusiastically to C.A. Chant that the enormous piece of glass 1.87 m in diameter and 34 cm thick, with a hole 23 cm in diameter in the middle of it to accommodate the Cassegrain focus, was perfect, with "hardly a bubble to be seen in it."[36] Brashear and Company immediately began rough grinding and shaping the disk to a cylinder 30 cm thick with plane, parallel faces and a circular cross section 1.85 m in diameter.

The Son and the Sun

By the summer of 1914 Harry Plaskett had completed two years of university and had learned a lot about his father's work by doing solar research during the summers at the Dominion Observatory (DO). The collaboration between father and son and their interaction with a co-worker at the DO, Ralph DeLury, is a good illustration of how scientists' time and effort can be eaten up in trying to track down the reasons for discrepancies in their results. The three men at the DO and astronomers at other institutions had been cooperating since the meeting of the Solar Union in 1910, trying to find the rotation rate of the Sun by measuring the Doppler shift of the lines in its spectrum, but with discordant outcomes.

Harry pursued the idea that the variation in the solar rotation measurements was a result of the habits of the person doing the measuring.[37] He found that some people, like his father, operated in an automatic mode and could make readings all day long without tiring. Others, like Ralph DeLury, did his work in what Harry called an attentive mode and tired after only an hour. He proposed that everyone involved in spectrographic solar rotation studies should measure specified lines on a common plate, thus providing a basis for standardization. The outbreak of war precluded the participation of European astronomers, but comparative studies already provided evidence of inconsistencies; even Harry acknowledged that people unconsciously exhibited ephemeral tendencies that could change during a day.

DeLury, on the other hand, attributed the divergent results to variable amounts of haze in the Earth's atmosphere. The absorption spectrum resulting from this haze blended with the solar spectrum, giving misleading measurements. He went so far as to claim that Harry, and even very experienced observers such as Walter Adams at Mount Wilson, had used observations made under hazy conditions.

Father and son each presented papers on solar rotation to the American Astronomical Society (AAS) in Evanston, Illinois, in August 1914, immediately after inspecting the newly arrived glass in Pittsburgh. Harry, just twenty-one years of age, was elected to membership in the society.[38] Another young man joining at the same time was Edwin Hubble, just beginning his graduate studies at the University of Chicago.

The really big news at the conference was Vesto Slipher's presentation of the record-breaking radial velocities he was finding for the spiral nebulae – those faint pin-wheel smudges in the sky.[39] Originally, he and his director, Percival Lowell, had thought that these swirls were solar systems in the making, but Slipher had found that some of these galaxies (as we now call them) were receding at over a thousand kilometres per second. Each elusive spectrum, requiring exposures lasting between twenty and forty hours, had to be accumulated on clear, moonless nights and fitted in with other telescopic work.[40] His assembled colleagues recognized his tremendous feat and gave him an extraordinary standing ovation – a tribute that made a lasting impression on the modest astronomer.[41] While most in the audience appreciated that Slipher's measurements of large speeds of recession showed that the spiral nebulae could not be part of the Milky Way, no one, including Slipher himself, could guess that they were seeing the first evidence for an expanding universe. He made no estimates of the distances of the nebulae but just stuck to what he knew – their speeds of recession. Fifteen years later, Edwin Hubble, who was present in Evanston, added the vital ingredient of distance to Slipher's data (which he used without acknowledgment) to produce what came to be known as Hubble's law, an empirical expression that the expansion rate was proportional to distance.[42]

Broader Interests

Though JSP could easily have become solely preoccupied with the new observatory project, he found time to reflect on the broader trends in astronomy. The address that he gave to the RASC at its annual meeting in Toronto at the beginning of the second year of his presidential term on 12 January 1915 is especially interesting and in its published form "The Sidereal Universe" attracted considerable interest. The editors of the widely read *Scientific American Supplement* recognized its general interest and decided to publish it in two instalments.[43] Plaskett covered many contemporaneous astrophysical topics, such as the role of spectroscopy, theories of stellar evolution, Ejnar Hertzsprung's recognition of "giants" and "dwarfs," and stellar masses.[44] In this address, Plaskett also spoke about the widely held idea that stars in the Milky Way seemed to move in two distinct streams, briefly mentioning the "recent" work of Herbert Turner, Savilian Professor of Astronomy at the University of Oxford. This attracted the attention of Turner, who then wrote, pointing out that Plaskett, like many other astronomers, failed to fully understand his simple explanation of the two-stream hypothesis.[45] Perhaps Turner's thoughts about the large-scale motion of stars in the Milky Way sowed some seeds in Plaskett's mind; in any event the topic would become JSP's ultimate contribution to astronomy.

Aside from his scientific work, JSP kept up his other interests in photography, in nature studies, and in his church. As a member of the Photographic Art Club of Ottawa, he continued to explore the aesthetic aspects of this avocation awakened over a decade earlier in Toronto.

The Ottawa club's small membership included two renowned scientists at the Experimental Farm, Charles E. Saunders and Frank T. Shutt, whom JSP counted among his good friends.[46] (Charles's brother, Frederick Saunders, had graduated from U of T in 1895, gone on to get his doctorate in physics under Henry Rowland at Johns Hopkins, and then had a distinguished career as a spectroscopist in the United States. He had used one of the DO's gratings to study the ultraviolet spectra of calcium, strontium, and barium.[47]) Another of their number, the professional photographer William J. Topley, spoke of Plaskett's "hearty laugh, wise counsel and cultured criticisms" at meetings of the club.[48] JSP had "particularly vivid recollections of the very enjoyable evenings we spent at the Topleys and how any of our outings that had W.J. Topley along were always most enjoyable."[49] The group held regular public exhibitions at the Sparks Street gallery of one of its members, James Wilson.[50] In 1915, when Plaskett was president of the club, it raised money for the Red Cross by charging twenty-five cents admission to such an exhibition and by selling little catalogues for ten cents.[51] Twenty of Plaskett's photos were on display, printed with a variety of techniques such as "toned bromide" or "sepia platinum." He gave some of the pictures evocative titles – "Shadows in the Snow," "Early Morning Mists," or "A Grey Day on the Rideau." Others, such as one taken in England in the summer of 1913, "The City Gate – Chester," may have been more representational. On one occasion, his "Sun-Tipped Cliff" (likely taken years earlier at Toronto's Scarborough Bluffs) was reproduced in the *Christian Science Monitor*, where it was described as "extremely impressive and lacking little in satisfying design."[52] Another of his compositions, entitled "A Heavy Load," was "a well-timed study of a team of horses bringing a heavy load down a steep country lane." Plaskett sometimes combined his professional and avocational interests; probably in response to a request for submissions of "American scientific photography," he sent some of his observatory photos of spectra, of lightning, and of the Moon, to the Royal Photographic Society in England, where they were exhibited in 1914.[53]

Plaskett continued his membership in the Ottawa Field-Naturalists too, and in 1916 he invited the group to the DO for a look through the telescope and to hear him give an illustrated talk entitled "A Journey through Space." The press coverage of this event apparently attracted the attention of the People's Forum, a broadly based quasi-religious group holding regular Sunday evening lectures and discussions. The group asked Plaskett to give a repeat performance. He spoke before an audience of 1400 people, enough to fill the Regent Theatre, the largest hall available in the city. He gave a clear idea of the vastness of space, stating unequivocally that there were countless other systems like the Milky Way out there, "illustrating the infinite power of the Creator."[54] While astronomers were coming to accept the mounting evidence for the first part of JSP's point, few would have spoken so forcibly about their religious beliefs. Lick astronomer Heber Curtis, speaking in San Francisco at almost the same time, took a somewhat more agnostic view: "It is very doubtful," he said, "if any of those who are seriously studying the heavens ever lose their feeling of reverence for this supremely wonderful universe and for Whoever or Whatever must be behind it all."[55]

Observatory Progress in 1914–1915

As the new telescope began to take shape, JSP had relatively little to do with the optical work on the mirrors but participated actively in the design and construction of the mounting.[56] The blueprints he had obtained of the 95 cm reflector at Ann Arbor, Michigan, gave him a good basis for proceeding.[57] With his own mechanical knowledge and training, he collaborated effectively with Ambrose Swasey and his team, which had extensive experience in building the non-optical parts of telescopes. On several occasions, Plaskett spent three or four days in Cleveland, discussing and modifying the detail drawings with Swasey and with two engineers on his staff, Edward P. Burrell and J. Walter Fecker. In spite of the fact that Fecker was only twenty-four and this was Burrell's first venture in large-telescope design, JSP was highly complimentary, saying that Burrell "showed no less than genius in developing the mechanism."[58] The firm made a one-tenth scale working model of the telescope, dome, and building.[59] It was destined for display at the 1915 Panama-Pacific International Exposition in San Francisco, although Plaskett also found that the replica was "extremely useful in settling complicated questions during construction."[60] Once the detail drawings were completed in August 1914, construction of the mounting began. Bethlehem Steel Company handled the casting and machining of the largest parts, which Warner & Swasey then used as they completely assembled the mounting in a special facility on their own premises. The scope of the project is evident from JSP's description of the mounting under the following headings: the tube, the declination axis, bush and housing, the polar axis, the pier heads,

6.2 The temporarily assembled telescope and the one-tenth scale model in the Warner & Swasey shop. (Courtesy National Research Council of Canada, DAO Historical Photo 001 (57))

the driving mechanism, quick and slow motions, the setting circles, the bearings, and electrical equipment.[61] He gave an idea of the challenges faced by the engineers when he described how "40 tonnes of the moving parts of this telescope must ... so accurately point to and follow the motion of the star that the deviation of a line 33 metres long [the focal length] from the true pointing is less than three hundredths of a millimetre."[62]

Out west, the B.C. government built the road up Little Saanich Mountain. In June 1915 the Vancouver engineering firm of McAlpine Robertson won the contract for erecting the building, including the observatory's foundation and concrete piers on which the telescope would rest.[63] In the days before ready-mixed concrete was delivered in trucks, cement and water had to be combined with crushed rock onsite. So when trouble was encountered in getting an adequate supply of water on the mountain, the concrete work had to be put off.[64] In the end, after a lot of negotiating with neighbouring property owners, the problem was solved and construction proceeded. Then there was a political glitch. Frank Shepherd, the local member of Parliament who had earlier been very helpful, received a letter from the secretary of the Saanich Central Conservative Association complaining that "city men and enemies [are] employed on [the] observatory."[65] Loyal to his party, Shepherd wrote the following letter to Gray Donald, a local engineer who had been doing satisfactory work:

> I beg to acknowledge receipt of your letter of the 4th inst. stating that you have been notified by Mr. Horwood, the Chief Architect that your services are no longer required on the Observatory at Little Saanich Mountain.
>
> My arrangement with the Department, and with Dr. Plaskett, was that you were to remain until the completion of the water system, but that the permanent appointment of the buildings, etc., rests upon the recommendation of the Central Conservative Association of the riding of Saanich, in which this work is situated. This is the usual practice and I do not propose to act counter to my political friends and supporters.[66]

Clearly, patronage was alive and well, and the unfortunate Mr Donald had had the temerity to support the wrong political party.

Construction advanced in spite of such setbacks and JSP had to take two more trips west – one to ensure that the piers were correctly aligned in the north-south direction and the second to set foundation

6.3 View of the Panama-Pacific Exposition grounds from the south, 1915, with San Francisco Bay and Marin County in the distance (Courtesy of the Bancroft Library, University of California, Berkeley, http://www.oac.cdlib.org/ark:/13030/tf609nb4qf/?brand=oac4)

bolts so the polar axis would be accurately tilted at 48°31 , the latitude of the observatory. On the first occasion he was able to attend his brother Frank's wedding, on 29 June 1915. Because the bride's father had died, JSP had the pleasant duty of "giving the bride away."[67] On the second trip, he went via California to discuss some questions about instruments with colleagues at Mount Wilson and to attend historic meetings of astronomers in the Bay area during the first week of August 1915.[68]

Canadians and Americans west of the Rockies often felt isolated from their compatriots in the east. As a result, associations grew up on the west coast, some affiliated with those in the east and some independent. One of long standing was the Astronomical Society of the Pacific, founded in 1889 by astronomers at California's then-new Lick Observatory. Its meeting in August 1915 was the first that JSP attended; on this occasion it was joined by over a dozen other societies, including the

6.4 A portion of a group photo of the joint meeting of the AAS and the ASP at Berkeley, 3 August 1915. In the front row, left to right, are JSP, Frank Schlesinger, Reba Plaskett, W.W. Campbell, an unidentified man, and George E. Hale. Behind Reba is Otto Klotz. Is he trying to avoid looking at the Plasketts? (AIP, Emilio Segrè Visual Archives)

American Association for the Advancement of Science (AAAS), holding the first congress of its newly formed Pacific Division, and the AAS, which was also meeting for the first time on the west coast.[69] One of the other groups was the Seismological Society of America, which attracted Otto Klotz. By coincidence, when Klotz boarded the train in Chicago after a stopover there, he found "Dr. J.S. Plaskett of our observatory on the train too."[70] Over the next three weeks the two men not only travelled together but also went to many of the same functions in California, though Klotz didn't mention JSP again in his diary entries about this trip. Because he wrote in detail about many other people he chatted with, it seems clear that the two had little interaction.

The scientists held sessions in San Francisco, at Berkeley, and at Stanford University. Some spent a day and an evening at Lick Observatory. All enjoyed fine banquets and luncheons and a visit to the fabulous ranch of Phoebe Hearst, who sent a fleet of automobiles to provide transportation. Whatever leisure they had left would likely have been taken up at the international exposition celebrating the first anniversary of the completion of the Panama Canal. At the meetings themselves Plaskett did not present any paper but he did move a lengthy vote of thanks to the many institutions and organizations that made the occasion so memorable. He then proceeded to Victoria, where he hosted Klotz for a whole day.[71] The two spent the morning at the observatory site looking at the water supply and admiring the fine vista. They had lunch together at the Empress Hotel followed by a drive out to Patricia Bay and Saanich Inlet – and there was no indication of dissension in Klotz's account.

By September Plaskett was on his way home to Ottawa, once more calling in at Cleveland to talk over details in the proposed contract with Warner & Swasey for constructing the dome and related equipment, including the observing bridge that would allow easy access to the telescope's prime and Newtonian focus.[72] They also discussed the silvering "car," which would eventually be used in Victoria to allow the 6 tonne mirror in its cell to be safely removed whenever its reflective surface needed replacing (at least once a year) and also to support the spectrograph whenever it was necessary to remove it from the end of the telescope (see figure 8.1). The contract for this part of the work was signed on 12 October 1915.

Although Plaskett's official reports give a very good idea of the scope of the work, only his correspondence reveals some of the frustrations typical of the civil service. Government accountants kept very close tabs on expenses, so he frequently had to explain why he arranged his travels the way he did, why he sometimes had to hire an automobile to take him to the observatory site rather than taking a tram, why he should continue to get $7.50 a day living allowance even after the mirror was complete but uninstalled, and so on.[73] He had to deal with the office of the chief architect, who wanted a say in the plans, and with the Department of Public Works, the overseer of all government building contracts. He found the architect's ideas "crazy" and offered his personal opinion, which could be "expressed in an expletive of one syllable."[74] And when he got a telegram requesting that plans for the dome be submitted to the department immediately, he wrote to King, "It is to laugh at the P.W. Dep. After dawdling away for months getting this contract ready it is so terribly urgent at the last."[75]

About the same time that the contract for the mechanical work was signed, Brashear had to suspend operations on the mirror, dashing

6.5 The massive piers that will support the telescope's polar axis, and the circular base from which the walls of the observatory will rise. (Courtesy National Research Council of Canada, DAO Historical Photo 001 (41))

Plaskett's hopes that the telescope would be ready by the end of 1915.[76] The glass disk had already been ground into its concave form but not accurately shaped or "figured" into a paraboloid. This process would involve polishing the surface to an accuracy of ten-millionths of a centimetre so that an incoming parallel beam of light would be reflected to a single focus regardless of where it impinged on the mirror. In this stage, progress should have been tested frequently using a large auxiliary flat mirror. The plan originally had been to get a flat 1.4 m in diameter from Saint-Gobain but it was destroyed along with the factory in the early days of the war.[77] Efforts to get a similar large flat in the United States proved futile, and so the work was put on hold for nearly a year.

In the meantime in Victoria, McAlpine Robertson, contracted under the supervision of the federal Department of Public Works, finished the piers by September 1915 and completed the surrounding steel building on which the dome would rest during the fall and winter.[78] There was a temporary scare when Prime Minister Borden suggested that the project be put on hold until the war was over but Minister Roche and Premier McBride held firm and construction went ahead.[79] Warner & Swasey finished the steel framework for the rotatable dome, 20 m in diameter, and temporarily erected it in Cleveland in February 1916.[80] Before the structure was dismantled and shipped to Victoria, JSP went to Cleveland again, this time to try out the assembled dome and related equipment. He told King that he had been at the factory each morning before eight and had not left until after six, having found plenty to occupy him. He had been up and down on the observing platform accessing the prime and Newtonian focus at the upper end of the telescope tube two or three times and found it worked very sweetly and smoothly. He continued:

> The whole structure is a splendid [piece] of high class engineering design and construction and I am exceedingly pleased with the workmanship and operation. Mr. Swasey told me confidentially that they stood to lose considerable on it and if a similar dome was needed now they would not undertake it for less than $50 000. So I think we have been exceedingly fortunate throughout. For one thing I think they have made it much more complete and finished than was proposed at the time the bid was made and they have certainly not stinted any expense or care to make it unexcelled anywhere.

The telescope is also making excellent progress but I think they do not expect to have all details completed for a month. All the big parts with the exception of the mirror [are in] place and [the] … shutter over the mirror is now being installed and appears to be very stiff, dust-tight, and easily operated by a hand wheel on the mirror cell. They have just completed the working out of the electric control system which is very complete containing all necessary interlocking & safety switches etc. to make it absolutely fool proof and very convenient in operation. It requires between 50 and 60 pairs of wires going down through the north end of the polar axis to provide for all the necessary connections.

The principal impression you get on looking at the assembled telescope is its bigness. It certainly is a monstrous looking machine and yet at the same time exceedingly well proportioned and effective in appearance.

Mr. Swasey is planning to hold a kind of private reception and inspection of the telescope after it is completed and before dismantling for shipment. It is his proposal to clean out the large room in which it is mounted, and have it in complete working operation. He will send out invitations to all the leading men of Cleveland and to many others in the country and is anxious that you should come down even if only for the day …

You will of course get a formal invitation later but I hope you will be able to manage it. They are planning an interesting programme and it will be quite an event.

I expect to go to Pittsburgh on Saturday [26 February] as there are some matters in connection with the optical account that require clearing up and will return home via Cleveland.

I sincerely hope that you have been feeling better since I left and with best regards. [81]

Invited guests did get a chance to see the complete telescope (except for the mirrors) at the Warner & Swasey reception but, sadly, King was not able to celebrate this important stage in a project he had thoroughly supported from the very beginning: he died on 23 April from cirrhosis of the liver. Plaskett took his death "with deep personal grief and loss" and wrote that "Canada has lost a man of surpassing intellect, high scientific attainments and the strictest integrity, and a pioneer in the establishment and advancement of scientific research. He was truly one of Canada's great men, and the loss to the country he so ably served and represented is irreparable."[82]

6.6 Some of the guests assembled in front of the Warner & Swasey plant on 25 May 1916 include (front row, left to right) Edward P. Burrell, Charles S. Hastings, James B. McDowell, JSP, Ambrose Swasey, Worcester R. Warner, and William J. Hussey. John A. Brashear is in the second row with his arms folded. (CWRU, box 26, folder 70)

6.7 The dome at the Warner & Swasey works with all parts numbered, ready to be disassembled for shipping to Victoria. (Courtesy National Research Council of Canada, DAO Historical Photo 001 (61))

Harrying Times

Another event at about this time was also very hard for Plaskett. His son Harry, with whom he was very close, enlisted in the army immediately on getting his BA at the University of Toronto in the spring of 1916. At this point Canada was still relying on volunteers – it was not until May 1917 that Sir Robert Borden's government enacted conscription for most men up to the age of forty-five.

According to Harry's student record, his results in math and physics were, at best, mediocre. As Harry himself later noted, "I got [my BA] by the skin of my teeth, because I had volunteered for the army." He did get an A standing in astronomy in his third year (usually a fourth-year option), although he was somewhat dismissive of the way that subject was taught, reflecting later that "astronomy at Toronto at that time was still that of the nineteenth century."[83]

Academic marks do not tell the whole story. With three published papers to his credit, including one in the *Astrophysical Journal*, Harry had probably learned far more from his father than he had from Chant.[84] With such an exceptional background, he had been chosen as president of the Mathematical and Physical Society, the same organization to which his father had given his first paper sixteen years earlier. The honorary president (staff advisor) for 1915–16 was mathematics professor Alfred T. DeLury, brother of Ralph E. DeLury, the solar astronomer at the DO. As solar scientist Vic Gaizauskas has recently commented, Harry's investigation of solar rotation revealed "the same tenacity and attention to detail in working through mountains of raw data that mark[ed] his later publications. His discussion and summation [were] products of an already rigorously disciplined mind."[85] And with the war, this promising talent was put on hold, even in jeopardy.

Students and faculty had been enlisting in the armed forces since war was declared and several had made the ultimate sacrifice for their country.[86] One who had interrupted his studies in 1915 was Joe Pearce, who a decade later would work alongside Harry and his father. Pearce was injured overseas and returned to Ontario just as Harry was signing up as a private with the Toronto University Battery.[87] By September 1916 Harry was commissioned as a lieutenant and was sent to Kingston for training in November and December.[88] While on leave he managed to assemble data, undoubtedly with his father's help, to rebut rather effectively Ralph DeLury's arguments about the effect of haze on solar rotation measurements.

6.8 Executive of the University of Toronto's Mathematical and Physical Society in 1915–16. Harry Plaskett, president, is the first figure on the left of the front row; Professor Alfred T. DeLury, honorary president, is the bald man in the middle. (UTA, Copied from *Torontonensis* 18 [1916]: 248)

At the end of an RASC meeting at the DO in December 1916, when DeLury spoke about solar rotation, JSP vociferously challenged his hypothesis. The gist of what he probably said is encapsulated in a letter he wrote at the time to his friend Walter Adams at Mount Wilson: "It is ridiculous to assume that any careful observer would make rotation plates under even 5% haze, and to my mind it was a piece of impertinence on [DeLury's] part to apply his half-baked theory to your observations. He is a nice fellow in many ways, but very unsatisfactory for true scientific work as he allows himself to become obsessed with theories such as this and neglects to carefully and fully examine them."[89] Klotz's take on the meeting was that

> Dr. J.S. Plaskett made a rather savage attack on DeLury, saying that he should not have rushed into print with his meager material & that Mt. Wilson's observations refuted DeLury's theory. – It was rather a spectacle. However DeLury replied and stood his ground quite well. – DeLury afterwards told me that P's son had a paper ready as I understood on the same subject & differing from DeLury's &c. – About "rushing into print" I should have liked to have said "I quite agree that it is reprehensible to rush into print"; this humour would I think have been much enjoyed by the staff, for if there is anyone that rushes into print & brags in & out of season that Canada has "the biggest telescope in the world" it is this same Plaskett, who is going to manipulate that instrument soon in Victoria, B.C., & we'll all be glad when he is gone.[90]

In the end neither haze nor psychology turned out to be the primary explanation for the inconsistent solar results.[91] DeLury ultimately found that the culprit was oil used in the micrometer, allowing a nut to slip relative to a screw by about 0.005 mm. Not only did this movement depend on the direction the screw turned but also on the properties of the oil, which changed with cleanliness, the speed with which the screw turned in the nut, and even the temperature. At last DeLury had the satisfaction of establishing the cause but only after fifteen years of hard labour, analysing and reanalysing the old data. But DeLury's explanation did not provide the whole story. There were real physical effects in the Sun's atmosphere, such as convection currents and oscillations, that were recognized only much later.[92] It would be Harry Plaskett who would show convincingly in 1936 that the so-called granulations appearing in high resolution images of the solar surface could be explained by convection currents bringing hotter gases to the surface in the central part of the granules and cooler gases sinking back near their periphery.[93]

Years of DeLury's career were taken up by this niggling problem. Sometimes it may be wise to cut one's losses and move on, though one never knows for sure if some important understanding lies concealed in the conflicting data. Fortunately, preparations for the new observatory diverted JSP from further involvement in such frustrating minutiae.

Publicizing the New Observatory

Each May JSP reported to the RSC on the progress of the giant telescope it had actively promoted. In addition, he kept members of the RASC well informed on its headway, a task made easier by his prominent role in the organization during this period. Plaskett's service to the Ottawa Centre and the national executive had culminated in a term as national president in 1914–15. Although this and other offices did not include onerous administrative duties, Plaskett certainly was far more than a figurehead. He wrote four papers for the RASC *Journal* about the telescope between December 1913 and 1918 and spoke about it to several RASC centres – Toronto and Guelph on 1 and 2 December 1914, Hamilton on 27 January 1915, and Ottawa on 24 October 1914 and 7 April 1916. At the society's annual "At Home" reception in Toronto, on 25 January 1916, he exhibited views of "the largest telescope in the world," as the Victoria telescope was then being called. He gave a public lecture in Ottawa's normal school on 16 February 1917, showing slides of the completed observatory building and telescope, without its mirrors.[94] He engaged the interest of others by writing about the facility for a variety of publications. Engineers were fascinated by the project because of its size, planned precision, and features such as the electrical control system, high-grade steel castings, and ball bearings.[95] The wider Canadian public followed its progress in a series of articles in *Saturday Night* magazine.[96] The world learned about the mammoth telescope through the RASC *Journal* with its respectable circulation among sister societies and foreign observatories and, more importantly, through *Nature*, whose articles were sometimes written by the periodical's own astronomical columnist after reading the RASC *Journal*, and sometimes by JSP or Chant directly.[97]

Campaigning to Be King's Replacement

King's death in 1916 left the country without a chief astronomer and the DO without a director. It also left Plaskett in an uncertain position. He had been appointed to take charge of the Victoria observatory on 31 December 1914 but had received no title to go along with his responsibilities. He had asked his superior for a promotion to a higher salary category, Class IA, but King died before considering it.[98] Plaskett felt entitled not only to more pay, but also to his late boss's title, though Klotz, having been assistant chief astronomer since 1911, believed that *he* was King's natural successor. The rivalry was inflamed when Warner & Swasey, planning the unveiling of the new telescope, sent out invitations to a luncheon for "Dr. J.S. Plaskett, Director of the Dominion Astronomical Observatory." When Klotz received the invitation he was furious, writing in his diary that "Plaskett is an adept at advertising, and should be corralled by some patent medicine concern."[99] JSP was likewise suspicious of Klotz: he wrote to George Hale at Mount Wilson and others (without naming Klotz) that he feared having "the work at Victoria interfered with by an envious or jealous man."[100] The public got wind of something amiss when the *Ottawa Journal* reported on 22 May 1916, "There is a likelihood that Dr. J.S. Plaskett will be appointed chief astronomer for the Dominion. He is in charge of the new observatory at Victoria, B.C.," and then on the following day stated, "The proposal to have Dr. Plaskett appointed chief astronomer in succession to the late Dr. King, is supported by several members of parliament. The report is, however, that a good deal of opposition to the appointment has developed."

Any appointments would be up to the minister of the interior, so JSP once more enlisted the aid of observatory directors and others whom he knew, including Chant, asking them to write to Roche.[101] He took an indirect approach by asking them to express their views on certain questions, though the answers he expected were pretty clear. He inquired:

1. To what department of the work of the Dominion Observatory and to whom personally is its reputation in the scientific world chiefly due?
2. What will be the relative importance of the astronomical and astrophysical research at Ottawa and at Victoria?
3. Who would be your choice among Canadians as best qualified by attainments and record to have charge of the astronomical work at Ottawa and at Victoria?[102]

Surprisingly, Plaskett also sent the circular letter to Minister Roche so that the respondents "need feel no hesitation in writing him about the matter." Even more boldly, JSP asked to receive copies of the letters. In this, he came across as too brassy even in the eyes of his good friend Frank Schlesinger, who replied to him immediately:

> Your letter arrived a few hours ago & has given me no small concern. I wish I could have had a talk with you before you had taken this step, and

> I reply at once in the hope that I may be able to make you see things in a little different light …
>
> If I were asked by the appointing powers for an opinion I should unhesitatingly say that no mistake would be made by app[ointing] you to the place & that I am sure your administration would be a highly successful one. But without being asked I cannot very well urge Min. Roche to appoint any individual, & if I were in his place I think I should look very much askance at such a letter from me.[103]

Schlesinger went on to explain that he thought he was being asked to judge between two men – Plaskett, whose standing and work he knew and admired, and Klotz, who worked in a field about which he, Schlesinger, knew very little.

Plaskett did not take Schlesinger's advice to heart. He wrote back to the Allegheny director that the other observatory chiefs, Pickering, Frost, and Campbell, gave no evidence of diffidence in *their* statements and that Schlesinger's "ideas as to the unwillingness of astronomers to make distinctions in favour of one claimant over another were not well founded."[104] Maybe Schlesinger had second thoughts, or perhaps Roche did solicit his opinion, but in any case he did exchange letters with the minister and sent copies to JSP. Roche also corresponded with Hale at Mount Wilson, saying, a little testily, that he had no intention of making the new Canadian observatory independent of Ottawa and that he would say nothing about King's successor at present.[105]

Over a year passed before any appointments were made. During the interval, acrimony between Klotz and JSP festered. It is easy to understand that making either man the superior of the other would have been a recipe for disaster. With the country in the midst of the First World War, Klotz faced other difficulties. Though he had many supportive friends and a long career of service to Canada, his German name was a liability and scurrilous rumours circulated of his supposedly unpatriotic activities.[106] Moreover his wife, Marie, was the daughter of the German consul in Ann Arbor, Michigan. Her sentiments were openly pro-German and anti-British.

Plaskett seemed to have no doubt of his own importance and flaunted it at every opportunity. Meanwhile, Klotz fumed in his diary. Wilbert Simpson, formerly the observatory's secretary and now acting director, muddled along with financial cutbacks, alienating just about everyone except his boss, deputy minister W.W. Cory, who held the view that scientists could not be trusted to manage their own affairs.[107]

Klotz's diary entry for 27 May 1916 gives a pretty good idea of the sad state at the DO at this time:

> At office. Mr. R.M. Motherwell was with me quite a while this morning. I was glad to hear from him (a son-in-law of the late Dr. King) that he wished to see me appointed the head of the observatory. He has no use for Dr. Plaskett – "Hot Air Plaskett," and still less for Simpson our Secretary who has on more than one occasion "knifed" him. It was interesting to hear from him his indignation at Simpson's autocratic doings in connection with the "Loving Cup" presented to Dr. King five years ago, & to which I refused to contribute & held aloof because I was not consulted until Simpson had every thing cut & dried. He narrated many instances of Simpson's tyrannical dealings towards the staff. I now know that the whole staff will hail the day that Plaskett & his two satellites, Harper & Young, leave for the observatory at Victoria.[108]

Progress Continues on the New Observatory

While all this intrigue was percolating, construction on the new observatory proceeded and other equipment was purchased, including the spectrograph from Warner & Swasey and clocks from E. Dent and Company in London.[109] Builders constructed a house as temporary sleeping quarters and office space; later, it would become the de facto director's residence.[110] They also put up a cottage for the observatory's engineer. Plaskett arrived in Victoria in July 1916 to find the observatory building, including the structural members of the dome, ready to receive the telescope. The telescope mounting that the astronomers had admired in Cleveland was on its way. It had been dismantled, packed into four freight cars, shipped across Lake Erie to Canada, then transported by train to Vancouver, and finally by ferry to Vancouver Island. It arrived in mid-August, and JSP was kept very busy supervising its transportation and installation over the coming months.[111] In the midst of this activity, his mother (from Penticton) and sister-in-law (from New Westminster) arrived for a short vacation.[112]

JSP wrote from the Empress Hotel, where he stayed while in Victoria, to Philip Fox, the AAS secretary, saying there was no hope of his attending the AAS meeting at Swarthmore College, near Philadelphia, but promising to send Harper in his place.[113] JSP had missed only one other AAS meeting – the meeting in Atlanta in December 1913 – since his first in 1908. Harper seemed delighted to go and wrote Plaskett with a full account of what transpired, including the news that the election of councillors, of which JSP was one, had required five ballots.[114]

6.9 The house at the DAO that became the director's residence once the observatory was completed; the dome structure is visible in the background. (Courtesy National Research Council of Canada, DAO Historical Photo 001 (57))

6.10(a) The telescope's polar axis on horse-drawn cart travelling along West Saanich Road en route to the observatory in August 1916. (Courtesy National Research Council of Canada, DAO Historical Photos 001 (110))

6.10(b) JSP standing by the dusty observatory road, watching the mid-section of the telescope arriving at the observatory. (Courtesy National Research Council of Canada, DAO Historical Photos 001 (118))

Plaskett engaged Skillings and Hamon, a local firm familiar with handling heavy equipment, to provide the man- and horsepower to haul the individual parts of the telescope, some weighing several tonnes, across town and up the 7 per cent grade of the new 2 km road to the summit. The moving operation took two weeks. A horse-powered winch and capstan arrangement lifted the massive parts off the wagons and into place in the observatory – a nerve-wracking experience, no doubt.[115] Once the equipment was inside the dome, work to enclose it could begin. Plaskett explained that the building

> is of steel construction with double sheet metal walls arranged to allow complete circulation of air, which enters at the bottom and passes up between the walls of building and dome, passing out through louvres at the top. The purpose of this construction is to keep the interior at the shade temperature during the day, and allow it to rapidly assume the night temperature. Only in this way can the conditions of atmospheric definition be kept at their best. [116]

Fortunately no rain fell during the three months of the cladding operation.[117] Warner & Swasey's man arrived on 5 September and supervised the assembly of the worm wheel, main driving gear, setting circles, and bearings into the 6 m long, 9 tonne polar axis. The axis was placed so exactly that it was less than a millimetre out of alignment. The other large parts followed suit and within ten days the declination axis, skeleton tube, and mirror cell were all in place. The electrical wiring and connections for manoeuvring the telescope were installed, followed by the electrically operated equipment for the dome, the shutters, the observing platform, and the car where the mirror would rest when it was being silvered.

Electric motors could move the telescope on both its axes at three different rates – 45 degrees per minute for quick pointing, ⅕ degree per minute for honing in on the target, and ½ degree per hour for guiding – that is, keeping the telescope pointing on the object during the period of observation.[118] An operator on the observing floor could control the motion of the telescope and dome from one of two switchboards on either side of the south pier, while an observer at either end of the telescope could use a portable hand-paddle to clamp or unclamp the telescope and move it at either of the two slower speeds controlling the telescope's motions. As Plaskett put it, "Everything had been designed to make the operation of the telescope convenient and efficient."[119]

He was delighted that "the whole work was completed in about six weeks without hitch or accident of any kind, a remarkably short time considering the magnitude of the undertaking, giving convincing evidence of the care used in the design, construction, and preliminary fitting and erecting of the installation."[120] Though the mirror was still not ready, at least Brashear had resumed work on it.

Members of the RASC Victoria Centre and their guests, 125 in all, had the privilege of a demonstration by Plaskett on a Saturday afternoon, 21 October 1916. The visitors made the journey from the city on the interurban train and walked to the summit of the hill. Arthur McCurdy, as president of the Victoria Centre and member-elect of the provincial legislature, introduced JSP, and Napier Denison, seconded by former Premier Edward Prior, thanked him for the splendid way he had explained the operation of the telescope and put all the moving parts through their paces.[121] Plaskett and the observatory certainly had friends in high places.

McCurdy was also a member of the Victoria's Board of Trade and saw that organization as a vehicle to publicize the new facility, asking JSP to write a piece for its annual report. Tailoring his writing to his readers, Plaskett began by saying that the new observatory "will undoubtedly increase the attractiveness of Victoria for the ordinary visitor, and especially for men of culture and science."[122] Once he returned to the east in November, JSP gave newspaper interviews to the *Toronto Globe* and the *Ottawa Journal*, eliciting a typical reaction from Klotz, "What bragging!!!" and "Bragging again."[123]

In Victoria, the DAO's newly appointed engineer, Tom Hutchison, was in charge and living on site when Brashear and Swasey came to inspect the installation on 29 November. Four members of the RASC centre executive, including President McCurdy, brought them to the site, and Hutchison demonstrated the mechanism to everyone's satisfaction.[124] To impress upon the American visitors that Canada was at war, McCurdy also took them to Esquimalt harbour to show them two submarines bought by the government of British Columbia to deter German attacks.[125] The resentment felt by many Canadians that the Americans were not "doing their bit" in the war effort ought to have been tempered, at least for McCurdy and his friends, by the realization that if the United States had been at war, its factories would in all probability have been unable to devote their resources to the manufacture of the telescope and observatory structure.

One consequence of the delayed arrival of the optics was the opportunity it gave JSP and Reba, still living in Ottawa, to travel to New York

City over the Christmas holidays of 1916. He attended joint meetings of the AAS and the AAAS at Columbia University and was especially pleased to be there for W.W. Campbell's address as retiring president of the "triple A-S" to over two thousand people gathered in the Museum of Natural History.[126] Also attending were Marie and Otto Klotz. The latter wrote extensively in his diary about New York City and the six-day convention, but the only comment he made concerning JSP was that "Plaskett and wife were on the train with us. For freaks of style, Mrs. P takes the buns."[127]

Director at Last

After the death of W.F. King in 1916, Plaskett had reasons to think that he was King's rightful successor as Canada's chief astronomer. At the beginning of 1917 the RASC seemed to endorse that idea by electing him as its honorary president, succeeding King, who had filled that role since 1906. Nearly a year had gone by following King's death and the government still had not named anyone as chief so Plaskett took a fallback position – that the directorships of the two observatories in Ottawa and Victoria would be considered of equal status, each reporting to the deputy minister of the interior. In March 1917 he prepared a memorandum outlining such an organization. Klotz and some of the Ottawa staff (Motherwell and DeLury, especially) saw this as a blatant power grab, contrary to King's vision of the Victoria observatory as a branch of the DO.[128] But the acting director, Wilbert Simpson, supported Plaskett's proposal, suggesting that the title "chief astronomer" no longer be used and that there should be a director of each facility, individually responsible to the deputy minister. Finally, with the support of the minister, William Roche (whom Klotz privately called "our imbecile pusillanimous Minister"), Plaskett was named director of the observatory in Victoria effective 1 April, three months before he actually moved there, with an annual salary of $3500, a hundred more than Klotz's.[129] About the same time, the name "Dominion Astrophysical Observatory," suggested by JSP six months previously, was designated by the federal government.[130] Klotz's fury at these developments was only fuelled by the delay in his own appointment as director of the DO and chief astronomer.[131] When that finally happened, in October 1917, Plaskett had the decency to go to Klotz's office to congratulate him, though Klotz thought he was a hypocrite for doing so.

At the same time as Plaskett became director of the DAO, Reynold K. Young was appointed as the only other astronomer on his staff. Young had been one of Chant's students and the RASC gold medalist upon graduating from the University of Toronto in 1909. He had gone on to Lick Observatory, married the astronomer Robert Aitken's daughter, Wilhelmina, and earned his PhD from the University of California at Berkeley in 1912. After an unhappy year of teaching at the University of Kansas, he returned to Canada and got a research position at the DO, a definite improvement as far as he was concerned.[132] From August 1913 until the summer of 1916, he had been the most productive observer there, gathering more spectrograms than anyone else.[133] Before he reported for duty in Victoria at the beginning of August 1917, he and his wife and two small children returned briefly to California.[134] The trip apparently marked the end of their relationship and they were divorced soon after.[135]

The Effect of War

The importance of pure research to the needs of the "real world" is often lost on the public, so it may seem astonishing in hindsight that the government would finance a major observatory devoted to such an arcane topic as stellar spectroscopy in wartime. Nevertheless, governments accepted in principle the key role that scientific exploration played in the war effort and in national prosperity.[136] (A direct path can be traced, for example, from the discovery in 1868 of unidentified lines in the solar spectrum to the use of helium gas in dirigibles during the First World War, though as far as I know that argument was not advanced at the time to promote astrophysics.[137])

During the War, the Canadian government was just beginning to develop a science policy. It set up the Honorary Advisory Council for Scientific and Industrial Research in December 1916 (it would become the National Research Council (NRC) in 1925).[138] Two of the nine unpaid advisers were well known to JSP: Toronto physicist J.C. McLennan and Robert A. Ross, JSP's boyhood friend, by then a consulting engineer in Montreal. Ross later chaired the council. Modelled on somewhat similar bodies in Britain, the United States, Australia, and New Zealand, the NRC was intended to promote inquiries in general areas of science not funded within individual departments of government. Very little opposition to establishing the NRC was voiced, though implementing its recommendations was another matter: its laboratory facilities did not materialize until 1932. During the war, there was a general recognition that pure research was the impetus behind German industrial strength and military might and that governments should do more to encourage scientific investigations.[139] As noted in Victoria's *Daily Colonist*, quoting from an English newspaper, the *London Standard*, "No nation can prosper in the long run that fails to encourage the spirit of inquiry which is the very breath of scientific progress."[140]

In spite of these noble aspirations, the Ottawa observatory, like many employers, was left short-staffed as its men enlisted in the armed forces.[141] Harold Parker was the first from the DO to sign up; he could not be replaced because the government imposed a hiring freeze on positions being held for overseas servicemen and no new positions were created. This meant that Plaskett could offer no prospects to even well-qualified Americans such as Harlan Stetson, who was seeking a position in 1915 after completing his PhD in astrophysics at the University of Chicago.[142] To staff the DAO, Plaskett would have to raid existing employees from Ottawa. In another wartime measure, appropriations for new buildings were cut off, so JSP would have to manage without an administration or office building at the new observatory.[143]

After Harry enlisted in April 1916, and the day approached when he would be called into combat, the Plasketts sent a photo of him to their friends, including the Russells in Princeton, who kept it on their mantelpiece, and to Annie Cannon at Harvard.[144] Cannon wrote to JSP, "I want to thank you for Harry's picture which is certainly excellent. I am so glad to have it, for often and often do I think of the dear boy. Oh, may naught of harm come to him in these dreadful times! Do let me know when he is ordered overseas, if he is, for I want to send him a 'comfort bag' or something he might use."[145] Harry sailed overseas with the 3rd Brigade, 4th Division, of the Canadian Field Artillery in April 1917. He was stationed first at Folkestone, Kent, where he enjoyed long walks to Canterbury and the Romney Marsh on his days off.[146] After further training in gunnery at Witley, Surrey, he left for Flanders in October, where his brigade participated in the brutal assault on Passchendaele ridge.[147] In that battle, more than four thousand Canadians lost their lives and almost twelve thousand were wounded.

The times must have been exceedingly worrisome for the Plasketts but the demands of the telescope project kept JSP fully occupied and his mind mercifully diverted from the war.

PART THREE

Career in Full Flower

7
This Is the House That Jack Built, 1917–1921

Plaskett, with persistence and ambition, had achieved his goals of getting agreement for a major Canadian observatory and, on April 1917, of being named its director. Because the Dominion Astrophysical Observatory (DAO) was a government observatory, JSP, unlike most of his American counterparts, would have no distracting teaching or external editorial duties and did not have to worry about fundraising from private benefactors or foundations.

The new observatory in Victoria was unique for a government observatory. Its major counterpart in the United States was the U.S. Naval Observatory in Washington, DC, but in its traditional and utilitarian roles the USNO was more akin to the Dominion Observatory in Ottawa than the DAO would be. By focusing solely on astrophysics, the Victoria observatory would also be unlike its government cousins overseas – in Greenwich and Potsdam for instance. With the poorer climate, smaller instruments, and different national traditions, their achievements were distinct from what the DAO would be able to accomplish.[1] The Europeans tended to shine in the traditional field of positional astronomy and in theoretical investigations. As JSP was reported to have said in a speech entitled "The Personality and Work of Astronomers To-day," "the emphasis of American astronomical work is observational ... [and is] not very strong theoretically and mathematically." On the other hand, he found "the treatment of the theoretical and mathematical viewpoints of astronomy by the present English School is unequalled in the world."[2] By this measure, the DAO would clearly fall in the American camp.

In Ottawa, geophysics was a major part of the DO's purview under Otto Klotz's direction. Plaskett assumed that once he left Ottawa, the mainstay of the astronomy program would revert to traditional positional work and that astrophysics would no longer be pursued. So he tried to have calculating machines and measuring equipment transferred to Victoria, including the Hartmann spectro-comparator, the Repsold measuring engine, and a Tait Arithmometer, all essential equipment at the new observatory. Eventually JSP got what he needed but Klotz, of course, was far from conceding anything. He took the position that the equipment belonged in Ottawa and would be used there to continue astrophysical research, resulting in quite a brouhaha involving even the deputy minister.[3]

Unfortunately, during the first year of Plaskett's directorship the Victoria telescope was still without its mirrors. Brashear and Company had been unable to find a large optically flat mirror to test the accuracy of its work as it progressed.[4] Eventually it decided to use a modified, more time-consuming procedure with a smaller (84 cm) flat already in the company's possession. Besides that, the figuring and polishing of the main mirror went much more slowly than anticipated when it was realized that the heat generated by the polishing action distorted the glass. Only a half-hour's polishing could be done at a time, after which the mirror had to stand for twelve hours before it could be tested. Much tedious effort was still required.

In addition, the war had threatened the completion of the spectrograph, the instrument at the heart of most astrophysical research. Its mechanical parts were built by Warner & Swasey, and the optical parts were to come from Brashear. Plaskett wanted to use flint glass because of its low absorption and high refractive index, thus wasting a minimum amount of light and yet spreading the spectrum out as much as possible.[5] But such glass would have had to come from Germany, an impossibility because of the war. Fortunately, C.A. Chant was able to come to the rescue by lending a prism of similar material from a laboratory spectrograph at the University of Toronto.[6]

In July 1917, the Plasketts – John, Reba, and twelve-year-old Stuart – left Ottawa, moved to British Columbia, and took up residence in the house on Saanich hill previously occupied by workers during construction of the dome.[7] Before long, the Plasketts made it comfortable, with Reba holding "at home" days twice a month – a social custom that would soon be a thing of the past.[8] Within a year of their arrival, the Victoria newspaper, the *Daily Colonist*, described "the cultivated gardens which surround their pretty home so near the clouds" and the "delicious spring water of crystal purity" pumped up from the foot of the mountain.[9] The Geographic Board of Canada, accepting a suggestion by JSP, renamed the little mountain "Observatory Hill."[10]

A suitable school had to be found for Stuart, something that was not perfectly straightforward, since, as JSP put it, his younger son was slow to develop.[11] There was a one-teacher school about three kilometres away but the Plasketts thought that he would receive more attention in a city school. Besides, JSP argued, the interurban railway stopped at the foot of Observatory Hill, making it more convenient for Stuart to get to Victoria than to the Prospect Lake school. The Victoria school trustees had to be petitioned to admit Stuart but apparently they did so.

The Plasketts were deeply worried about their older son, Harry, especially once he began active duty in Belgium in October 1917. Fortunately he managed to keep in touch with his family and his Ottawa girlfriend, who all longed for his safe return to Canada.[12] The only other person to get a letter from Harry that winter was the Harvard astronomer Annie Cannon, who had knit him a scarf that he treasured. She wrote to JSP in March 1918, "I had a good letter from [Harry] about two weeks ago. And now the terrible drive is on, I think of him constantly. To-day is a very anxious one. I cannot bear to think of the sacrifices that must be made. My eyes are full of tears when it all comes over me." About the same time, JSP wrote to Brashear, "We hear from Harry every week or so and he writes in good spirits, though he is at the front. He was for a time in brigade headquarters taking the place of an officer on leave, and we only wish he could have stayed there where he would have been comparatively safe. But we must hope and pray that he may be brought home safe and sound."[13]

Almost every day the papers listed those killed in action – grim news for affected families and fearsome fuel for the anxiety of loved ones serving abroad. Canadians at home were also making monetary sacrifices. By popular demand, in 1917 the wealthiest started paying federal income tax, and in 1918 citizens bought $660 million in Victory Bonds, a large sum indeed for a small population.[14] JSP contributed a hundred dollars each year, roughly 2 per cent of his salary, to the Canadian Patriotic Fund, created to provide support and comfort to wives and other dependants of those who had volunteered their services to their country.[15]

In its brash way the immense war effort bolstered patriotism, at least among Anglo-Canadians, and strengthened Canada's progress towards mature nationhood through its full-member status in the League of Nations and in nascent attempts to develop a home-grown foreign policy. The parallel growth of political and scientific maturity, as evinced in the DAO and National Research Council (NRC), were two sides of the same coin.

During his first two years as director of the DAO, JSP's only astronomer assistant was Reynold Young from the DO. Ed Harper remained in Ottawa, with the difficult task of trying to do Plaskett's bidding without alienating his colleagues.[16] Even with a mirrorless telescope, Plaskett managed to find plenty of details to keep him busy. He spent two months in the fall of 1917 in the east, mainly overseeing progress at the Brashear works; as a result, he likely missed a courtesy call to the new observatory by the governor general, the Duke of Devonshire, on 20 November 1917.[17] When JSP was in Victoria, he and Young took every step they could think of to ensure that, once the mirror arrived, everything else would be in place so that observing could begin without delay. They adjusted the telescope's balance accurately by weighting the cell that would contain the mirror and, after the spectrograph arrived in January 1918, they carefully tested its dispersion and properties. Young wrote to Chant, "I really think the spectroscope is beautifully designed and reflects credit on Dr. Plaskett who is almost wholly responsible."[18]

Relying on the generosity of colleagues elsewhere, JSP began building up a library – essential for any research institution – and asking advice as he considered what investigations the DAO should undertake.[19] George E. Hale told JSP that, on the basis of his experience, he knew "of no task more difficult than to prepare a satisfactory scheme of research for a large Observatory."[20] Edward C. Pickering, whom JSP consulted at Harvard prior to moving to Victoria, took seven pages to outline his ideal plans.[21] His central proposal was to take direct photographs of star fields through a coarse diffraction grating (wires spaced about a centimeter apart) to estimate the apparent brightness (magnitudes) of very faint stars.[22]

Most significantly, further advice came from Dutch astronomer Jacobus Kapteyn, who had been striving for years to understand the structure of the Milky Way, a project requiring the cooperation of observatories around the world in gathering a vast amount of information about the brightness, types, and motions of the stars. JSP was sympathetic to Kapteyn's suggestion to him: "There is no doubt in my mind but that we must study the radial velocities in connection with the astronomical proper motions. I am no less strongly convinced that, in order to study to the best advantage the arrangement of the stars in space and their systematic motions, we will have to study the several spectral classes separately. On the programme of even the best equipped observatory, it will be safe not to put more stars (for radial velocity) than some 1500 or 2000 stars *at most*."[23]

While JSP was thinking through the various options of an observing program, the Brashear shop was finally finishing the telescope's main mirror. Plaskett was called to Pittsburgh on 22 March 1918 to check it over.[24] A few days later he telegraphed the Victoria newspaper to say the testing had been a big success. The opticians had to make only a very small correction near the centre of the mirror. Otherwise JSP was delighted to find that the surface was correct within an eighth of a wavelength (0.0006 mm) – twice as good as the specifications stipulated. Plaskett gave a great deal of credit to "Jimmy" McDowell and Fred Hageman of Brashear and Company for their skill in this stage of the work.[25]

The mirror was silvered, carefully packed, and shipped to Victoria, where it arrived in perfect shape on 26 April, three days after JSP got back from Pittsburgh and Ottawa.[26] The director of the Lick Observatory, William W. Campbell, supportive since Plaskett's first forays into stellar spectroscopy, was able to spend 30 April at the DAO during the alignment of the mirror. He had been invited to Vancouver to give the commencement address at the University of British Columbia on 2 May, and JSP took a break from fine tuning the telescope to accompany him to the event.[27] This diversion had little effect on JSP's ambitious schedule. Only a week after the mirror's arrival, he began observing, and on the night of 6 May 1918 he captured the first stellar spectrum.[28] Plaskett later exulted, "There was no delay or hitch from any cause and ... the telescope operated perfectly from the first without any alteration or further adjustment being required."[29] Atmospheric conditions at Victoria also met or exceeded expectations, allowing the giant instrument to live up to its potential.

Brashear and Company naturally supplied several oculars (eyepieces) of different power, as well as adapters and prisms to allow convenient viewing of star images formed at the prime, Newtonian, or Cassegrain focus. Public viewing on Saturday evenings was always part of the plan; to minimize interference with the research program, a special optical arrangement was made to divert the image close to the Cassegrain focus so that viewing could take place at the side of the telescope without needing to remove the spectrograph.

An observer cannot rely entirely on the telescope's setting circles to point it at any particular star being studied, as atmospheric refraction

7.1(a) The original spectrograph attached to the 1.83 m telescope at the DAO. The temperature case has been removed to give an idea of the internal structure. The arms marked I, II, and III denote the positions where the spectrogram would be obtained under low, medium, or high dispersion, with one, two, or three prisms. (Reproduced from W.H. Christie, "The Spectroscope and Its Work," *JRASC* 23 (1929): 247)

β Can Ven $\alpha = 12^h\ 29^m.0$
1 $\delta = +41^\circ\ 54'$

7.1(b) The first spectrogram obtained with the DAO telescope as it appears on the original negative. The two identical iron-arc spectra serve as standards for measuring the tiny stellar spectrum between them, originally about 4 cm in length. JSP has written the name of the star and its celestial coordinates on the plate. (Courtesy DAO)

and other smaller effects make the star's apparent position different from its catalogued position. To fine tune the pointing, an observer looks through a finder telescope attached to the main tube and centres the desired star on the crosshairs. Brashear provided the DAO instrument with three finders but in practice two were found to be sufficient. They were f/15 refractors of 10 cm aperture mounted on opposite sides of the main tube for easy access no matter what the telescope's orientation.[30]

As we have seen, JSP had years of experience and experimentation with spectrographs in Ottawa. He had decided as early as 1909 that the universal instrument supplied by Brashear was not suited to radial velocity work. He had instead designed two separate spectrographs – a single-prism one used for O and B stars with few lines in their spectra and a three-prism instrument used for other stars with so many lines that only a relatively small region of the spectrum needed to be studied. He had also found, in 1913, that a plane grating had some advantages over a prism. So it seems surprising at first that he designed the Victoria spectrograph with no grating and of a type that could be used with one, two, or three prisms. Plaskett cited cost as the reason for opting for a single, multipurpose instrument as opposed to separate spectrographs but he took pains to ensure that it would be rigid and yet easily and rapidly alterable from one configuration to another.[31] The spectrograph box was enclosed in an electrically heated and well-insulated case.

Opening Ceremonies

The inauguration of the world's largest operational telescope certainly demanded some sort of ceremony, but who would be able to come? Many were doing war work; moreover, finances were tight and federal government officials were not going to take days to travel across the country for an official opening, even for one as important as this. However, Plaskett knew that many American astronomers would be gathering in nearby Washington State to witness a total solar eclipse on 8 June and so, to encourage their attendance at the DAO opening, he planned the ceremony to take place three days later.

A total solar eclipse visible in one's neighbourhood is a very rare event and, to satisfy public curiosity, JSP delivered a lecture, arranged by the Royal Astronomical Society of Canada (RASC), to a full house at the Girls' Central School on 13 May.[32] As he explained, the eclipse's path of totality would miss Victoria but would sweep across the United States, from Washington through Colorado and on to Florida. He recalled his own experience thirteen years earlier in Labrador and gave a number of reasons why astronomers go to a great deal of trouble and expense to observe a solar eclipse. One of the new motives he discussed was the possibility that stars seen near the Sun during the darkness of totality would appear slightly displaced from their normal positions because of a bending of their light path by the Sun's gravitational influence or, to put it in relativistic terms, by the distortion of space-time near the Sun. Plaskett would have known that the amount of deflection was predicted in 1916 by Albert Einstein as an outcome of his general theory of relativity, though whether he actually said so was not mentioned by the newspaper reporter covering his lecture.[33] Some papers, especially the *New York Times*, ran a story on the eclipse by Campbell, who did name Einstein, giving North Americans what was likely their first introduction to the unique man and his controversial ideas.[34] Before long, the public would think of Einstein as the epitome of twentieth-century science.

Astronomers from the Lick Observatory selected a site close to Goldendale, Washington, to obtain photographs to test Einstein's prediction. JSP, Reba Plaskett, and Reynold Young all made a quick trip to join them.[35] The DAO team did not plan to make scientific observations, as they had been preoccupied with the observatory start-up. Besides, JSP had left solar studies behind when he departed from Ottawa. But he and Young could hardly have passed up the chance to see an awe-inspiring

7.2(a) The solar eclipse of 8 June 1918, photographed at Goldendale, WA, by the Lick Observatory staff. The stars that were measured for the relativistic shift predicted by Einstein are highlighted on this photo. (UCSC, ua0036_glp_0979)

solar eclipse close to home. They were not disappointed: an extremely lucky break in the clouds just at the crucial moment allowed Plaskett to make a quick sketch of the coronal streamers, though the time was very short and the darkness made drawing difficult. He and Young also noted in some detail the shadow bands – those strange ripples of light and dark appearing on the ground just as totality is about to begin. It would be the only occasion in four attempts during his life, that JSP would see a total solar eclipse. As for the Lick results, the star images on the photographs were poor and the measurements showed almost no deflection, a setback for Einstein but no great surprise, as the astronomers had to use makeshift equipment. Their best eclipse instruments had been shipped

7.2(b) Observers and guests of the Lick Observatory solar eclipse expedition to Goldendale, 8 June 1918. Starting near the centre of the photo, and going from right to left: the leader of the expedition and director of Lick Observatory W.W. Campbell (wearing a bow tie), Samuel Boothroyd (University of Washington), and Reba Plaskett. Further to the left is JSP, looking over the shoulder of John Brashear, and R. K. Young. For a complete key, see *PASP* 30 (1918): 222. (UCSC, ua0036_glp_0336)

The Director of the

Dominion Astrophysical Observatory,

Victoria, B.C.

requests the pleasure of the presence of

Dr Geo. E. and Mrs Hale

at a

Reception and Informal Opening

of the Observatory

Tuesday, 11th June, 1918, from 8 to 11 p.m.

7.3(a) Invitation to the opening ceremonies of the DAO to director of the Mount Wilson Observatory and his wife (HL, Hale papers, Plaskett file)

to Russia for a similar attempt in 1914, but because of the war and then the Bolshevik Revolution, they did not make it back to the United States for the 1918 eclipse.

By coincidence, at the same time as the solar eclipse was dazzling daytime observers, a much rarer event was sparkling at night – the brightest nova in three hundred years. At its peak on 8 June, Nova Aquilae was the brightest star in the evening sky and eighty times brighter than Nova Geminorum, which Plaskett had studied six years earlier. Young, who hurried back to Victoria from Goldendale, captured the first DAO spectrogram of it on 10 June.[36] The next night – which also marked the opening ceremonies – a second spectrum was taken. What an auspicious beginning for the observatory!

Campbell, Ambrose Swasey, and John Brashear were among those who made their way to Victoria after the eclipse. They took the ferry from Seattle on the 10th and addressed the Canadian Club of Victoria at a luncheon meeting in the Empress Hotel the next day. Members of both the men's and women's branches of the club attended, as did many Americans visiting Victoria. They were all charmed by "Uncle

7.3(b) Guests at the opening ceremonies of the DAO. Standing (from left to right): William W. Campbell and his wife, Elizabeth, John Brashear, Reba and Jack Plaskett, and Ambrose Swasey. The Hales were unable to come. (Courtesy National Research Council of Canada, DAO hist-175)

John" Brashear's personal reminiscences of how astronomy caught his fancy as a boy.[37]

Brashear wrote about the week's events to a friend: "After the eclipse was over, Dr. Campbell and his wife, Dr. Plaskett and his wife, Mr. Swasey and I went directly to Victoria, British Columbia, to attend the opening exercises and dedication of the Dominion Observatory there, where we had a right royal time."[38]

The next night, 11 June, marked the so-called unofficial opening of the DAO. Aside from the absence of federal government officials, it hardly seemed "unofficial." As reported in a front-page story in the *Colonist*, the guests included BC Premier John Oliver, provincial cabinet ministers, the chief justice and other judges, Victoria mayor Bert Todd, ministers of various churches, representatives of every imaginable organization, and their spouses.[39] They were greeted by the Plasketts at the door and shown upstairs to the dome, where Young explained the operation of the telescope. About 10 p.m., when twilight was beginning to fade, the proceedings got under way with a welcome from JSP and speeches by Brashear, Swasey, and Campbell. As the director of the Lick Observatory, Campbell spoke for professional astronomers when he said that a large telescope, like the one being dedicated, "cannot … be devoted to searching for comets, new stars, satellites, or planets. Its function is the study of the motions and physical properties of objects already known."[40] Then the province's lieutenant governor, Sir Frank Stillman Barnard, declared the observatory open and the assembled guests sang "God Save the King." Light refreshment and ices and a peek through the telescope rounded out the festivities. One can imagine the immense pride that Brashear and Swasey felt as they spoke at the culmination of the project on which their firms had laboured for so long. Brashear recalled:

> We worked on this great glass for nearly four years, under very trying conditions toward the last. Indeed, it was practically perfect, humanly speaking, on two occasions, but due to temperature changes, which we could not control owing to a scarcity of gas fuel and a desire to get just a little nearer to our ideal, the "figure" was lost; but the final corrections over that great surface exceeded by half the specifications – the greatest error found by nearly a week's study under the best conditions not being greater than one eighth light wave, equal to one four-hundred-thousandth inch or one sixteen-thousandth millimeter, and there were some parts of the surface that did not deviate one twentieth light wave,

> or one millionth inch. I had a royal view of the cluster Messier 13 (given as thirty-five thousand light-years distant from the sun) in the great glass, and the photo taken of it reveals a mighty universe or cluster of sixty thousand suns.[41]

If Brashear and Swasey were well satisfied, Plaskett must have burst his buttons. Nonetheless, he gave full credit to those who had helped to bring the project to fruition, naming the federal member for Nanaimo, Frank Shepherd, and the former provincial premier, Sir Richard McBride, and especially the former minister of the interior, William Roche.[42] At the time of the dedication, the minister was Arthur Meighen, who would hold the post until he became prime minister following Robert Borden's resignation in 1920.[43] They were all Conservatives – perhaps a bit embarrassing to the provincial Liberal premier and his ministers as they listened to Plaskett's speech. But as historian George Webb has observed, Victoria residents generally "welcomed the new facility warmly, recognizing that its presence promised a new cultural foundation and would convey a more sophisticated image of the provincial capital."[44]

One astronomer whom the Plasketts especially hoped could come was Annie Cannon from the Harvard College Observatory. JSP had submitted a paper on the new telescope for the American Astronomical Society (AAS) gathering there in August, but was not able to attend in person. Cannon wrote JSP, saying that he was missed and regretting that she had not been able to go to Victoria for the opening.[45] She had been in Colorado for the solar eclipse and afterwards went on for a holiday in Glacier National Park, where she enjoyed riding horseback and admiring the magnificent scenery. It would have been convenient for her to continue on to Victoria but she did not get JSP's invitation to the opening ceremonies until her return to Massachusetts. The closing paragraph of her letter shows what warm feelings she had for the Plasketts:

> Remember me lovingly to Mrs. Plaskett, and tell her my thoughts are with her very, very, often. Thanking you for the invitation and truly hoping to see your Observatory some day, and with every good wish for your successful work and for the dear boy [Harry] in France.
>
> Yours very cordially,
> Annie J. Cannon.

Given her fondness for the elder Plaskett son, Cannon would have been pleased to learn that, in June, Harry had once again been assigned to brigade headquarters staff in France, as the tide turned in favour of the Allies.[46]

Despite the observatory being fully operational, JSP still had to get by with Young as his only astronomical assistant. There were experienced astronomers who would have loved to work at the DAO but funding was unavailable to pay their salaries. One potential candidate was Ralph E. Wilson, son of Herbert C. Wilson, editor of *Popular Astronomy*.[47] The father wrote JSP to say that Ralph would be returning in the summer from Chile, where he had been in charge of the Mills expedition of the Lick Observatory for the past five years and would be looking for suitable employment. His experience with Director Campbell and his extensive radial velocity work would have made him an ideal fit for the DAO. Another interesting possibility was Antonia Maury, a niece of the pioneer spectroscopist Henry Draper, and one of a group of women associated with Cannon at Harvard. She had a BA in physics and astronomy and demonstrated remarkable insight in analysing stellar spectra. Towards the end of 1918 she applied for work at the DAO. JSP would have been happy to hire her to analyse plates. He wrote to the Harvard director, Edward Pickering, "It just occurred to me on reading your letter that perhaps the difficulties [of paying her] might be overcome by your making application on her behalf for some specific research here, such as quantitative classification of the spectra we have."[48] JSP's suggestion did not materialize, as money available for such investigations in the United States could not be applied to research in Canada. Given Maury's independent spirit, it may have been for the best.[49]

As soon as the regular routine began at the observatory, JSP knew he would need a secretary. He wrote to the Civil Service Commission explaining why he preferred a woman for the job. Although women began filling the lower ranks of the civil service during the war, only men had ever been employed at the DO, so JSP's attitude was somewhat unusual.[50] He said, "Other things being equal, I am convinced that a young girl would most acceptably perform the work, as the experience of other observatories has shown that they more rapidly pick up the details of measuring and computational work, are more easily trained and adaptable, and are content to remain at routine work, where a man would chafe and want promotion. Moreover the salary is so small that no man would long be content, and would be continually worrying to get increases."[51] JSP got his way and hired Helena Keay in August 1918 as secretary at $1000 per annum, slightly more than a quarter of his salary.[52]

Whose Is the Biggest of Them All?

Plaskett had been calling the Victoria telescope the "world's largest" for a couple of years, but was he justified in doing so? There were a couple of competitors for the title. The oldest instrument that might have been in the running matched the DAO telescope in diameter. Astonishingly, "the Leviathan of Parsonstown" as it was known, was completed seventy years earlier in Ireland by the third Earl of Rosse at his own expense. This remarkable telescope reached stars far fainter than any previously observed and revealed for the first time the spiral structure of some nebulae. But its metallic mirror tarnished quickly, the instrument could only see a portion of the sky near the meridian, and Ireland's climate was a dreadful one for astronomy. By 1878, no one was making observations with it, and in 1908 it was dismantled.[53]

A much more viable contender was the long-planned 2.54 m reflector at the Mount Wilson Observatory, founded and directed by George Hale. His plan had begun to take shape after Los Angeles hardware magnate John Hooker donated funds allowing the huge 4 tonne glass disk to be ordered from the Saint-Gobain firm in France in 1906. The glass was delivered two years later to the observatory's optical shop in Pasadena, California, to be ground and polished there by the outstanding optician George Ritchey.[54] Though he initially considered the glass to be fatally flawed, he eventually used it after attempts to cast a replacement failed. In 1911, Andrew Carnegie himself assured the ultimate success of the endeavour with a gift of ten million dollars to his institution which owned and operated the observatory. Plaskett, we have noted, had seen Ritchey's work in progress in the spring of 1913, but many challenges ensued and years passed before the mirror was mounted in the tube and housed in the observatory. By November 1917, the telescope could be used visually but it was not until 23 December 1918 that Francis Pease secured the first spectrograms. Several more months elapsed before it was fully functional.[55] Financial problems and personal clashes were partly responsible for the delay, but a crucial factor was Ritchey's diversion to oversee the production of prisms and lenses for rangefinders urgently needed by the U.S. Army during the war.[56]

Thus, although the Mount Wilson 2.54 m telescope was larger than the DAO 1.83 m reflector, it became completely operational more than a year after the Victoria instrument. So Canada's claim to the world's largest working telescope was justified, if only for a short time. And for the remaining sixteen years of JSP's directorship, the Pacific coast facilities – Lick, Mount Wilson, and the DAO – were the world's pre-eminent research observatories.

The DAO telescope remained the second biggest in the world until the University of Toronto's David Dunlap Observatory edged it out by 5 cm in 1935. The DAO telescope was also reputed to be more than twice the size of that in any other national government observatory.[57] It would remain the largest government telescope until the Crimean Astrophysical Observatory inaugurated its 2.6 m reflector in 1961. For the federal government to have seen it to completion at a time of enormous financial sacrifice and horrendous human loss was almost unbelievable. It was quite extraordinary that the federal government had, to that point, invested in only two large capital projects in pure science, and both were observatories, the DO and DAO, which were undertaken only a decade apart and at considerable expense. Plaskett recognized that, even though Canada lagged behind older countries in the advancement of astronomical science by amateurs and in study at universities, the nation was well ahead of others in government support.[58]

While it was an amazing achievement for Canada, a country with fewer than eight million people, we should not over-exaggerate the government's generosity. The total cost of the observatory, $155 000, was spread out over a number of years and was less than 0.1 per cent of typical annual federal expenditures at the time, even excluding the costs of war.[59] As a point of comparison, the government appropriation in 1919 for the Honorary Advisory Council for Scientific and Industrial Research (the forerunner of the NRC) was $500 000 to $600 000, an amount its advocate, J.C. McLennan, described as "a mere bagatelle."[60] By observatory standards, the DAO was certainly a bargain, considering that the 2.5 m telescope and dome on Mount Wilson cost about $600 000, even with much of the job done "in house."[61] Furthermore, by 1924, Swasey estimated the cost for a telescope identical to the Victoria one would have been almost three times as much.[62]

The End of the War

When the war ended on 11 November 1918, there were over a quarter million Canadian troops overseas. For a couple of years, the government had realized that there could be problems with a mass demobilization, so it had recruited Henry Marshall Tory, the visionary first president of the University of Alberta, to plan and organize study courses for Canadian soldiers in England. This educational section of the army became known as the Khaki University.[63] Fifty thousand men took courses and about a thousand received credit for a year of regular college study. Harry Plaskett was among those benefiting from this enlightened opportunity. He chose to work with Alfred Fowler, whose laboratory at Imperial College, London, had impressed JSP when he was overseas in 1913.[64] Though Fowler had earned a reputation in solar spectroscopy under Norman Lockyer, he had become more of a physicist following the removal of astronomical equipment from London to Cambridge in 1911. Harry modestly recalled, many years later, "There was nothing dramatic about Fowler at all. Just honest, self critical, persistent work – the only kind that I can really admire ... He allowed me to take some laboratory spectra. I didn't do anything, I'm a damn poor experimentalist, and I just absorbed the atmosphere."[65] In these early postwar months, Harry enjoyed associating with a number of British astronomers. He lived with the family of the Astronomer Royal, Frank Dyson, at Greenwich, a privilege he may have enjoyed even earlier when he was on leave during active service. The Dysons' teenage daughter, Margaret, was quite taken with him. She later recalled that he "became almost like a son of the house."[66] Over Easter, Harry had three weeks at Cambridge, where Arthur Eddington was in charge of the observatory; in Oxford he visited Herbert Hall Turner; and in London he joined the Royal Astronomical Society (RAS) on 9 May 1919, seven months before his father.[67] (Coincidentally, in somewhat similar circumstances, Edwin Hubble was mingling with some of the same British astronomers that spring after serving overseas for a short time.[68] Hubble was proposed for fellowship in the RAS at the same meeting at which Harry was formally elected.)

In Canada, in the aftermath of the war, labour strife, a soaring cost of living, which undercut the security of the middle class, and the deadly influenza epidemic of 1919 placed a strain on the population at large.[69] Probably this general malaise did not help the festering relationship between the observatories in Ottawa and Victoria.[70]

One rather petty aspect of the continuing feud involved the RASC. After the society had elected Plaskett as its honorary president at the beginning of 1917 and the government had named Klotz as chief astronomer in October, trouble arose. The executive of the society's Ottawa

Centre, thinking that the country's chief astronomer should be the honorary president of the RASC, wrote in Klotz's name in place of Plaskett's on the ballot for 1918 and suggested that others should follow their lead.[71] It is hard not to think that they were lobbied by Klotz or his friends. Plaskett remained as honorary president for 1918 but evidently was asked by Chant to consider standing aside at the end of the year, if only to keep the peace. Plaskett shot back:

> I have not the slightest intention of stepping out in favor of Dr. Klotz ... Dr. King was Chief Astronomer of Canada by virtue of title and duties. [The] work of which he was the head was divided last year into four branches, each equal in rank in the Department of the Interior – the Dominion Astronomical Observatory, Dr. Klotz Director – the Dominion Astrophysical Observatory, Dr. Plaskett Director – The Geodetic Survey of Canada, Noel Ogilvie Superintendent – The International Boundary Commission, J.J. McArthur, Commissioner ... Since Dr. King's death and the division of his work, the title of Chief Astronomer is only an empty one and probably given to Dr. Klotz as a matter of course on his promotion from Assistant Chief Astronomer and Director.
>
> So far as astronomical work in Canada is concerned, I can leave it to the good sense of the Council [of the RASC] to judge which observatory, the one with a 15 inch or the one with a 72 inch telescope, is likely to accomplish the most, or which as soon as the latter is properly organized and staffed, will occupy the most prominent place in the astronomical world.[72]

In truth "honorary president of the RASC" was a rather empty title too, but it was not in JSP's nature to simply take pleasure in the fact that he had been chosen to succeed King in that capacity and had served for two years. Could he not have graciously let Klotz, a sixty-five-year-old man with many friends and a strong scientific standing in geophysics, have his turn? Once Plaskett was assured that the position was not a permanent one and that the RASC council understood his opinion about his status relative to Klotz, he simmered down and accepted the situation.[73] And the society's Victoria Centre came to the rescue and named him as *its* honorary president, starting in 1919.[74] As things turned out, Klotz did serve as honorary president of the national RASC from 1919 to 1921. He would probably have relished the insult to Plaskett more than the position, as he had no great interest in popularizing astronomy, which was above all the society's forte.[75]

The Observing Program

In the public's mind, astronomers are associated with beautiful pictures of spectacular objects in the far reaches of the universe but JSP, like most of his colleagues at other major observatories, took few direct photographs of celestial objects. Those that he did secure soon after operations began were intended for publicity or to display the superiority of the telescope's optics. He sent four photographs via Sir Frank Dyson to the Royal Astronomical Society in London in 1919 and to the Royal Photographic Society in London for their exhibition in 1922.[76] The RAS was quite taken with the separation of the stellar images at the centre of the globular cluster M13, and with two faint streaks just visible across the middle of the Ring Nebula, M57. Subsequently, they reproduced these two images and offered them for sale as part of their slide sets.[77] But pretty pictures were never a major part of Plaskett's plan.

Kapteyn's advice, quoted earlier, seems to have given Plaskett the idea of instituting a program to obtain the radial velocities of all stars in the *Preliminary General Catalogue* compiled in 1910 by Lewis Boss of the Dudley Observatory in Albany, New York. The Boss catalogue was chosen because it had the best data on proper motions of stars (i.e., perpendicular to the line of sight), data which could be combined with radial velocities and the stars' distances to yield information on the actual three-dimensional velocities of the stars in space. Proper motion was also useful as a rough indicator of distance; the most distant stars showed no perceptible proper motion, while the nearest ones tended to have the largest proper motions. (Compare what you would see if someone beside you wagged a finger with what you would see if someone on the far side of an auditorium did so.) Big surveys were part of JSP's vision of the usefulness of astronomical research, especially for a publicly funded institution such as the DAO.[78] As he replied to Kapteyn, "Dr. Hale in his letter emphasizes the desirability of opportunity for individual initiative and to avoid becoming a mere cataloguer. But at the same time, I think the greatest advances in Astronomy have been the direct result of the unselfish labours of cataloguers, and I believe a plan for observation here can be evolved which will combine the greatest usefulness to the science with sufficient scope and spur to individual initiative and enthusiasm."[79] The only catch was that others might already be involved in a similar endeavour. Indeed, the Lick Observatory had by this time determined the radial velocities of all stars brighter than fifth magnitude, and Mount Wilson with its

1.5 m telescope was beginning to extend its investigations to fainter stars. JSP consulted carefully with Walter Adams, the acting director there, to avoid duplicate effort.[80] They agreed on a division of labour and JSP was left with about 800 stars on his list – enough, he reckoned, for a program lasting four or five years.[81] His plan was to obtain six spectra of each star and somewhat more if the lines were poorly defined. Because random errors in radial velocity from a good plate were about a kilometre per second, he hoped to reduce the error to less than half that amount by averaging several results.[82] The idea was fine as long as the star did not have variable velocity, and to ensure that was not the case, observations had to be judiciously spaced. Naturally, Plaskett knew that in the course of surveying the radial velocities, some would be found to be variable. Such stars would be potential new spectroscopic binaries.

It took time, of course, to secure enough spectra to contemplate calculating binary orbits. The first to be completed was H.R. 8170 (number 8170 in the *Harvard Revised Catalog*) using plates taken from August to December 1918.[83] Such a brief interval was sufficient in this case because, it turned out, the period was unusually short – just 3.24 days. JSP wanted everyone to know how well the telescope and spectroscope were working, so when he published his orbit for H.R. 8170 in the *Journal of the RASC*, he asked Chant, as editor, to send him 150 reprints of the paper, saying "I need that many for my correspondents."[84]

In a series of four announcements in the RASC *Journal*, JSP published data on the stars that showed variable velocities. By September 1919 the DAO had seventy-five new suspected spectroscopic binaries to its credit, half of them discovered by JSP and half by Young.[85] There were advantages in publicizing results as they became available. For instance, Adams at Mount Wilson pointed out that he had already recognized three of the stars as possible spectroscopic binaries.[86] Investigators elsewhere found some of the DAO stars to be quite interesting for additional reasons, such as EW Lacertae, described as "the star with disappearing bright lines," and Struve 3050, a long-period visual binary whose two components JSP measured separately even though the angle between them was a mere two arcseconds (0.0006 degrees).[87] Eventually, in 1920, Plaskett included all his previous entries in a compilation he called "One Hundred Spectroscopic Binaries."[88]

In hindsight, the title cannot be taken at face value. JSP assumed that any star whose radial velocity changed by considerably more than his perceived margin of error was a spectroscopic binary. In fact fifty-five at most would now be classified this way. Unknown to Plaskett, there were reasons why the spectra of a few of his stars showed measurable, even periodic, changes in their radial velocities – some were single pulsating variables or rotating ellipsoids, for instance. But even taking all physical explanations into account, probably a third of the stars in his list were there because he mistakenly thought the range of radial velocities exceeded what he believed to be normal statistical scatter. He even went so far as to say "there can be no doubt of the reality of the variation in all the stars published."[89] A decade later, he would write, "There are many stars previously announced [not only at the DAO] as having variable velocity, which later knowledge and greater experience with these spectra seem to indicate as having constant velocity."[90] We can almost hear Klotz saying, "Just what I said about rushing into print!"

Another set of observations, unrelated to the program of Boss stars, was initiated the night before the observatory's opening ceremonies – the spectrograms of Nova Aquilae.[91] To ensure quick dissemination of his findings, JSP followed normal protocol for discoveries and sent a telegram to Pickering at Harvard describing the spectrum on 15 June and noting the shift in the main absorption lines in the nova's spectrum towards the blue, indicating approach.[92] He also pointed out the pair of very sharp unshifted lines – the ones labelled H and K by Fraunhofer in the solar spectrum a century earlier. In 1861, Gustav Kirchhoff had identified them with two lines in laboratory spectra at wavelengths of 396.8 and 393.3 nm respectively. He recognized that they were produced by ionized calcium (Ca+), an atom that, having lost one of its electrons, is left with a net positive charge. So Plaskett was able to explain the stationary lines in the nova's spectrum as an indication that the "the light from the star is shining through a relatively cooler calcium vapour which is undisturbed by the tremendous cataclysm occurring in the star."[93] It was foretaste of important investigations on the interstellar medium that he would carry out a few years later. From 10 June to 5 August 1918, Plaskett and Young photographed the nova's spectrum sixty times. From measurements of dozens of absorption lines, including an amazing nineteen consecutive lines in the Balmer series of hydrogen, Plaskett found the velocities of approach ranging from 1400 km/s on 10 June to 1750 km/s on 24 June. He thought these immense speeds were very hard to accept, just as he had after looking into Nova Geminorum in 1912, but they were in fact legitimate indications that the star was explosively shedding huge gaseous layers from its atmosphere. With the large light-gathering capability of the new telescope, he was able to follow the nova to a later stage in its development when its spectrum sprouted emission lines characteristic of the nebula growing around it. His extensive series of observations drew widespread attention. John Evershed, director of the Kodaikanal Observatory in India, commented that Plaskett's queries about why the dark absorption lines were not broadened like the bright emission lines could be answered by assuming the absorption originated in a part of the star that had become detached from the main body, "like solar eruptions, but on a vastly greater scale."[94] Frank Baxandall of the Solar Physics Observatory in Cambridge, England, used copies of spectrograms that JSP had sent him to identify a pair of erratically behaving absorption lines with nitrogen.[95]

Staffing and Administration

An observatory director, of course, has many duties besides research. Personnel management inevitably brings some friction, and JSP soon found that that was the case with Tom Hutchison, the man who had been providing engineering oversight during construction of the observatory and now was doing night work as an observing assistant along with mechanical and maintenance chores. JSP gave him full credit for the "very efficient assistance in the setting of the telescope and the guiding."[96] But to see if his duties were in line with what others expected of their employees, JSP wrote to Walter Adams at Mount Wilson and to Edwin Frost at Yerkes:

> We have an assistant, Mr. Hutchison, for the night observing. He is a first rate mechanic and electrician and is supposed in addition to keep the equipment in working condition. He is however, none too fond of work and thinks too much is expected of him in the way of night work and day duties. When he is on duty at night he is not expected to work the next day but I think when there is no night work he should be on duty in the day time for Civil Service hours, 9 to 5. He has Saturday, Saturday night and Sunday off but has to be on hand on Sunday night for observing. I want to do what is fair by him and if I could find out what is the practice … it would be a great help to me.[97]

Both astronomers replied with helpful information. In thanking Frost, Plaskett added:

> The difficulty in a Government institution is, as you can imagine, as soon as the average person gets a position in the service, they seem to imagine that they do not need to hustle around any more, their job and pay are secure. We had a great deal of trouble at Ottawa and I hoped to get away from it here, but human nature and civil service notions are much the same everywhere.[98]

Whatever difficulties there were, they were resolved; Hutchison remained on at least until JSP's retirement, "thoroughly competent in maintaining the telescope and equipment in good working order, and ... cheerfully and ably assisting the observers six nights a week." He also became a source of observatory lore. When a younger astronomer asked him some years later how JSP got his crooked forefinger, Hutchison, in his Irish brogue, told the story of the night when, "as we were closing up, we had some trouble with the shutters. The gears were emitting a horrible growl. We went up on the platform together to investigate. I threw the switch; the gears growled in the gloomy darkness and JSP said, 'I think the trouble is right in there ...' as he stuck his forefinger between the moving gears. You should have heard him yell!"[99]

For most of the telescope's first year of operation, JSP got by with Young as the only other astronomer on staff. The other Ottawa astrophysicists were surprisingly uninterested in moving to Victoria. John Cannon and Harold Parker, back from the war, both asked to be transferred to the Geodetic Survey. The one man from the DO whom JSP most wanted was Ed Harper. But as the observatories' historian, John Hodgson, pointed out,

> Harper was somewhat reluctant to transfer. He had a fine new house on Fairmont Avenue, a young family, and a chance to become number one in Ottawa astrophysics. On the other hand he acknowledged the opportunities for important work with the new telescope. Pressure was applied. The Deputy Minister received a letter from the Member for Esquimalt urging the transfer. "The letter was undoubtedly written by Dr. Plaskett, for the details about radial velocity work could have been written by none other than Plaskett." Harper felt that he should have an increase in salary and wanted to be named Assistant Director. Plaskett said that his salary was

> adequate and that there was no justification for an Assistant Director in a three-man establishment.[100]

Klotz tried his best to keep Harper in Ottawa, telling him that "his future prospects were more favorable here, reminding him of the return of Plaskett's son from overseas to the Observatory, & the prospect of Harper's eventual subordination to Harry Plaskett, also consideration should be given to his family, his wife & two girls, their isolation, their education &c."[101] Klotz's persuasive tactics may seem to refute his earlier anticipation of the happy day when Plaskett and his acolytes, Young and Harper, would be gone from Ottawa but perhaps Klotz now realized that Harper was his best hope for continuing an astrophysics program in Ottawa. At any rate, his arguments were to no avail. An agreement with the deputy minister that Harper should be paid a housing allowance of $50 a month clinched the deal and Harper commenced his duties at Victoria on 22 April 1919 – one grade above Young on the salary scale. (Harper had five more years' seniority, though Young had his PhD and Harper did not.)

Naturally, JSP was also counting on Harry joining the DAO staff. The only possible hitch was that his appointment might appear to be a case of blatant nepotism. The Civil Service Commission would have to approve his appointment, and JSP took steps to help his son get in on the basis of merit. He ensured that the advertisement posted on 30 January 1919 for "Observer, Astrophysical Observatory, Victoria, Salary $1700" listed qualifications matching Harry's exactly, stipulating that "candidates must be graduates in Arts of a recognized University and must have taken an honours course in Astronomy and Mathematics. They should have practical experience in observational and measuring work in some observatory."[102]

Harry's military service was also a strong point in his favour, and it did not hurt that William Roche, the minister of the interior whom JSP had publicly credited for approving the DAO project, was now chair of the commission handling applications. JSP corresponded with him about Harry and also got Frank Schlesinger and C.A. Chant to provide references.[103] Character referees were two Ottawa ministers (Rev. Canon Mackay, the first rector of All Saints, Sandy Hill, and Rev. Thomas J. Stiles of St Albans) and JSP's friend Dr Frank T. Shutt, Dominion chemist. All went well, and by April JSP reported to Schlesinger that "Harry has been appointed but is having such a profitable time in London with Fowler, Dyson, Stratton, etc., that I hardly expect he will be home before

July or August. Naturally we are longing to see him but now that the war is over we can wait without anxiety and feel that he is being repaid for some of the very strenuous times he had."[104]

Harry got back to Canada in July 1919. At last he was able to be with Edith Smith, the girl from next door on Fairmont Avenue, Ottawa, who had waited anxiously for his return. Of necessity, their courtship soon became a distant one when Harry reported for duty at the DAO in October.[105] For his $1700 annual salary, he was allotted 2 nights per week of observing, dusk to dawn (7 to 14 hours depending on the season), while during a typical day he spent 4 hours measuring, reducing, and discussing spectrograms, and another 3 on other duties, 16–25 days per month, depending on how much night observing was done.[106] He and Harper both had to make their way from their homes in town to the foot of Observatory Hill by electric tram and then walk the 2.5 km uphill to the dome. It was an inconvenient arrangement at the best of times and hopeless late at night when the trams did not run. The astronomer on duty, not knowing if a change in the weather would necessitate opening up or closing down, was stuck on site. Young fared better, as he had a car, for which he received an allowance.

In addition to his permanent staff – Harry, Harper, and Young – JSP had occasional temporary assistants. A professor from the University of Washington in Seattle, Samuel Boothroyd, who had met W.F. King while employed on the Alaska Boundary survey some years earlier, was eager to do some astronomical research.[107] He came to the DAO for the summer of 1919 as a volunteer assistant, an arrangement that released JSP from night duty for a while and turned out so well that it would be repeated in future summers. Though Plaskett had discovered the binary nature of the star H.R. 8803 in 1918, he gave Boothroyd the opportunity to obtain more spectrograms and eventually to publish an orbit.[108]

Having handled the telescope for a year, JSP reported that it was even more accurate, satisfactory, and convenient in operation than had been anticipated:

> The quality of the optical parts is well shown in the results of the Hartmann tests, in the short exposures required for the spectrograph, and in the remarkable smallness and crispness of the star images in the direct photographs ... I have yet to find any part of the design where any improvement could be suggested. In making exposures on star spectra the average time ... from the end of the exposure on one star to the beginning

7.4 The completed DAO telescope, with JSP standing by the electrical panel. The "bridge" that allowed access to the Newtonian focus near the upper end of the tube is clearly shown in the photo. (Courtesy National Research Council of Canada, DAO hist-161)

> of the next is less than three minutes and if the stars are not very far apart is generally only two minutes. I do not believe that record is excelled by even very small telescopes and, when we consider that the moving parts weigh 45 tons, the ease of handling is a remarkable evidence of the perfection of design and workmanship.[109]

In spite of this glowing report, some final tweaking was desirable. When JSP had gone to Pittsburgh to take part in testing the telescope's optics in 1918, it had not been feasible to do a full Hartmann test on the combined primary and secondary mirrors. When he did get an opportunity to do so in 1919, he found little cause for complaint, though he did recognize that the secondary (Cassegrain) mirror could be improved with refiguring. The contract with Brashear had provided for this eventuality, and so McDowell, the company's optician, came to Victoria in August as part of a holiday. In just three sessions on two nights, he worked his magic and the small defects were reduced to an imperceptible level.[110]

Though Plaskett had the finest observational equipment and an increased staff, he had no adequate workspace for them. The original vision for the DAO had called for an office building but it was quashed because the government was strapped for funds as a result of war debts. So everyone had to make do with the space partitioned off in the dome underneath the observing floor. There was, of course, a darkroom so the plates could be developed right after they were exposed at the telescope. And there was an office for the director, another room with the library and four desks for astronomers, a small room for the secretary, and a lavatory. This whole area was noisy because of the sheet metal cladding of the building and the windows were too high to permit an outside view; all in all it was a very unsatisfactory, confining arrangement.[111] Presumably the heating needed in the winter was not good for observing, with the telescope just above. JSP spared no effort to rectify the situation but had to continue living with it for years.

A New Direction

Plaskett had just published the first spectroscopic binary orbit based on DAO data when an exciting development diverted his attention in the spring of 1919. In February he had written to his friend the theoretical astronomer H.N. Russell at Princeton, saying that with Harper soon joining the staff, he hoped to expand his area of research a little. He wondered about eclipsing variables and asked if Russell had any suggestions. Russell was thrilled, saying "Your letter [is] … a bonanza. To have someone offer to observe some of the things I have wanted to see observed for several years past is remarkably satisfactory – especially when the observer is as competent for the job as you are!"[112] He appended a list of seventeen eclipsing binaries whose photometric orbits had previously been published. These were pairs of stars in orbit around one another and, because the plane of the orbit happened to be almost edge-on, the combined brightness varied with periodic regularity as one star eclipsed the other. Even though the components were too close to each other to be separately resolved, by examining the way their combined light varied, the system's orbital parameters, including inclination, could be calculated.[113] As we have already noted, the inclination was a crucial orbital element that could not be found from spectroscopic data alone; however, by using such data along with the photometric results, actual physical data such as the ratio of the two stars' masses, radii, and luminosities could be figured out. The real payoff came for those systems having double lines in their spectra – one set of lines for each star in the pair. In such rare and valuable cases, astronomers could calculate the stars' individual masses, radii, densities, and their actual separation in kilometres. These were the reasons why Russell was so pleased to have Plaskett study them.[114] Such data were grist for the mill of any theoretician trying to explain how the laws of physics could maintain a huge ball of gas and produce prodigious amounts of energy over eons. About the same time, the English theoretical astrophysicist Arthur Eddington also saw the great need for data on stellar diameters in his investigation of the internal constitution of stars, "in order to make sure that our theoretical deductions are starting on the right lines."[115]

Plaskett immediately began looking into the matter and started observing one of the stars on Russell's list, U Ophiuchi, on 19 March. Observations on two others, RS Vulpeculae and TW Draconis, soon followed, and by 6 August 1919 JSP had calculated the orbits of all three as well as their physical properties.[116] His findings have stood the test of time fairly well in spite of tremendous advances in instrumentation and theory. The following list shows the masses he found, with currently accepted values in parentheses, all in terms of the Sun's mass:

U Oph1 = 5.36 (4.93)	RS Vul2 = 1.69 (1.76)
U Oph 2 = 4.71 (4.56)	TW Dra1 = 3.21 (2.16)
RS Vul1 = 5.40 (6.59)	TW Dra2 = 1.19 (0.93)[117]

Over the next three years he completed a similar analysis of five more systems from Russell's list, making an impressive contribution to what was known about the stars. An instructive table, published by Robert Aitken in 1935, listed twenty-two binary systems for which complete spectrographic and photometric observations were available.[118] Plaskett's research published between 1919 and 1922 provided more than a quarter of the information – more than that of any other individual.

Most significantly, Eddington used JSP's data in establishing the important relation between stellar mass and absolute luminosity (not to be confused with apparent luminosity, which depends on how far away the star is).[119] The conclusion – that (in the great majority of cases) the more luminous a star is the more massive it must be – was of great importance in relation to the question of stellar evolution.[120] Russell's theory had stars contracting from cool, red, puffed-up giants of type M, through the spectral sequence K, G, F, A, B, to the hottest O-class until they reached the density of a liquid or solid and then cooling down through the main spectral sequence in reverse order to end up as red M-type dwarfs. But, as we shall see, Plaskett found O-stars with a density that was only 1 per cent of that of water. His data suggested that if stars evolved according to Russell's ideas, they would have had to gain and lose mass, a highly unlikely scenario. Plaskett did not venture to say that his data killed Russell's theory, but it was certainly a nail in the coffin. Better explanations evolved in which the stars kept their initial mass for most of their lifetime. Those in pairs or larger clusters, as they were presumably formed at the same time, would ultimately shed light on how stars of different luminosity and mass evolve.

The diversion to eclipsing binaries did not seriously slow down the regular program of Boss stars, in which all the DAO staff participated. JSP allotted the stars fairly, taking a quarter of them himself, and giving each of the other astronomers the responsibility for observing, measuring, and reducing the spectra assigned to him. As he noted, "This added interest to the work, prevented confusion in credit for binaries or other interesting objects discovered and gave the observers greater incentive to effectively follow up their stars."[121] Hutchison kept the telescope in good order and assisted with most of the observing; the secretary, Helena Keay, looked after record keeping and proofreading,

	Star	P_1	*SP*	*rb*	*rf*	*mb*	*mf*	*pb*	*pf*	Dist. bet. centers	Authority
		d.		$\odot = 1$		$\odot = 1$		$\odot = 1$		10^6 *km* ×	
1	*Boss* 46	3.52	B0	23.8	15.5	36.3	33.8	0.0027	0.0091	27.92	Pearce, *Publ. D.A.O.* **3**, 275, 1926
2	V *Puppis*	1.45	B1	8.45	7.70	19.4	19.4	0.042	0.055	8.83	Shapley, *Ap. J.* **38**, 169, 1913
3	Y *Cygni*	3.00	B2	4.7	4.7	17.4	17.6	0.16	0.16	19.90	Redman, *Publ. D.A.O.* **4**, 341, 1930
4	AG *Persei*	2.03	B3	3.75	2.62	5.18	4.57	0.10	0.25	9.92	Huffer, *Publ.* Washburn 0, 15, 192, 1931
5	U *Cor. Bor.*	3.45	B3	2.90	4.74	4.27	1.63	0.175	0.015	12.08	Plaskett, *Publ. D.A.O.* **1**, 187, 1921
6	u *Herculis*	2.05	B3	4.56	5.35	7.66	2.93	0.095	0.022	10.29	Shapley, *Ap. J.* **38**, 169, 1913
7	Z *Vulp.*	2.45	B3	4.23	4.46	5.24	2.36	0.085	0.033	10.47	Plaskett, *Publ. D.A.O.* **1**, 251, 1920
8	σ *Aquilae*	1.95	B3	3.56	3.56	6.19	5.14	0.15	0.12	10.22	Wylie, *Ap. J.* **56**, 232, 1922
9	TT *Aurigae*	1.33	B5	4.5	4.0	6.7	5.3	0.11	0.12	8.10	Joy and Sitterly, *Ap. J.* **73**, 77, 1931
10	U *Ophiuchi*	1.68	B8	3.23	3.23	5.36	4.71	0.18	0.16	8.92	Plaskett, *Publ. D.A.O.* **1**, 138, 1919
11	RS *Vulp.*	4.48	B8	2.05	10.25	5.40	1.69	0.63	0.0016	15.30	Plaskett, *Publ. D.A.O.* **1**, 141, 1919
12	TV *Cass.*	1.81	B9	2.50	2.83	1.83	1.01	0.118	0.044	6.16	Plaskett, *Publ. D.A.O.* **2**, 141, 1922
13	U *Sagittae*	3.38	B9	3.4	5.7	6.7	2.0	0.171	0.011	13.63	Joy, *Ap. J.* **71**, 336, 1929
14	β *Aurigae*	3.96	A0p	2.81	2.81	2.40	2.36	0.11	0.11	12.31	Shapley, *Ap. J.* **38**, 169, 1913
15	RX *Herculis*	1.78	A0	1.54	1.38	0.89	0.89	0.25	0.34	5.20	Shapley, *Ap. J.* **40**, 399, 1914
16	TX *Herculis*	2.06	A5	1.33	1.33	2.04	1.77	0.87	0.75	7.41	Plaskett, *Publ. D.A.O.* **1**, 207, 1920
17	S *Antliae*	0.648	F0	1.67	1.29	0.75	0.42	0.31	0.38	2.30	Joy, *Ap. J.* **64**, 287, 1926
18	Z *Herculis*	3.99	F5p	1.77	3.29	1.6	1.3	0.3	0.04	10.52	Adams & Joy, *Ap. J.* **49**, 192, 1919
19	RS *Can. Ven.*	4.80	F8	1.6	5.3	1.85	1.71	0.45	0.012	12.78	Joy, *Ap. J.* **72**, 41, 1930
20	W *Urs. Maj.*	0.334	G	0.78	0.78	0.69	0.49	2.8	1.9	1.50	Adams & Joy, *Ap. J.* **49**, 189, 1919
21	RT *Lacertae*	5.07	G5	5.0	5.0	1.0	1.9	0.01	0.02	12.45	Joy, *Ap. J.* **74**, 101, 1931
22	α *Gem.* c	0.814	M1e	0.76	0.68	0.63	0.57	1.4	1.8	2.70	Joy and Sanford, *Ap. J.* **64**, 250, 1926

7.5 Table of eclipsing binary stars published in 1935 by R.G. Aitken. The first two columns identify the star, column 3 gives the period (i.e., the time in days for one star to orbit the other), and column 4 shows the spectral class. The next six columns give physical information about each component of the binary system – the brighter (b) and fainter (f), in comparison to the Sun, where *r* is the radius, *m* is the mass, and *p* is the density. The last two columns give the separation of the two components, measured in millions of kilometers, and finally, the source of the information. (R.G. Aitken, *The Binary Stars* (1935), 200) assistance that JSP did not fail to acknowledge. Impressive statistics for 1920 showed that 2119 spectra were obtained on 212 nights, each night averaging six hours of clear conditions.[122]

Societies and Meetings

In earlier chapters we saw that JSP had gone to important meetings of the International Union for Co-operation in Solar Research in 1910 at Mount Wilson and in 1913 in Bonn, Germany. The Solar Union's intention to hold triennial assemblies was thwarted by the war; the next meeting was scheduled for Brussels in July 1919, as part of a new umbrella organization called the International Council of Scientific Unions.[123] It was at this assembly that formal approval was given to a desire, first voiced in 1910, to broaden the Solar Union's mandate to include astrophysics generally and to work as part of a new group, the International Astronomical Union (IAU), with the Solar Union itself formally disbanding. The IAU charter specified that each of the adhering countries have a national committee to promote and coordinate the study of astronomy in their respective countries, especially in relation to international requirements. Earlier in 1919, the U.S. national committee had chosen Plaskett as one of three members of the IAU Radial Velocity Commission and one of nine members on the Commission on the Spectroscopic Classification of Stars.[124] He was the only non-American to be appointed by the U.S. astronomers.

The reasons why Canada did not appoint him and why he did not go to the assembly in Brussels are a bit convoluted. Although Canada was one of the seven original members of the IAU approved in 1919, the country had no apparent mechanism to appoint its delegates. The IAU charter required that each national committee "be formed under the responsibility of the principal Academy of the country concerned, or of its National Research Council, or of some other national institution or association of institutions, or of its Government."[125] In the United States the AAS effectively took on that role but there was some confusion at first as to what body would choose Canada's two official delegates. Would it be the new Honorary Advisory Committee (later the NRC)? The RASC? The Royal Society of Canada? When Plaskett attended meetings of the RSC in May 1919, he got the impression that nothing was happening with respect to appointments to Canada's national committee and that it was too late to plan a trip to Brussels in any case.[126] In typical Canadian fashion, a compromise was eventually reached whereby the DO chose five representatives, the DAO chose three, the RASC three, and McGill University one to the national committee, which then selected two of its number as voting delegates at the IAU.[127]

Austerity measures in the postwar period made the major expense for international travel or even trips across the continent hard to justify. Plaskett did manage to get to the RSC meetings each May in Ottawa (except 1918), perhaps combining them with obligatory consultations with his DO counterpart, Klotz, and with government officials. In 1919 he delivered three papers to the RSC – two concerning the performance of the new telescope and its mirror and the other on the eclipsing variable U Ophiuchi. He missed all the American Astronomical Society sessions held in the east during the period 1918–21, though he prepared papers to be read by others in attendance. His absence from the AAS was offset by his presence at meetings in the western United States. In June 1919 the Pacific Division of the American Association for the Advancement of Science held its annual assembly in Pasadena, as usual in conjunction with several other organizations, including the Astronomical Society of the Pacific (ASP), where Plaskett spoke on the same topics he had just addressed in Ottawa. The highlight of the meeting was the unveiling of the 2.5 m telescope on Mount Wilson, about to surpass the DAO as the world's largest. To make their trip into an extended holiday, JSP and Reba took an eleven-day cruise to California.[128] Reba was greatly taken with Pasadena on this her first time there and enjoyed visiting the Hales at their home.[129]

The following June there was the usual joint meeting, this time in Seattle, and Plaskett became a member of the ASP on this occasion. Boothroyd, who was about to take up a second summer at the DAO, was the main organizer of the event and, with JSP's concurrence, included an optional trip to Victoria in the plans. Though general relativity was intended to be a key topic, it did not turn out that way. The Lick results from the Goldendale eclipse two years earlier were still not available, although Joseph Moore, an astronomer at that observatory, did speak briefly on the topic.[130] The four DAO astronomers all presented papers on their own research. JSP's was once again a repeat of an RSC paper of a month earlier. It was on the radii, masses, densities, and distance of the stars in an eclipsing system, this time U Coronae Borealis. He hurried back to Victoria to welcome the conference delegates for their planned excursion.[131] The DAO astronomers met their guests at the dock and drove them around the city and up to the observatory, where all enjoyed a picnic supper under the trees before an evening at the telescope. Boothroyd stayed on for six more weeks and produced a paper on the orbit of spectroscopic binary Boss 4602, another example of many discovered at the DAO in the course of the regular program of finding radial velocities.[132]

Following the Seattle/DAO meetings, the Plasketts repeated their trip of the previous summer, mainly as a holiday but giving JSP the opportunity to exchange ideas with the Mount Wilson astronomers.[133] Again in 1921, for the third year in a row, the Plasketts sailed to California, where JSP delivered an endowed lecture on the dimensions of stars to the ASP in San Francisco on 11 November, followed by more discussions at Lick and at the Mount Wilson office in Pasadena.[134] The Plasketts had another lovely visit with the Hales at their home, and JSP was touched that George Hale, who was not well, had travelled to San Francisco from Pasadena to hear him speak.[135]

8
Challenges and Rewards, 1921–1923

Ongoing friction between the observatories in Ottawa and Victoria escalated in 1920 after Arthur Meighen, formerly minister of the interior, succeeded Robert Borden as prime minister for the remaining year of the Conservative mandate. The new minister, Sir James Lougheed, called for a reorganization of his department to consolidate a number of branches.[1] In particular he supported long-standing Deputy Minister W.W. Cory, who did not like doing business with a head (Plaskett) 4600 km away. They formed an advisory technical board, composed of branch directors, which in due course recommended that the Dominion Astrophysical Observatory be placed under the direction of the Dominion Observatory (that is, under Otto Klotz). Plaskett was livid and insisted that any decision should be deferred until he could present his case in person. Because he was planning to be in Ottawa for the meetings of the Royal Society of Canada (RSC) in May 1921, the board agreed to schedule a special sitting for that time. As Klotz recorded in his diary for 25 May,

> [Plaskett] came fortified with a very enlarged photograph of the interior of the Victoria Observatory, many reports – all in fact so far issued. When the meeting opened the Chairman read the letter of the Deputy calling the meeting, also three letters – from Prof. G.E. Hale of Mount Wilson, Prof. S.A. Mitchell of Leander McCormack Obs., & one from Prof. J.C. McLennan of Toronto – written at the request of Plaskett. Hale's was sensible without palaver like Mitchell's and especially McLennan's. Plaskett then made a half hour's speech, full of eulogy of himself & his work.[2]

The board dealt fairly with Plaskett, even allowing him to vote on the question, but his objections were to no avail. The resolution was upheld by a vote of 9 to 2. According to Klotz, Plaskett "was crimson & perspiring with rage." A few days later, however, the two men met and came to an understanding, both signing a memorandum for the deputy minister. They agreed that the administrative functions, apart from the scientific investigations, would be placed in the hands of the director in Ottawa. Klotz gloated in his diary that "all official correspondence will now be through me – all accounts, all estimates, all requisitions of whatsoever nature go through my hands. In short I have the administration of the Victoria observatory, while he [JSP] plans the scientific work & sees to its carrying out."[3] While still in Ottawa, Plaskett had his first experience with the new arrangement. He discussed with Klotz various projects that he would like to have carried out. One item Klotz could not recommend, "with the present stringency of money," was a tennis court for the DAO. It would be hard to fault Klotz on that score.

In spite of the latent hostility, the two men were remarkably civil to one another whenever they met. Klotz made an official trip to Victoria that same summer, primarily to seek improvements in the time signal service radioed to shipping in the Pacific.[4] He thought the city's promotion of the observatory as a tourist attraction was rather derogatory. He was not impressed to see a taxi company advertising a "Saturday night special" to the observatory including "free lecture and observations." Not that he had to take a taxi himself – Plaskett came in to town to get him and drive him out to the DAO. The next morning the two men had a good talk on a wide range of administrative issues and then went over to Plaskett's house at noon, where, Klotz says, "a light lunch was served; there were only Dr. Plaskett, Mrs. Plaskett, their boy 'Laddie' (a little off) and myself. Since leaving Ottawa, Plaskett has learned to smoke cigarettes which he produced after luncheon. We met in the Observatory Hill road several automobiles with visitors to the Observatory. After luncheon we climbed up stairs to see the big 72-inch reflector, which I saw for the first time." He also saw "probably a dozen people, men, women & children, wandering about in the dome without guidance. This seemed rather odd to me. They come and go without let or hindrance."[5]

It was the only time that Klotz would see the completed DAO. Emphasizing that he had the best interests of the observatory at heart, he wrote to a colleague,

> In the latter part of July I paid my first official visit to the Victoria Observatory as its administrator, and things are moving very smoothly, and

> I think that Dr. Plaskett finds that it [the new administrative arrangement] is of great advantage to his Observatory. I recently secured some fifty-five acres additional land for the Observatory to provide for all times against encroachment. A microphotometer was also obtained and I had Mr. Harper sent to the San Francisco meeting and also on a visit to Mt. Wilson. Being at headquarters here and in touch with the Department more can be accomplished than by one who is 3000 miles away.[6]

After he returned to Ottawa Klotz even urged the deputy minister to make provision for a DAO office building (though not residences for the astronomers).[7]

Observatory Research

The microphotometer to which Klotz referred was used mainly by Harry Plaskett to make carefully calibrated measurements of stellar spectrograms. He used Planck's theoretical distribution of radiation with wavelength to confirm that O and B types were the hottest stars but rightly found that one of them (Gamma Cassiopeiae) had a spectral distribution showing significant departures from the theory. Harry's research brought a detailed and theoretical basis to his father's observations of the hottest stars and was described as "painstaking" and "brilliant."[8]

JSP wrapped up his original observing program of photographing the spectra of northern hemisphere stars between fifth and ninth magnitude (fainter than those he had been able to record in Ottawa) in July 1921, well ahead of schedule. The results appeared as a large table of the radial velocities in the second volume of the *Publications of the DAO*. With similar commitments at Mount Wilson and Lick, the radial velocities of all the stars in the Boss catalogue had been determined. Such a complete data set was the ticket for future investigations of the motion of stars in the Milky Way. Plaskett remembered the role Jacobus Kapteyn had played in motivating him to undertake the program and sent the DAO data to him in Holland even before it appeared in print. Kapteyn's young student Jan Oort was grateful to be able to use this pre-published material for a paper he completed in 1922, the first of many important contributions he would make to the understanding of the motion of stars in our home galaxy.[9]

Naturally, the data collected suggested many further areas to investigate, especially for the stars not making the final cut. Indeed, a couple of hundred of the original eight hundred on the list proved unsuitable for the catalogue, mostly because their radial velocities were not constant.[10] This variability, when sufficiently large, made the stars candidates for further study as potential spectroscopic binaries; for the most part, though, Plaskett left the orbit calculations to others.

Sometimes these others were visiting professor-assistants – such as Samuel Boothroyd, whom we met in the previous chapter. John W. Campbell came from the University of Alberta in the summers of 1921 and 1922 and Daniel Buchanan of the University of British Columbia joined him in the summer of 1922.[11] Both men were mathematicians with PhDs from the University of Chicago. They were interested in computing orbits but were willing to get some practical hands-on experience as well. The extra help was all to the good, as JSP had a couple of setbacks that laid him low for a while and he also had to manage without Reynold Young for several months, for reasons to be explained shortly.

Late in 1921 and again in 1922, Plaskett suffered serious falls from a ladder – an easy thing to do while observing in the dark.[12] The second tumble came about when his glasses fell from his nose onto the generator supplying power to the telescope, causing an electric arc across the 220-volt comparison switch and knocking him off the ladder onto the steel floor beneath.[13] He was totally disabled for four weeks and partially for five more but used the time to write a thirty-seven-page booklet about the observatory, intended mainly for visitors.[14] Its lasting value can be found in the concluding paragraphs, where Plaskett discussed the economic and "elevating influence" of astronomy and pure science in general. Another blow came on 27 March 1922, when JSP learned that his eighty-one-year-old mother, Annie, had died of myocarditis and nephritis.[15] For the previous fifteen years she had been living with her only daughter, Josephine (Fraser) at Osoyoos, BC. JSP attended the funeral at St Saviour's Church in Penticton, where Annie was buried.[16] She left an estate of only $950 to her seven surviving children.[17]

But as the cycles of nature follow their inevitable course, the sad departure of the old generation is paralleled with happy new arrivals. Harry had married Edith Smith in Ottawa on 4 January 1921 and they moved to Victoria in February, where Harry was supremely happy, telling Annie Cannon that he "lived in the seventh heaven of bliss."[18] On 9 June 1922, they had their first child, Barbara Rochester.[19]

Starting in the summer of 1922, Young was away from the observatory. His departure originated with a telegram sent to him by his old University of Toronto professor, C.A. Chant, who was laying plans to observe the solar eclipse in Australia later that year.[20] Chant's intention was to verify the minuscule deflection in apparent position of stars predicted by Einstein's general theory of relativity just as the Lick astronomers had attempted (unsuccessfully) in 1918. Even though two British teams had verified the deflection of starlight during the 1919 eclipse to the satisfaction of some, there were still plenty of scientific sceptics. Further tests were clearly necessary and the eclipse of 21 September 1922 would provide an opportunity to conduct them. Chant, with absolutely no experience in making the very fine observations and measurements required, blithely planned to do so.[21] Although he saw the eclipse as an opportunity to make a major scientific contribution, he was probably even more aware that the press would latch on to an appealing story of Canadian astronomers participating in such a crucial experiment, and the publicity would boost his chances for a large telescope at the university.[22] On his request, his old friend W.W. Campbell allowed him (and others that he could bring along) to join the Lick team at the remote site of Wallal on the west coast of Australia. There the prospects for clear skies were very good and the eclipse would be seen high in the sky. Campbell even supervised the construction of special lenses that Chant required and arranged for comparison plates to be taken months in advance. Moreover, Chant and his group would simply piggy-back on all the arrangements that Campbell made for transportation in Australia.

As Chant was still the sole astronomer at the University of Toronto, he desperately needed some expert assistance in making the observations and measuring the plates afterwards. He had written to Yerkes to see if the director there could recommend anyone for an astronomy position at the university. Frost replied that he thought it would be desirable to look for someone from Canada, apparently giving Chant the impetus to try to recruit Young.[23] Barely a month before sailing for Australia, Chant telegraphed Young to see if he was interested in going. Young was enthusiastic, though Plaskett was reluctant to let him go, realizing that several other parties were attacking the same problem. In the end, JSP and the Department of the Interior did agree to excuse Young from his duties at the DAO for several months, and so the Chants – the professor, his wife, and daughter – arrived in Victoria to meet up with Young in June 1922.[24] The Harpers invited the Chants to stay at their house, and the Plasketts drove them around and had them for dinner.[25] Chant had his first look at the DAO telescope and viewed Jupiter and Saturn with it using a 600-power eyepiece. The RASC Victoria Centre arranged a luncheon at which he spoke, in a general way, about Einstein's theory of relativity and the observations needed to test its validity.[26]

The great adventure took place. The astronomers were blessed with clear skies in Australia, and Young returned to Victoria in November to measure the plates that he and Chant had secured. Chant told the story of their "experience of a lifetime" in a four-part illustrated series in the RASC *Journal*.[27]

The RASC and the Public

Though Chant will always be thought of as the great popularizer in Canadian astronomy of the first half of the twentieth century, Plaskett also knew the importance of public outreach and seemed to enjoy any opportunity to promote astronomy, the work of the DAO, and his own research. His roster of engagements, listed in Appendix B, shows that he spoke to associations holding conventions in Victoria, to church groups, and to service organizations including the YWCA. He may have been encouraged to speak to the "Y" by Reba, who was very active in the organization.[28] Naturally, he spoke to the Victoria Centre of the RASC at least once each year, often more frequently, and went to their meetings regularly.[29] His presentations were well received and sometimes widely noticed, especially when he ventured into more sensational subjects.[30] In one instance, a number of newspapers (via Associated Press) picked up on a lecture in which he rightly dismissed as nonsense the idea that a lineup of planets would cause violent disturbances on Earth.[31] Another talk he gave, to the local Rotary Club, was cited in many American dailies under headlines such as "Feels Stars Will Supply Energy for the Entire World." Plaskett was quoted as saying, "Our stores of coal and oil are rapidly being depleted and in 200 or 300 years they will be exhausted, if not before. Long before that we will be seriously seeking a means of obtaining energy from the stars. We know that there is untold energy in atoms of matter and that the stars send out tremendous energy, sufficient to meet our needs for uncounted aeons. So far we have been unable to harness that energy but such a process will come in time."[32] The media seemed to think that Plaskett was saying that the actual energy from the stars would be collected rather than that energy would someday be produced on Earth in the same way that it is in stars. Of course JSP had the story right. In fact, several months earlier he had concluded his DAO booklet by saying that "the enormous supply of energy, which has been radiated into space for aeons of time from these bodies, can only be maintained undiminished by the energy released by the transformation of atoms in the interior of stars, where conditions of temperature and pressure prevail at present unattainable in terrestrial laboratories."[33] That process, which we now call fusion, had been described by Arthur Eddington as early as 1920.[34]

Occasionally JSP was able to speak to RASC centres in other parts of Canada, usually when he was attending professional conferences. When he came east in May 1921, he took the opportunity to deliver a talk entitled "The Work of the DAO" to the Toronto Centre and "Modern Ideas of the Universe" to the Montreal chapter. In the latter city, he told the audience of nearly 150 that within the previous two or three years the dimensions of the stellar universe had been magnified about twenty times, mainly due to the research on globular clusters carried out by Harlow Shapley at Mount Wilson.[35] This trip east, which lasted a month, also gave JSP and Reba a chance to visit many friends.[36]

Though there was no RASC chapter in Vancouver until 1931, Plaskett was often asked to speak in that city, just a few hours away from Victoria by ferry. Different groups met at the University of British Columbia, such as the BC Academy of Sciences, which he spoke to on 27 February 1922 and the Vancouver Institute, a free public education forum, which he addressed on five occasions between 1919 and 1933.[37]

Even Plaskett's papers in the RASC *Journal* after 1921 were written with a broader audience in mind. Initially he had been very eager to use the *Journal* for professional papers and indeed he often did so while he was still working in Ottawa. He hoped to continue the practice in Victoria, writing in 1918 to Chant, the founding editor, "I want to publish practically everything of interest from the Observatory in the Journal and I presume this will soon make it indispensable for all astronomers."[38] Although it is true that during Plaskett's lifetime almost every astronomer in Canada belonged to the RASC and that most of the professionals at one time or another led the organization, its membership was composed overwhelmingly of amateurs – those keen on astronomy, but of limited expertise. So the *Journal* could not be aimed primarily at a professional readership and Plaskett soon accepted that. After 1921 he was choosing other places for his research and sending only non-technical articles to Chant.

That left the only Canadian venue for JSP's scholarly papers as the elite RSC, but it met only once each year and published only some of the work presented at these meetings in its *Transactions*. Though JSP continued to attend these meetings and make presentations each May, he may have felt a particular obligation to do so as vice-president (1922–3) and as president (1923–4) of the society's section III (the division for astronomers, mathematicians, physicists, and chemists).

More Frustration

While Klotz found the administrative arrangement between Ottawa and Victoria was working well, Plaskett was less than pleased. By the summer of 1922 he was once more fulminating against it. He appealed to his old ally, William Roche, who had supported the observatory project and Harry's appointment to the staff. Plaskett wrote him:

> You of course remember clearly that after Dr. King's death and the establishing of this observatory you, as Minister of the Interior, appointed me as its director and gave it an organization independent of the observatory at Ottawa, as a separate branch of the Department.
>
> This condition of affairs worked admirably and this observatory under its independent control has already established an enviable record of scientific work of the highest caliber and of hitherto unrivalled quantity and is doing a great deal towards placing Canada on the scientific map of the world.
>
> Unfortunately, however, this independent status of the observatory was changed about a year ago in a reorganization of some branches of the Department and Dr. Klotz director of the observatory at Ottawa, against my protest, was given a certain amount of jurisdiction over the administration of this institution. You must realize that it is absolutely impossible, at a distance of 3000 miles, for such an arrangement to work effectively, that it is likely to result in a loss of efficiency and morale. Further there is a danger of the work of the observatory being cramped or belittled if the director of another observatory, who may perhaps consider this a rival institution, has any jurisdiction over the work or administration.[39]

Roche, as head of the Civil Service Commission, couldn't overrule a policy of the new Liberal government under Prime Minister William Lyon Mackenzie King. All that JSP could realistically hope for was to forestall any permanent changes to the job descriptions and classifications for which the commission might be responsible.

A further glimpse of Plaskett's opinion of the Ottawa bureaucracy comes in a letter he wrote in January 1923 to Harlow Shapley, now Pickering's successor as director at Harvard College Observatory. Shapley had lectured at the Montreal, Ottawa, and Toronto centres of the RASC on a tour the previous November and had prefaced his Ottawa speech with a tribute to Canada. He said the government of no other country was so effectively and so successfully carrying on the work of astronomy as Canada.[40] Plaskett wrote thanking Shapley for his kind remarks and added, "When one depends upon a government department whose head officials know absolutely nothing and care little about astronomical research ... unsolicited favourable comments such as yours from so prominent an astronomer are a great help."[41] Clearly, Plaskett had become used to dealing with Meighen and Roche and was not too pleased with the new Minister of the Interior, Charles Stewart, who did not share their advantage of a university education (or their political affiliation).

JSP's negative feelings notwithstanding, Mackenzie King's Liberals accepted the urgent need for office space at the DAO. The problem was that, like all government construction, the project had to make its way through the bureaucracy.[42] JSP tried to enlist support from all sources, from the Chamber of Commerce to the prime minister.[43] He described the situation to Chant in 1923:

> I have sent requisitions and full arguments through the regular channels and I have no doubt that the estimate will get into the Public Works list. It has been there before several times but has always been out when the Treasury Board prunes the department estimates.
>
> I understand that you know the Premier [Prime Minister King] very well, you know the needs out here and what a fine record the observatory has made even under disadvantageous conditions. You are the premier and practically the only professor of astronomy in Canada and can speak with authority on the subject to the Government. If you could write soon a more or less personal letter to the Premier emphasizing the record and needs of the observatory and asking his influence towards *ensuring money being voted*, by speaking to the Finance Minister, for example, who is Chairman of the Treasury Board, I would take it as a great kindness and I believe it would have great weight towards the desired end.[44]

Chant could hardly have refused to do what he could, as Plaskett was, at the same time, striving to get him elected to the RSC.[45] Exactly what Chant did is not known but within a few weeks JSP thanked him and mentioned that Ontario Premier Ernest Drury had written to King on the matter. JSP himself wrote to the prime minister on 27 March 1923, to ask his support for approval of $24 000, already in the supplementary estimates for construction of the office building.[46] Persistence would eventually pay off.

Instrumentation and New Projects

What the Victoria astronomers achieved in their poor workplace was amazing – almost eight thousand spectra photographed, developed, measured, and reduced by the end of 1923. The measuring alone, Plaskett estimated, took three times as long as the observing itself.[47] The astronomers originally examined the plates on one of the two machines transferred from Ottawa and, after 1919, also on a new Gaertner instrument.[48] Calculations had been carried out with an old Tait Arithmometer transferred from Ottawa in 1917 but a more modern Monroe calculator replaced it in 1921.[49]

Plaskett was always very conscious of the importance of temperature stability in spectroscopic work. Soon after the DAO telescope was put into service, he had carried out a number of tests to check for aberrations in the main mirror under different temperature conditions and found these small defects could be almost eliminated by covering it with a thick insulating cover when it was not in use, thus keeping the mirror at nearly constant temperature.[50] Eventually this precaution was forgotten, and astronomers using the telescope decades later found the mirror's figure was bad.[51] Plaskett would have known better. Moreover, he had always intended to have a system of accurate temperature control built into the spectrograph. The Callendar recorder, a major component of the system, was to come from England, but war production initially prevented the Cambridge Scientific Instrument Company from supplying it. When it did arrive in 1921, the observatory technician, Tom Hutchison, installed it very capably and efficiently in spite of the finicky threading of wires through both axes of the telescope. The system recorded the temperature in the spectrograph and kept it practically constant. Simultaneously a recorder charted the temperature of the dome and main mirror.[52] At a cost of $2391 these two instruments constituted one of the biggest equipment outlays during the early years at the DAO. JSP made a description of the way the recorders, thermostats, and heating coils operated the subject of a paper "Temperature Regulation of Spectrograph and Mirror."[53]

For the first five years of the DAO, Plaskett continued to modify and test his spectrograph, always striving for an instrument that would provide sharp images of the whole spectrum, whether one, two, or three prisms were used.[54] Decades later a testament to his success was expressed by Ira Bowen, director of the Palomar observatory in California.[55] Bowen recalled that when he and his colleagues were designing the spectrograph for their 5 metre telescope in 1948, they faced many of the same problems that Plaskett had dealt with thirty years earlier and were scarcely able to improve upon the design of his fastest spectrograph camera.

JSP also continuously strove to improve the efficiency of the telescope. Keeping its mirror bright was one aspect of regular maintenance, so every three or four months he had it removed for re-silvering.[56] Plaskett and each member of staff had a role to play in the two-day exercise, which involved making a large amount of solution and pouring it gently onto the dish-shaped mirror.[57] Two of the components in the ancient formula of the Brashear Company were rock candy and ethanol (grain alcohol), ingredients that no doubt raised a few eyebrows in Ottawa and required a liquor permit from the province during the prohibition era.[58]

With the Boss program complete and the eclipsing binaries project nearly finished, JSP took aim at another target – all the O-type stars with absorption lines in their spectra. It was an undertaking that occupied him from 1921 until 1924.[59] Previous studies, including his own, covered only seventeen such stars, and Plaskett knew there were probably three or four times as many within reach of the DAO telescope. He thought (correctly, as was later established) that stars of this rare class were likely the most massive and that any physical data he could gather about them would be useful. To select those of interest, JSP enlisted the assistance of Annie Cannon at Harvard.[60] She was in the midst of producing a mammoth compilation, known as the *Henry Draper Catalog*, in which she and her assistants listed the spectral types of thousands of stars. From those volumes already published, JSP selected O stars suitable for the DAO, and Cannon suggested several others from as yet unpublished lists.

The spectra of O stars are notoriously difficult to measure and interpret. Some have bright emission lines, but JSP set those aside and concentrated on the ones with dark absorption lines, as they seemed to form a continuum with the B stars. The few absorption lines that do show up in O stars are broad and diffuse, and plagued with problems of blends where the lines of two elements fall so close together that they merge on the photographic plate. This was the case, for example, with

8.1 JSP watches as the primary mirror (1.83 m in diameter) is gently lowered into its cell on the silvering car. (Courtesy National Research Council of Canada, DAO hist-146)

the Balmer series of hydrogen lines and the Pickering series of ionized helium whose wavelengths were less than 0.2 nm apart, as we saw in chapter 5. By choosing such a challenging field, JSP did not have much competition – not that that figured into his motivation, but it was a factor working to his advantage. Rather, he realized that, with relatively few stars of the type, he could get the required observations in a reasonable length of time.

Among the O stars, several were spectroscopic binaries. By studying these, Plaskett helped to establish that the O class contained the most intrinsically luminous and most massive stars. His published orbits of three of these systems in 1922–3 contained some startling results.[61]

Plaskett's Star and the Honours It Brought

None of the O-type spectroscopic binaries that Plaskett studied were visual or eclipsing binaries, so the inclinations of their orbits were unknown. However, one of the three had double lines in its spectra, lines arising from each component of the pair, which meant that the mass ratio of the individual components could be estimated. It was a star in Monoceros, barely visible to the naked eye at sixth magnitude, and listed in the *Henry Draper Catalog* as HD 47129.[62] It turned out to be a lulu.

Plaskett obtained thirty spectrograms of this star between 16 December 1921 and 5 April 1922.[63] He measured each plate twice in opposite orientations to reduce errors and calculated the radial velocities for both sets of lines. He was astonished to see the orbital velocities at times exceeded 200 km/s, but when he tried to reconcile that with a period of fourteen days, he thought there must be a mistake. Stars moving so rapidly around one another would normally be very close together and would complete an orbit in a much shorter period. To check, he took some extra spectra just hours apart but they showed no difference in velocity, so short periods had to be ruled out. The only remaining possibility was that the stars were more massive than those in any known system.[64]

As he usually did, JSP determined the preliminary elements of the orbit graphically and then refined them using Frank Schlesinger's least-squares method. He calculated the minimum separation of the stars to be 129 solar radii (a bit less than the distance from the Sun to Venus) and their minimum masses to be 75.6 and 63.3 solar masses. They were minimum values because of the unknown inclination of the orbit. However, as the light from the system showed no variation, the inclination could not be so great as to make one star eclipse the other, and so Plaskett figured that the inclination could not exceed 73°. Consequently, he reasoned, the actual masses were at least 14 per cent greater than the values just given, making the total mass of the binary about 158 times the Sun – about four times that of any previously known system. Using empirical results for surface brightness derived by H.N. Russell and theoretical considerations by Eddington, Plaskett went on to discuss what the stars' actual luminosities or absolute magnitudes might be and from this deduced the distance of the system to be about ten thousand light years, the greatest then known for any individual star in the disk of the Milky Way Galaxy. (Four years earlier, Shapley had found some variable stars in globular clusters to be at greater distances.) JSP also discussed the two sharp, strong lines (known as H and K) that did not share the motion shown by the other lines in the stellar spectrum and he convincingly argued that they did not originate in clouds surrounding the binary system but must originate in clouds that are stationary with respect to the stars of the Milky Way. He also pointed out that, in spite of the great mass of the stars, the gravitational redshift predicted by relativity was too small to explain the difference between the radial velocities of these H and K lines and the other lines in the stellar spectrum.

Plaskett probably had no trouble deciding where to publish his findings. He wrote Frank Dyson at Greenwich, "We would much prefer to have our papers presented [to the Royal Astronomical Society] rather than anywhere else. Again there is the very strong patriotic and sentimental feeling of having our work presented and discussed at home, at the centre of the Empire, rather than in a foreign country such as the United States."[65] Though American journals had served him well, Harry's wonderful experience in England after the war may have bolstered JSP's preference for Britain. Of course his ancestral roots were English and, like most residents of Victoria, he was unfaltering in his support of the empire. Besides, since 1920 he had been submitting annual reports of the DAO to the *Monthly Notices* of the RAS, where they appeared alongside similar summaries from other observatories in the British Empire. So, after introducing his obese binary system to the world at the RSC in Ottawa in May 1922, he submitted his paper to the RAS, where it was discussed at a meeting of that society in London on 9 June. Arthur Eddington, as president, was in the chair while Hugh Newall summarized Plaskett's work and fielded questions about it afterwards.[66] (JSP and Newall, the director of the Solar Physics Observatory at Cambridge, had been correspondents since Plaskett's Ottawa

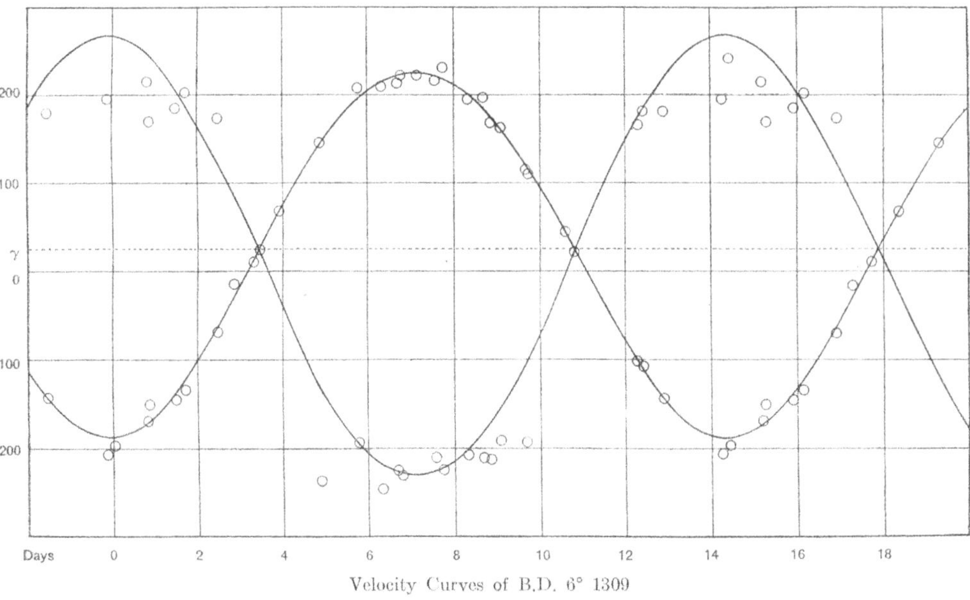

8.2 The velocity curves for Plaskett's massive binary star. The agreement between the individual points and the curve is much better for the primary component than for the secondary. (From *PDAO* 2 (1924): 157)

days.) Eddington noted that in the case of the previous massive record holder, V Puppis, its components were almost touching, distorting the actual stars and putting the computed orbit in doubt. However, he said, "In Plaskett's binary star, the components are much more separated than those of V Puppis, and I regard the mass in this case as more certain." With Eddington's endorsement and his reference to "Plaskett's star," HD 47129 was well on its way to the record books. *Nature* picked up on the story in its issue of 17 June, and, of course, JSP's full paper was published in the *Monthly Notices of the RAS*.[67] In addition, he wrote a complete account in the *Publications of the DAO* and a more accessible version for the RASC *Journal*.[68] International societies would soon be honouring him with recognition.

Plaskett's Star, as the massive pair came to be known, caused quite a stir in the press. The media, always attracted to the "biggest" of anything, really played up the discovery. Select Pictures Corporation wanted to make a movie about the achievements of the DAO, and the newspapers were agog.[69] William Loudon, JSP's former associate in the

8.3 Full page from *Current Opinion* 73 (1922), an American periodical that advertised itself as skimming the cream of all the current magazines. (Image courtesy of Toronto Public Library)

University of Toronto's Physics Department, wrote to say that the discovery stirred the imagination of one of his students who now wanted, more than anything else, to study astronomy. On a broader scale, he said, JSP's description of this immense star was "so inspiring that it deserves to be presented to all those who walk about upon this earth of ours from day to day with their eyes cast down upon the ground."[70]

Besides getting coverage in the Canadian press, the story got huge international attention. In Australia, for example, where the news arrived by cablegram, nearly a hundred papers carried reports of the discovery.[71] Sydney's *Morning Herald* snapped up a personal interview with Reynold Young about the discovery and about the DAO, as he and Chant happened to be staying in the city for a few days en route to Wallal to photograph the solar eclipse.[72] The *New York Times* reported "the suns have been named Plaskett for their discoverer," and quoted Professor Harold Jacoby of Columbia University as saying it was "the most outstanding of recent astronomical developments."[73] (Plaskett's Star retained its record-breaking status for a very long time, though recent developments have somewhat reduced estimates of its mass.[74])

With the postwar economy finally picking up and funding to attend conferences improving, JSP not only went to Ottawa in May 1922 to speak about his massive star, but he and Reba travelled to the American Astronomical Society (AAS) meeting at Yerkes Observatory near Chicago in September.[75] There, surprisingly, he did not speak about his famous binary but did give a talk on the DAO's new spectrograph, purchased from the English instrument maker Hilger, at a cost of $2515.[76] It transmitted near-ultraviolet light, an especially useful feature in the investigation of the spectra of the hottest blue-white stars – those of type O. After the meeting, the Plasketts went on to Pittsburgh and Rochester. They then returned to Chicago, travelled west on the Santa Fe railroad, and stayed a while with Vesto Slipher and his wife at the Lowell Observatory in Flagstaff, Arizona, before heading back to Victoria.[77] Perhaps Plaskett considered this trip somewhat of a consolation, as he had wanted to go to Rome in May for the triennial general assembly of the International Astronomical Union (the first under this name). To understand why he could not go, we need to take a small step back.

The IAU's National Committee for Canada, chaired by Klotz at the time, had the responsibility for naming the Canadian delegates to the IAU. The committee members arranged to get together and make their decision over the Christmas holidays in 1921, when they would be gathered at the American Association for the Advancement of Science (AAAS) sessions in Toronto. JSP and the DAO astronomers were not present at the meeting but the DO staff was well represented. As might have been expected under the circumstances, the national committee passed over Plaskett, selecting Klotz and Young. Plaskett was greatly annoyed and wrote to Chant, "I have no doubt it was a conspiracy by DeLury and others at Ottawa, who have apparently never forgiven my success, to put me in my place. I have no objection to Klotz' election but I certainly do to Young, who is not a member of any of the committees and who certainly should not have been chosen instead of his director. It seems to me it will create considerable unfavourable comment abroad, where I have good reason to know I am undoubtedly considered the foremost astronomer in Canada."[78]

In a similar vein, he felt it necessary to explain to the most prominent astronomers, such as Hale and Dyson, that he was the victim of jealous intrigue.[79] Though he did not get to Rome, JSP did make a number of recommendations for the IAU through the national committee. These entailed certain improvements to the system of stellar classification, the criteria for selecting stars for radial velocity determinations, support in publishing tables drawn up by Young to convert coordinates, and (jointly with Young) the need for a program for velocities of spectroscopic binaries whose orbits had not been found.[80] Most of the IAU's work was done by standing committees (or commissions, as they were called); scientific papers by individuals were not part of the IAU sessions. Commission members would discuss matters of international importance ahead of the general assembly – items dealing with standards, cooperative efforts, and urgent needs. They would hammer out a report and, once the members agreed, would forward their recommendations to the IAU as a whole for ratification.

Meanwhile Plaskett's Star continued to excite attention. Generally speaking, astronomers pooh-pooh the widespread notion that discovering a new object in the sky is their raison d'être, but Plaskett's Star was the only O-type star pair whose individual masses (or at least their lower limits) were known and they were far outside the bounds of what had become accepted as normal. This was of real significance to astronomers, as it allowed them to feel confident that this class of stars was the most massive and most luminous of all the spectral types. Scientific societies and universities seemed eager to follow the lead of the popular press to recognize Plaskett's achievements, and not just his spectacular discovery, as worthy of their highest accolades.

In March 1923 he received word that the Royal Society of London, probably the world's most prestigious scientific body and certainly one of the oldest, had decided to elect him a fellow.[81] JSP was "very proud and pleased." He said, "I know full well my own limitations and consequently never expected to receive so high a distinction. That it has come so soon I must ascribe to the kindly offices of my friends in England and especially I believe to Newall who initiated the idea and has apparently so soon succeeded in completing it."[82] To be formally admitted to the Royal Society, JSP would have to be present at one of its meetings, a requirement that he wasn't able to fulfil for over two years. Becoming an FRS was (and still is) a triumph for any scientist, especially for a Canadian – only three others had that distinction at the time: geologist Frank Dawson Adams, mathematician John Charles Fields, and physicist John C. McLennan, JSP's associate from his University of Toronto days.[83] It was an honour worthy of prime ministerial congratulations, as Mackenzie King recognized when he wrote to Plaskett, noting that "your distinguished services have helped to bring not less fame and credit to the Dominion than to yourself."[84] In the same letter King voiced his support for the long-awaited office building at the DAO.

Perhaps McLennan was working behind the scenes when JSP's alma mater awarded him a DSc on 7 June 1923.[85] As Plaskett travelled east in May to attend the RSC meetings in Ottawa and to receive his honorary degree in Toronto, he stopped off in Edmonton to deliver the convocation address to the University of Alberta. Dean R.W. ("Billy") Boyle of Applied Science had suggested this idea but JSP initially told him, "Attempting to preach a sermon to the graduating class … is decidedly out of my line … I certainly cannot talk platitudes for 20 minutes."[86] Boyle would not take no for an answer, and so Plaskett reluctantly agreed to give the convocation address as well as a talk to the Science Association.[87] On the social side, he enjoyed some early spring golf in Edmonton – perhaps his first experience with the game. He wrote to Boyle later that year to say, "Probably my visit to Edmonton was the deciding factor in making me take membership in the Victoria Golf Club. I could not help but see what fun you all seemed to get out of it. I am now in the throes of learning the A.B.C. but next time I meet you I may be able to have a game if I manage to get anywhere near your class."[88]

After the Ottawa meetings JSP arrived in Toronto, where Chant hosted a dinner for him at the faculty club, thoughtfully including a couple of his former classmates among the guests.[89] Chant probably felt like celebrating too, as he had just been named an FRSC after years of lobbying by Plaskett and others.[90] Following dinner they proceeded to Convocation Hall, where fifteen distinguished recipients, most notably the university's proud graduate of 1895, Prime Minister Mackenzie King, accepted honorary degrees.[91] When it came Plaskett's turn, university president Sir Robert Falconer not only praised him but also referred to his three assistants, all "able graduates of this University."

8.4 Photos from an album of the 1923 Yerkes eclipse expedition (UCL Photographic Archive)
8.4(a) above. Two of five photos of the site taken by JSP.

8.4(b) top. Canadians assemble for the solar eclipse on 10 September 1923, on Catalina Island, California. Standing (left to right) JSP, H.R. Kingston, Elizabeth Laird, N.B. McLean, Annie Laird, J. W. Campbell. Seated in front: Mrs Kingston and Reba Plaskett. (Key from *JRASC* 17 (1923): 325)
8.4(c) bottom. E.B. Frost, director of the Yerkes eclipse expedition, and JSP with a high-speed camera for photographing shadow bands. (Photo by B.W. Harris)

The *Star* interviewed JSP, characterizing his life story as "one of sheer grit and bulldog determination."[92] He emphasized that his early engineering and mechanical training was largely responsible for his success, describing the modern telescope as "just a huge machine."

If the AAAS western division had met in June as usual, Plaskett might have gone directly to the United States, but he was able to return to Victoria as the convention date had been postponed by three months to follow right after the solar eclipse of 10 September, visible in southwestern California and Mexico. Combining this event with the AAS and Astronomical Society of the Pacific joint gathering at Mount Wilson was very attractive for many astronomers. It was just the push Annie Cannon needed to make her long-anticipated visit to the DAO just prior to the west coast meetings.[93] Harry had just returned from five weeks at Mount Wilson, so she was delighted to spend some time with the younger Plasketts too. After her time in Victoria she, along with JSP and Reba, headed for Catalina Island off the coast of California to see the eclipse.[94] There they joined the Yerkes team, along with many others, including several Canadians. Harper, quarantined with smallpox, was unable to go.[95]

In spite of very favourable odds in sunny California, the day of the eclipse was cloudy. The astronomers whose elaborate plans had been foiled were dejected but for Plaskett, who was there more or less as an interested spectator, the disappointment was short lived. He put the gloom behind him and went on to enjoy the triple meetings.[96] He did not seem to mind that, owing to a pressing schedule, many papers were not read, including his two, "The Absorption Lines in O-Type Stars" and "Two New Camera Lenses for Spectrographs." Instead he was happy to present Harry's contribution, "A Possible Origin of the Nebular Lines."[97]

JSP found a presentation by Charles St John, a solar astronomer at Mount Wilson, to be especially important.[98] In spite of earlier doubts and in the face of enormous difficulties, St John had successfully measured the gravitational redshift in the Sun's spectrum as predicted by Einstein. Plaskett, of course, knew well how difficult it was to separate the various potential causes of such a tiny shift of only 0.1 nm in the spectral lines. He also made a trenchant comment about inadequate theories not completely disappearing until their older adherents were no longer active: "Though present generations of physicists and astronomers may find difficulty in accepting and using this great generalization [i.e., the general theory of relativity], this difficulty will disappear with time."[99] This was exactly an attribute of scientific revolutions to which the influential philosopher Thomas Kuhn drew attention in 1962.[100]

Between the Stars

Plaskett's paper on absorption lines in O-type stars (which went unread in Pasadena in 1923) is an indication that he was extending his research to an exploration of the space between the stars, though that was not entirely clear at the time, even to him. Until the 1920s astronomers generally operated on the assumption that space was transparent to starlight. The implications of the opposite assumption were too far-reaching to consider: if there really were absorption of light, the stars would appear dimmer than they actually are and their presumed distances would be overestimated. The astronomers were not engaging in willful neglect; there was just too little information to make allowance for interstellar material. Some evidence of patchy absorption had been accumulating since 1895, when Edward E. Barnard began taking wide-angle photographs showing dark areas in the Milky Way, but those "coal-sack" regions appeared to be localized and of no general significance. Kapteyn had carried out pioneering work on general absorption in 1909 but there was a lack of certainty on the subject for another twenty years.[101] Obscuration of this type is now understood to be caused by clouds of dust particles ranging from a few molecules to a few microns in size that scatter the incident light. The effect is akin to the dimming and reddening of sunlight by Earth's atmosphere, so familiar at sunrise and sunset.

A different type of evidence that some material existed in interstellar space came from the selective absorption in certain stellar spectrograms. The first instance was noted in 1904 by Johannes Hartmann, the astronomer whose optical equipment and testing procedures were so widely used. He saw that most of the lines in the spectroscopic binary Delta Orionis shifted back and forth as expected but that two prominent, sharply defined lines did not. These were the H and K lines produced by ionized calcium (Ca+). It was certainly not evident from Hartmann's observations of Delta Orionis whether the Ca+ was situated in a cloud surrounding the binary system or somewhere else in space between the star and Earth, though it was obvious that the absorption did not arise in Earth's atmosphere, as the lines appeared only in some stellar spectra.

Before long, major observatories with sufficiently large telescopes began to capture the attenuated light of fainter and presumably more distant stars showing the sharp H and K lines from ionized calcium and sometimes similar D lines from neutral sodium as well. The first astronomer to present convincing arguments that these apparently stationary absorption lines originated in interstellar regions had been Vesto Slipher in 1909. Initially he published his findings in the *Bulletin* of the Lowell Observatory in Arizona where he worked (and where JSP visited in 1906 and 1910), but he later reprinted them in *Astronomisches Nachrichten*, where they would have received wider circulation and credibility.[102] Plaskett was probably referring to this version when he told Slipher, "I read your article on calcium absorption with a great deal of interest."[103] Though Slipher couched his conclusions rather tentatively, saying "It might, then, for the present be assumed that the calcium absorption has its origin in an interposing cloud covering at least certain extensive regions of the sky," it seems odd that Plaskett (and astronomers generally) ignored the significance of Slipher's work for many years. As mentioned in chapter 6, this also happened with his radial velocity measurements of spiral galaxies; his biographer attributed this lack of recognition to Slipher's reticence and to an "aura of skepticism" that surrounded anything coming from the institution founded and directed by Percival Lowell, who had promoted the fanciful idea of intelligent life on Mars.[104] But long after Lowell's death in 1916 and even after Slipher became director in 1920, his accomplishments were overlooked, notably at the DAO by JSP and Young in their comprehensive examination of calcium lines in O- and B-type stars.[105]

Young favoured the hypothesis of a cloud containing ionized calcium surrounding many binaries of this sort and he concluded that the cloud moved with the stars and was sometimes set into revolution, of smaller amplitude but in the same period, by the motion of the stars. Young's analysis appeared in the *Publications* of the DAO immediately after a paper by JSP on the eclipsing binary Y Cygni. Plaskett showed that the velocity of that system as a whole, as determined from its diffuse hydrogen and helium lines, differed from the velocity given by the sharp H and K lines by 40 km/s, implying that the source of the latter lines could not possibly have been in the star system. Young did not think the 40 km/s difference was real and attributed it to the difficulty in measuring the diffuse and asymmetrical lines. As Plaskett himself put it, "If you ever saw a spectrum of Y Cygni, which I think is without exception the worst I ever saw, you would wonder how the faint washed-out slight weakenings of the continuous spectrum which pass muster for lines could ever be set on by a micrometer wire."[106] He conceded that neither of the two hypotheses appeared "to fit all the observed facts, and objections can be adduced to each."[107] Another problem facing JSP was his mistaken understanding that the interstellar calcium could be ionized only if high-energy, hot O and B stars were nearby. In spite of these difficulties he still thought that the overall evidence favoured the theory of a widespread distribution of interstellar calcium.[108]

Plaskett gathered more and more spectra of O stars distributed in different regions of the Galactic disk. All of them, not just the binaries, showed the Ca+ lines. With the larger sample, Plaskett was able to move beyond mere collection to interpretation of the data at hand. By the time of the Pasadena meeting in September 1923, he had substantiated what Slipher had earlier suggested on much scantier evidence: the radial velocities given by the H and K lines were opposite to the Sun's velocity relative to the local standard of rest.[109]

To understand the meaning of that claim, it might help to imagine an analogy.[110] We are on the upper deck of a ship at night, on an ocean where nothing is visible except the lights of other vessels, apparently moving in all directions and at different speeds. Careful examination, however, shows that on average they have a tendency to move in a direction towards the stern of our ship. In fact this tendency is just a reflection of our own forward motion. Similarly we, as passengers in the solar system, zip along with the Sun at about 20 km/s in a direction towards an apex near the star Vega, though the only evidence is the apparent motion of our neighbour stars in the opposite direction towards the antapex in the constellation Columba.[111]

Plaskett realized that the velocities determined from the H and K lines showed that the Ca+ gas where the lines originated also seemed to share in this motion towards the antapex. He elaborated on his findings based on forty stars and submitted a paper on the subject to the RAS in November 1923. Frederick Stratton, on the society's council, gave an account of it at the meeting of 14 December and later included Plaskett's conclusion in his textbook, *Astronomical Physics*.[112] During discussion following the RAS meeting, various objections were raised to Plaskett's ideas. "Some thought that the clouds would be luminous and show bright lines; another difficulty was the practically perfect transparency of the stellar spaces which Dr. Harlow Shapley deduced from his work on globular clusters."[113] Herbert H. Turner, the Savilian professor of astronomy at Oxford, wondered if the interstellar medium, since it did not seem to share in the motion of the stars, could provide a stationary reference framework to which all other velocities could be referred. In this sense it reminded him of the theory of ether that had been held by scientists until it was expunged by the experiments of Michelson and Morley in 1887. If the interstellar medium had been truly at rest, there would have been implications for the theory of relativity, in which there is no such thing as absolute motion or absolute rest. As it turned out, the interstellar medium did move but not at the same rate as the more distant stars whose light passed through it.[114] Turner acknowledged, rather wryly, that "the subject is still somewhat obscure."[115] What Plaskett's research clearly showed was that interstellar gas that gave rise to H and K absorption lines pervaded the disk of the Milky Way. Definitive evidence that interstellar dust was similarly widespread had to wait until the investigations of Lick astronomer Robert Trumpler in 1930.[116]

Looking Up

The extent to which Plaskett orchestrated media coverage of his discoveries, especially of the "Plaskett twins," is not clear, but Klotz and R. Meldrum Stewart, the man who would soon succeed Klotz, were sure they knew. Klotz wrote in his diary:

> Stewart is disgusted with Dr. Plaskett's self-glorification in the press. He has had the grand double star named after himself, an unheard of scientific glorification, especially in the heavens. Canada has been asked to send some typical photographs to Paris [presumably for the International Exposition of 1925]. Mr. Stewart & Plaskett agreed upon a few from each observatory, but later Plaskett on his own initiative furnished several plates, visibly drawn, of the great "Plaskett binary" etc. Plaskett is the outstanding braggart and as such, a disgrace to Canada, although he has done much to give Canada recognition in the astronomic world, but he loves newspaper notoriety.[117]

Plaskett disavowed any role in press publicity as "entirely unsought."[118] Even so, he did often curry the favour of the media and sometimes seemed to reap the benefit of their attention. An example can be seen in the letter that he wrote to the prominent editor of the Ottawa *Citizen*, Charles Bowman, on 16 January 1924:

> I enclose this clipping from the editorial column of yesterday's Colonist which apparently refers to an interview of Turner in London in reference to recent work of mine sent to the Royal Astronomical Society, London.

> I recall with gratitude your reference last year to our work out here and the value it was in my campaign for an additional building which I am glad to say, is now under construction. If you can find an opportunity for using the enclosed reference in your columns it would be of very great value with the Government. "A prophet is not without honour" is often a very true proverb.[119]

Bowman replied eight days later: "I shall run it with pleasure in *The Citizen* and at any time if you have articles or press references which you would like printed please do not hesitate to send them along. I often think with pleasure of my visit to Victoria last year."

Plaskett was not only bolstered by the recognition of his achievements but by new leadership at the DO that seemed to promise an improvement in relations between Ottawa and Victoria. Otto Klotz had not been well since his IAU trip overseas in 1922, and he died at age seventy-one on 28 December 1923. Soon thereafter, JSP wrote to the man who would become the new director in Ottawa, R. M. Stewart:

> The news of Dr. Klotz' death came as a shock to me as I had no expectation but that he would get around to the office again after a rest at home. I regret very much his untimely death, for he was still very mentally alert, and consider it a great loss to Canada, to the observatory, and to the community in which he lived. I do not know of any one who can quite take his peculiar place in Canadian science.
>
> The main purpose of this personal letter is, however, to express the hope that there will be no obstacles to your prompt promotion to the directorship of the observatory. Certainly no one else in the observatory or the department should even enter into serious consideration for the position and I hope the Department will view the matter in the same light.
>
> If I can do anything to further your interests in obtaining the position which your ability and service so fully merit, please let me know.[120]

At last it looked as if Plaskett's administrative problems might be eased. Perhaps, he hoped, old animosities would be forgotten and his latest honours would bring him the credit in Ottawa that he deserved.

9
The Farthest Stars, 1924–1926

Early in February 1924, JSP made one of his occasional trips to mainland British Columbia to give a public lecture under the auspices of the Vancouver Institute.[1] As he spoke about the life and death of stars, he may also have reflected on human mortality. Otto Klotz, whom he had known so well for twenty years, had recently passed on, and in three days' time, on Sunday 10 February, the Plaskett family would gather at a special service to honour Annie Plaskett, JSP's mother, who had died in 1922.[2] Her children had arranged to have a beautiful double stained glass window placed in her memory in the baptistery of St Mary the Virgin Church, Sapperton, New Westminster, BC, where her son, the Reverend Canon Frank Plaskett, had been rector since 1912. Other family members at the dedication service would have included the rector's wife, Mary, and their three children, John, Mary, and Joe. Joe would later become a highly regarded artist in Canada and France. His revealing autobiography sheds no light on his eminent uncle but paints a relevant picture of conditions in British Columbia in the 1920s:

> The wireless was just beginning, television was not even dreamed of. My father had a motor car, a model-T Ford, but hardly any of his parishioners had one. There were not even buses. It was the age of the electric tram, the trolley. Chinese with baskets suspended from a pole on the shoulder sold fish from door to door. The breadman came each day to sell bread. Food was cooked on wood-fed kitchen ranges. Refrigeration was unknown – there was a cooler in the cellar. Life was primitive by today's standards, but comfortable ...
>
> What seems curious today was the outpost of empire mentality that I was part of in remote British Columbia. I certainly grew up without realizing that we were witnessing the sunset of the British Empire.[3]

Double Stained Glass Window in the Baptistry: The Home in Bethany

pg 28

Manufactured by: N. T. Lyon Co., Toronto.

(2 Light Gothic with Kite)

Dedicated: February 10, 1924 by the Rt. Rev. Adam de Pencier, the Rev. Frank Plaskett, Rector.

Damaged by the fire on December 6, 1932 and restored to its original beauty by Bogardus and Wickens before the restoration service March 29, 1933.

9.1 Memorial window depicting the biblical scene of the house in Bethany, dedicated to JSP's mother, Annie Plaskett. (Reproduced with permission from Crichlow and Turney, *Stained Glass*, 28)

Without doubt thousands of Victoria residents were sure that Britannia ruled the waves, when, in June 1924, they cheered the arrival of a special squadron of seven Royal Navy warships, including the famous HMS *Hood*, on a round-the-world tour. The federal government hosted a dinner at the Empress Hotel for Vice-Admiral Sir Frederick Field and his officers. JSP was there along with the lieutenant governor, the premier, the mayor, members of Parliament, bishops, and others who represented the social elite of the city.[4]

JSP's love of the old country was clear even in his professional correspondence. When he was thinking of augmenting his staff in 1923, he had written to Sir Frank Dyson, "I would prefer to have some astronomer from England rather than from the United States, not only on account of national feelings but because I believe the English training is far superior to the American. Our propinquity is bound to give us enough of the American style in our work and I can feel that we naturally follow American methods which have a decided tendency to superficiality as compared with the work in England."[5] He was not just buttering up the Astronomer Royal; his words reflected a genuine admiration of all things British. A few years later in an exchange of letters with Frank Schlesinger, he said how pleasant it was to hear an American expressing admiration for the make-up of the "Britisher" as a good sport under the most trying circumstances, an opinion, "needless to say," that Plaskett agreed with.[6]

New Colleagues in 1924

Though Plaskett had extended an olive branch to R.M. Stewart after Klotz's death, he wasted no time in letting the new director of the Dominion Observatory (DO) know in no uncertain terms that he would not tolerate any interference or delay from Ottawa. Within a month of Klotz's death, JSP wrote two long and carefully reasoned letters to Stewart, one outlining why the prescribed procedure for requisitions was often flawed, and the other explaining that JSP alone should approve the publications of the Dominion Astrophysical Observatory (DAO). And just in case Stewart did not get the message, he was very clear that he would appeal higher up in the chain of command when necessary. It is hard to disagree with anything he wrote, but equally hard to imagine that the bureaucracy could be as flexible as he would have liked. He wrote in part: "It must be remembered that we have a big and complicated machine to maintain in efficient working order, that troubles may develop quickly and must be cured quickly ... Much of the work is experimental and requires special supplies on short notice if any progress is to be made and if the investigator is not to be discouraged by long waits."[7]

He proposed a reasonable compromise that only expenses over $50 require prior authorizations from Ottawa and that emergencies, such as the recent failure of the opening mechanism on the dome shutters, should be exempted. In the other letter to Stewart, he protested a delay of over a month in getting approval for the latest DAO *Publications* and offered no conciliation: "To have this work passed on by anyone outside ourselves or for this highly specialized technical work to be edited or criticized by any departmental officer is the height of absurdity ... We have already had one instance of work being partly forestalled and some of the credit due the observatory and department lost by similar results elsewhere and the importance of cutting out any unnecessary delay is obvious."[8] Plaskett's concerns were well founded. Ottawa bureaucrats with no knowledge of astronomy occasionally did get their way, even long after Plaskett's time, introducing silly changes with sometimes ludicrous results.[9]

To keep the government accountants happy, JSP was responsible for annually submitting an inventory of all material in the possession of the DAO. Filling out these multi-page returns must have tried his patience, as they included even the most trivial items such as individual screwdrivers, forks, and knives, valued at just a few cents each. There was also a ten-page section listing all the maintenance costs, item by item, for the year.[10] Plaskett found all this annoying and inefficient. As he said to Stewart, "I want to make out these accounts in the way that will give least trouble to you, the accounts branch and us, but if they get much more complicated and involved they will have to send out an interpreter and we will have to employ additional assistance."[11]

Those irritations aside, the Liberal government under Mackenzie King, including his minister of the interior, Charles Stewart, were supportive of Plaskett and the work of the observatory.[12] It was the King government that approved the long-awaited office building at the DAO. After years in a cramped, unsuitable space under the dome, in October 1924 the astronomers finally moved into a building with individual offices, photographic facilities, and a library. The caretaker had living quarters in its attic.[13] But Plaskett still wanted more. He had always maintained that the original vision for the observatory included houses on the property for the director and the astronomers.[14] Always

9.2 The DAO office building (Courtesy National Research Council of Canada, DAO hist-173)

relentless, he stepped up his campaign for these residences just as soon as the office building was assured.[15] When he got nowhere through normal channels, he once again appealed directly to the prime minister, a tactic bringing gentle rebuke from Deputy Minister Cory.[16]

The house which he, Reba, and Stuart occupied was never intended for the director, he protested. The government, he argued, should build a proper residence, one befitting his status and in keeping with the entertaining he was expected to do, and in recognition of the distinction that the DAO had brought to the country. He suggested that the present house, as well as additional ones, could then be used for the other astronomers. The Ottawa astronomers thought their own need for a library extension was a higher priority, and Charles Stewart could not understand Plaskett's legitimate argument that it was virtually impossible for astronomers on night duty to respond to changing weather conditions at the observatory if they were miles away, at home in the city.

Part of the trouble with any proposed construction was that the Department of the Interior had to get the Department of Public Works to agree, and Public Works, in turn, had to rely on the recommendation of its resident architect in Victoria. He was a man approaching ninety years of age, and any action proceeded at a leisurely pace.[17] Another factor, though unspoken, may have been that the astronomers might not have shared their boss's enthusiasm for living so close to work and to each other.[18] In any case, further housing was never built, though (as we shall see in the next chapter) JSP did succeed in getting approval for a one-room addition to his residence in 1930.

Around the same time that the office opened, human help also arrived, when "assistant computer" Sherwood N. Hill transferred from Ottawa.[19] He had been in the Surveyor General's office and knew nothing about astronomy but Plaskett helped him develop. "I am glad when he takes up and carries through an orbit as it is good for him to do it. Of course, he needs some help and guidance but that we are glad to give."[20] Plaskett also went to bat for Hill to get his salary increased and his classification as an "astronomical computer" recognized in Ottawa.[21]

There were few departures of personnel during JSP's directorate, surely a sign of happy relations among the staff. The first took place in 1924 when Reynold Young decided to leave permanently. He had originally been away for five months in 1922, observing the solar eclipse in Australia with C.A. Chant. They had obtained a couple of good photographs showing about twenty star images around the eclipsed Sun, and once Young got back to Victoria he measured the plates with some help from Ed Harper. (Chant did not have any suitable measuring equipment at the University of Toronto.) The results for the individual stars showed a lot of statistical scatter, but the general trend suggested a mean value only about 20 per cent higher than relativity theory predicted. Though Chant and Young's findings were the first to be announced after the 1922 eclipse, they were later overshadowed by the more accurate and long-anticipated conclusions of the Lick astronomers, which demonstrated a very close agreement with the deflection predicted by Einstein's theory.[22] Another aspect of the Canadians' eclipse efforts was an attempt to look for polarization in the coronal light. This had been the subject of Young's PhD thesis a decade earlier and now he was able to use new equipment at the DAO, a microphotometer and wedge, just received from the Topley Company in Ottawa.[23] JSP had ordered it for Harry's research on the overall profile of stellar spectra, essential to estimating stellar temperatures, and for studying intensities of spectral lines, crucial in distinguishing giant stars from dwarfs and thus to finding their luminosities and distances.

Because Young was still on the DAO staff and had used the observatory equipment for his measurements, JSP rightly requested that Chant and Young's paper "Evidence of the Bending of the Rays of Light on Passing the Sun," appear in the DAO *Publications*.[24] The paper did appear in this volume, but Young also wrote a simpler version for the RASC *Journal* to complement Chant's series of articles on the expedition itself. Soon after this project was completed, Young accepted Chant's offer to join him on the faculty at the University of Toronto, beginning in September 1924.[25] Plaskett was displeased at Chant's failure to contact him before openly recruiting Young but forgave the slight, bearing in mind all that Chant had done to promote astronomy in Canada.[26] Young confided to Chant that he suspected Plaskett was "glad I am going as it gives Harry a boost upward."[27]

After Young's departure, JSP considered a few astronomers for the vacant position. The only professional in Canada capable and interested was Richard McDiarmid from Ottawa. He had completed his PhD under H.N. Russell at Princeton in 1915, writing his thesis on four eclipsing binaries.[28] In fact Plaskett had recently studied one of these systems, TV Cassiopeiae, spectroscopically and, by combining his data with McDiarmid's, had found the masses and radii of the individual stars.[29] McDiarmid, though, could not make up his mind whether or not to move from Ottawa and dithered for so long that Plaskett was almost relieved "that a man so uncertain and so pernickety about minor details is finally not coming."[30] Two possible American candidates were also protégés of Russell, just completing their PhDs based on research at the Harvard College Observatory: Donald Menzel and Cecilia Payne. JSP naturally asked Harlow Shapley, the director there, for his opinion, and Shapley replied on 29 October 1924: "I cannot unreservedly recommend Menzel for a place on your staff. He is a pretty good man, especially in some ways, and is ambitious ... Some of his immediate ancestors, I suspect, were Austrian Jews – a confidential statement, of course, for he does not show Hebraic characteristics very conspicuously, and he would not admit such connection."[31]

JSP gave no hint of how he felt about Shapley's anti-Semitic smear, though such feeling was widespread at the time. He merely thanked the Harvard director for the opinions expressed in his "frank letter in regard to Menzel."[32] (Shapley's biases could not have run too deep, for he himself hired Menzel in 1932. On Shapley's retirement in 1952, Menzel succeeded him as director.) As for the other candidate, Plaskett told Shapley that Russell had "referred to Miss Payne very highly indeed but of course I do not know what position she occupies with you nor whether she would be available and suitable. All our scientific staff take part in the observing and I doubt the desirability or possibility of woman observers at this rather inaccessible place. Unless she observed it would be rather difficult to get observational material for her to work upon."[33] The thought of a woman observing alone at night – or, possibly even worse, accompanied by the male night assistant in the darkened dome – was almost inconceivable. Within a few years the director might have been aghast to know that "some of the young astronomers had out-of-town girl friends who would occasionally come up to the dome and spend the night with them."[34]

Another slim possibility to replace Young was W.F.H. Waterfield. A great-grandson of the famous astronomer Sir John William Frederic Herschel (hence the initials), he was a talented amateur astronomer in 1924 living in Nakusp in southeastern British Columbia. Shapley described him as "easily one of the best observers" in the American Association of Variable Star Observers.[35] The DAO would have moved in very different directions if Plaskett had opted for a dedicated observer such as Waterfield or a budding theorist such as Menzel or Payne. In the end, he hired none of the above. Instead he chose Joseph Pearce, a University of Toronto graduate then proceeding towards his doctorate on a fellowship at the University of California's Lick Observatory.[36]

Pearce was keen to return to Canada and willingly put his doctoral studies on hold to take up the opportunity. R.M. Stewart supported Pearce, telling the deputy minister, "It is men of this enthusiastic type that we want in Canada for our scientific institutions and I believe it is worth some concession to prevent such men from being snatched up by our neighbours to the south."[37] Even with such an endorsement and with war service to Pearce's credit, securing him was something of a coup, because American observatory directors were eager to hire able doctoral graduates, who were in very short supply.[38]

JSP's choice (approved by the Civil Service Commission) turned out remarkably well, as Pearce seemed perfectly willing to collaborate on the director's projects for the next decade. In 1923 Plaskett had decided to expand his program to include stars of spectral classes from B0 to

9.3 Joseph Pearce (far left) as a graduate student at Lick Observatory in 1922; standing next to him is R.G. Aitken, and in the middle is the director, W.W. Campbell. For a full key, see http://digitalcollections.ucsc.edu/ (Courtesy Special Collections and Archives, Lick Observatory Records, UCSC ua0036_glp_0375)

B5 because many stars classified among the early Bs were in reality O-type.[39] (In the spectral sequence from O through B, A, F, G, to K, those nearer the O end were called "early" and those nearer the K end were called "late." Furthermore, each class was divided into ten sections, B0, B1, B2, ... to B9, for example.) Over the next seven years Plaskett and Pearce would photograph and measure the lion's share of three thousand plates of these early B stars brighter than magnitude 7.5.[40]

In 1924, just as Pearce was getting started at the DAO, Plaskett was wrapping up his study of the O stars. It had not been easy, as he explained to Benjamin Boss, the astronomer son of Lewis Boss whose catalogue provided the impetus for the initial DAO program: "I have been observing the absorption line O-type stars for radial velocity and it is some job I assure you as nine-tenths of them have diffuse lines and three-fifths of them have variable velocity."[41] The paper that marked the culmination of his investigations was based on 528 spectra of 80 stars, of which 47 were type O.[42] Plaskett established conclusively that this class of stars was the most luminous and the most massive and he was able to use criteria established by his son Harry to categorize the stars into their subclasses from O5 to O9 as well as Wolf-Rayet stars with broad emission lines in their spectra. Edward Arthur Milne, a brilliant young theorist, then Hugh Newall's assistant at Cambridge, cited this work as "one of the most important memoirs published during the year."[43]

In light of JSP's significant findings, he was asked to contribute an article to a collection – a *Festschrift* marking Hugo von Seeliger's seventy-fifth birthday.[44] JSP would have known Seeliger at least since his visit to Germany in 1913, as Seeliger was president of the Astronomische Gesellschaft at that time and for many years before and after. He was the director of the Munich observatory for most of his life and was, like J.C. Kapteyn, renowned for his studies of the distribution of stars in the Milky Way Galaxy. The *Festschrift* papers were to bear upon this life's work of Seeliger, so it was appropriate that Plaskett's contribution should be on problems of the O-type stars, as they were the most distant markers of the Galactic disk. It was an honour for Plaskett to be one of only four North Americans among the thirty-eight contributors (the others were his younger colleagues Schlesinger, Shapley, and Stebbins).

JSP's publications on the O stars were lauded not only for their solid addition to the knowledge of the hottest and most massive stars but also for the stimulation and suggestions they provided for further investigation.[45] Why did the O stars, as a group, seem to have high radial velocities while the B stars, next in order of mass and temperature, have the lowest? Why did some O stars have emission lines and others not? Why were some associated with surrounding nebulae when others seemed only to hint at such a feature through the absorption lines in their spectra? Much more remained to be done, but Plaskett did not pursue these problems.

Interstellar Stuff

Instead he followed up another key aspect of his research on O-type stars – the convincing difference between the radial velocities he found from the stars' few broad absorption lines and found from the sharp H and K lines. By this time, he had examples where the difference was great, as much as 60 km/s. He also had evidence confirmed by the sodium D line, whose shifts gave very similar results. As he put it, "It is beyond the bounds of credibility to assume … that any cause but difference in velocity" can explain these shifts. Furthermore, he said, "If it is noted … that the calcium velocities in any one region are nearly the same and agree approximately with the solar component while the stellar velocities wander all over, it can no longer be assumed that the calcium and sodium vapour is connected with and moving with these stars and the hypothesis of an individual surrounding cloud belonging to each star must also be abandoned."[46] He expressed similar ideas even more clearly in a speech he gave on 21 May 1924 in Quebec City as president of section III of the Royal Society of Canada (RSC):

> The sharpness and stationary position of these lines as contrasted with the diffuseness and shifting position of the star lines unmistakably points to a stationary origin outside the stellar atmospheres. The only possible explanation seems to lie in the presence of very widely, almost universally, distributed but exceedingly tenuous clouds of matter throughout interstellar space; practically at rest with respect to the stellar system. All the stars are moving through these clouds with varying speeds but only the very high temperature stars such as the O's or early B's have sufficient exciting power to produce the sharp calcium and sodium absorption lines.[47]

This latter idea was Harry's.[48] JSP said he was "greatly indebted to his son … for many helpful ideas and valuable assistance on the physical side of the question," and was "glad to acknowledge the stimulating assistance provided by our discussions on the phenomenon and its origin."[49] Unfortunately the idea was wrong. Arthur Eddington soon explained that calcium ions *could* exist in the very cold environment of interstellar space. He also succeeded in dispelling another thorny problem – starlight did not seem to be dimmed by these calcium clouds; they were much too rarefied. (Dust grains were responsible for that.)

It may seem curious that calcium and sodium are the only culprits leaving their fingerprints at the crime scene. Surely they are not the only constituents of the interstellar medium. In fact, as Eddington showed in 1926, they are not.[50] By pure chance, Ca+ and Na are the only ions or atoms abundant enough and at the same time with spectral features in the visual part of the spectrum observable with detectors on Earth at the base of the terrestrial atmosphere, which filters out much of the ultraviolet and other radiation from space. (We know now that nearly all the interstellar gas, like the stars themselves, is hydrogen and helium.)

It may also seem puzzling that the "detached" H and K lines (a much better descriptor than the misleading term "stationary" and so-called because they did not have the same radial velocity as the star itself) were visible only in O and the "early" B stars (i.e., those whose spectra were more like O than A stars). An obvious reason was the sharpness and darkness of the H and K lines that made them immediately evident in comparison with the few broad and diffuse stellar absorption lines. Also, in the later B stars and in the rest of the spectral classes, there were increasing numbers of spectral lines, including H and K lines originating in the stellar atmosphere itself, masking any interstellar lines that might have been present. More significantly, O stars, being highly luminous, could be seen at great distances and therefore their light passed through a greater amount of interstellar material.

Theory and observation went hand in hand in advancing the understanding of the interstellar medium. Without question Eddington did more than anyone on the theoretical side. This was evident in his Bakerian lecture "Diffuse Matter in Interstellar Space," delivered in 1926.[51] (The Bakerian lecture, instituted in 1775, was the Royal Society's highest distinction awarded in the physical sciences.) In his address Eddington explained that proximity to hot O stars not only was unnecessary for the ionization of calcium but that calcium atoms in this environment would lose two of their electrons, and Ca++ had a very different spectrum from Ca+. He also clarified how ionized calcium could exist in "cold" interstellar space. First he established the density of gaseous matter diffused throughout the Galaxy as about 10^{-24} grams per cubic centimeter or less than one atom per cubic centimetre. By laboratory standards that is empty space! Even with huge astronomical distances, there would be only a milligram of material in a 1 cm square column 1000 light years long. Eddington recognized that the interstellar gas is so tenuous that interactions are rare between atoms or between ions and the free electrons left over after the ionization. So, interstellar Ca+ is likely to stay in its ionized state much longer than it would in a denser cloud. Paradoxically the light from a distant star encounters many more Ca+ ions as it passes through this rarified gas than it would in a denser cloud. The formation of an absorption line, as usual, occurs when starlight of the right wavelength, 393.3 nm, happens to encounter the ions, raising their other outer electrons to a higher energy level. When those excited electrons return to their original state, light of the same wavelength is re-emitted but, because it is sent out in all directions, only an exceedingly tiny fraction comes our way, and so our spectroscopes detect an absence of light at 393.3 nm.

Probably for reasons that were outlined in the previous chapter, Eddington overlooked Vesto Slipher's early contribution and consistently credited Plaskett with the first intimation of widespread interstellar material in 1924.[52] For his part Plaskett was "proud of having provided the observational material and given a preliminary hypothesis on which this important contribution of Eddington's may have been based and would like to believe that other work of ours may prove equally useful."[53] It was a great boost for JSP to have his role advanced by a man widely acknowledged as "the most distinguished astrophysicist of his time," and as an author who was greatly admired by the public.[54] Often an authority figure is needed to champion a cause before it gains general recognition and Eddington's Bakerian Lecture can be viewed in that light, as a watershed in promoting the idea of a uniform and pervasive interstellar gas. Of course there have been many developments in the decades since. The uniform distribution is only a first approximation; clumpy clouds is a much more apt description.[55] Even Eddington was not always right. A more modern take on his famous lecture is provided by astronomer and author Gerrit Verschuur: "Few have been willing to suggest that the paper had serious flaws, perhaps because Eddington has since been elevated to a stature larger than life … It reveals how a great mind like Eddington's makes progress through mistakes and an occasional chase up a blind alley which leads nowhere."[56]

Meetings in 1924

The RSC meeting in Quebec in May 1924 was special for Plaskett. Not only was he there as president of the society's section III, but he had chosen to speak on a topic that he hoped would be sufficiently comprehensive to include all the sciences represented in the section: the O-type stars and their relation to the stellar evolutionary sequence.[57]

Another special occasion of 1924 was an infrequent Canadian gathering of the British Association for the Advancement of Science, known simply as the BA, scheduled for August. The BA had been founded in 1831 as a less aristocratic alternative to the Royal Society. In an effort to encourage scientific development outside the metropolitan centres, the BA moved its annual conventions around to smaller cities, mainly in Britain but sometimes in other parts of the empire. Canadian meetings had been hosted in Montreal in 1884, Toronto in 1897, and Winnipeg in 1909, and would once again be held in Toronto, on 6–13 August 1924. The BA was a large organization of almost three thousand members covering an extensive range of pure and applied sciences.[58] Among the 577 BA members making the overseas voyage in 1924 was Arthur Eddington. Immensely popular for his writing and public speaking, he was asked to give "citizen lectures" in Toronto and elsewhere on the subject of relativity and the interiors of stars. Other astronomical presentations were given by Arthur Milne from England, Ludwik Silberstein (recently emigrated from Europe to the United States), and by a good contingent of Canadians: Harper, Harry Plaskett, and Young representing the DAO, and Judson Henderson, François Henroteau, and Meldrum Stewart from the DO. J.C. McLennan and his wife hosted a large garden party at the York Club, and Chant presided at a lunch for forty-three astronomers at the University of Toronto's Hart House, which Eddington described as "the most enjoyable function" of all.[59] Plaskett was at both.

Other bodies sought to benefit from the presence of scientists from abroad, and the International Mathematical Congress planned its first North American gathering for 11–16 August 1924, also in Toronto.[60] JSP was a member of its organizing committee by virtue of his presidency of the RSC section III. American astronomers arranged their annual AAS meeting for the first week of August at Dartmouth College in Hanover, New Hampshire, allowing astronomers the option of going to the conventions consecutively.[61]

A special attraction taken by 360 of the BA delegates was a train excursion (actually two, by CPR and the newly formed CNR) from Toronto to

C.A. Chant
A.S. Eddington
Henry Norris Russell
C.A. French
G. Lemaître
W.E. Harper
J.L. Chant
C.D. Perrine
Cecilia H. Payne
James Stokley
Margaret Harwood.
H.C. Plummer
R. Meldrum Stewart
Reynold K. Young.
W.J. Humphreys.
Mary B. Jackson
Oliver Justin Lee
[illegible]
R.J. McDiarmid
Aline Henroteau
François C. Henroteau
W.E.W. Jackson
F.H. Safford
[illegible]
F.M. da Costa Lobo
Louis V. King.
[illegible]
[illegible]
Sydney Chapman

A. Fowler.
F. Lowater
Jean Mascart
Mme Jean Mascart
G. Van Biesbroeck
J.S. Plaskett
Ernest W. Brown
Ernest G. Hodgson
J. Jackson
A.L. Cortie.
Chas. E. St. John
J.A. Carroll
F.R. Moulton
R.H. Fowler

Astronomical Luncheon in Hart House, University of Toronto, at the close of the B.A. Meeting, August 13, 1924

9.4 Signatures on the menu card of the lunch given for astronomers at the British Association meeting in Toronto, 1924 (RASCA)

Vancouver and Victoria and back, with stops along the way.[62] The hope was that the illustrious and influential visitors would be impressed by Canadian engineering projects as well as points of scenic beauty and scientific interest. To encourage this opportunity J.C. McLennan, in concert with other scientists across the country, including JSP, successfully lobbied the federal and provincial governments to provide a subsidy of nearly $100 000, leaving delegates to pay only $100 return fare (Toronto–Victoria), exclusive of meals, but including sleeping car berths.[63] When the party arrived in Victoria on 25 August, JSP, as the vice-president of the BA's Mathematics and Physics Section, joined the lieutenant governor and the mayor in welcoming the delegates. Those with interests embracing astronomy, about a hundred in number, went to the observatory for a tour and a look through the telescope at the globular cluster M13. Other delegates arriving the next day also included the DAO in their tour. Eddington stayed with the Plasketts before heading off to California to give more lectures.

A sideshow to the scientific assemblage received substantial press coverage, under the banner "Messages from Mars."[64] Each morning, at precisely 7:14 a.m., radio operators at the Point Grey radio station in Vancouver reported receiving four groups of four dashes which some seemed eager to characterize as messages from Mars. With radio still something of a novelty, and the public (though not the scientists) still eager to believe that there was intelligent life on our neighbouring planet (at the time about as close to us as it ever gets), there was enough credulity to give life to the story. Eddington, asked for his opinion, immediately denounced the idea of Martian signals as "too absurd for words." JSP was dismissive because of the fifty million kilometres between Mars and Earth and was more inclined to think the signals arose from some effect in our own atmosphere.[65] Another newspaper more bluntly headlined the reports as "ridiculous," quoting the astronomer "B.S. Plaskett."[66] In the end it seemed likely that some radio buff was either testing equipment or intentionally playing a prank.

Return to England

JSP enjoyed hosting the distinguished British guests and eagerly anticipated his own turn to travel overseas again the following summer. The triennial assembly of the International Astronomical Union (IAU) was to be held in Cambridge, England, on 14–22 July 1925. It was essential for Plaskett to attend, especially because he was on three IAU commissions, but getting Ottawa's approval was, as usual, a challenge. Another pressing reason for him to go abroad was a requirement of the Royal Society. As noted in the preceding chapter, he had been informed of his FRS in 1923 but, in accordance with the society's statutes, his attendance in London was needed, and not later than the fourth meeting after election, for him to be formally admitted. As he told Deputy Minister Cory in January 1925, "I feel that this obligation is also shared by the Department as this honour, the highest scientific distinction in the Empire, was given solely for my astronomical work for the Department. Further it is the only F.R.S. in the Government service and it would ill requite the special honour thus paid to us not to fulfill the necessary requirements."[67] Unspoken but equally true was the irresistible desire to see many of his astronomer friends, such as Annie Cannon, Frank Schlesinger, and Frank Dyson again.

Cory's reply held out a glimmer of hope though he cautioned, "We have been warned that no more extended trips are to be considered for some years to come."[68] JSP sent two more letters reiterating the reasons why he should go, outlining the tributes he had received in the past, and, eschewing any false modesty, stating "that men of science the world over look upon me as the foremost astronomer of Canada and I believe will almost unanimously expect me to be an official representative of Canada at the Cambridge meeting." (He qualified all this by saying how distasteful it was to discuss his personal reputation but those comments somehow seem a bit disingenuous.) Cory's impatience began to show when he wrote to Plaskett that the director of the Ottawa observatory had volunteered to pay his own travel expenses and "from what I have seen of Mr. R. Meldrum Stewart, I feel confident that in the matter of ability he can hold his own with any engaged in his chosen sphere of activity. It may be that his diffident manner has prevented him from getting the advertising that has accrued to others, but I am proud indeed to have him at the head of our Observatory here." Cory's kind words about Stewart were meant as a check on Plaskett's sense of self-importance but were likely sincere. He would not have worked closely enough with Stewart to recognize that his leadership was "uninspiring and lax."[69]

Plaskett finally received permission to travel abroad at government expense less than two months before he needed to leave. Also attending the IAU were R.M. Stewart and François Henroteau from Ottawa along with Chant and McLennan from the University of Toronto.[70] It may seem strange that McLennan, a physicist, would be chosen to represent Canada at the Astronomical Union but he and his graduate student Gordon Shrum had just solved the long-standing puzzle of how to explain the prominent green line in the spectrum of the aurora borealis. A presentation on their results, attributing its origin to ionized oxygen, had been a memorable part of the RSC meeting in Ottawa that May, doubly so for McLennan, who was just concluding his term as president of the society.[71] JSP was able to hear about their discovery first-hand, as he was on hand to present three DAO papers on spectroscopic binaries – the only truly astronomical contributions of the session. A week later, on 29 May, he and Reba set sail for England aboard the Canadian Pacific steamship *Montcalm*.[72] They arrived in London on 10 June and, over the next few days and weeks, were indeed treated royally. On the 12th, JSP participated in a meeting of the Royal Astronomical Society (RAS) at which his friend, Astronomer Royal Sir Frank Dyson, received the society's gold medal.[73] The president, James Jeans, paid tribute to Dyson's years of labour on proper motions: "Such observations, systematic, complete in the sense of covering the whole of the field they set out to cover, perfect in the sense of being as refined as modern appliances and technique can make them, form the foundation on which astronomical progress is built."[74] It is no stretch to say that JSP's radial velocities could have been similarly described. After the formalities, members gathered at the RAS Club, where JSP was toasted and invited by Dyson to his home: "Can you & Mrs. Plaskett spend Sunday June 14 with us? I have asked Miss Cannon to come & suggested bus 53 from Charing X to Greenwich Church. If you come in time for Church, you will see the church where Wolfe was buried. But in any case I hope you will come in time for lunch. It will be a great pleasure to see you both again after these many years."[75] A few days after these happy events, on 18 June 1925, at a well-attended session of the Mathematical and Physical Section of the Royal Society, JSP signed the roll admitting him formally as a fellow, followed by a dinner hosted by Dyson at the Royal Society Club.[76]

The Plasketts were also included in some functions that had nothing to do with astronomy. On the 29th they witnessed the king and queen officially opening Canada House on Trafalgar Square and were invited guests at the reception that followed. They attended a special Dominion Day dinner (with Alfred Fowler as their guest) and a garden party at Buckingham Palace. And they went to the British Empire Exhibition at Wembley, where they found the Canadian displays of agriculture and industry impressive. In between times JSP visited the Royal

9.5 Two photos at the Royal Observatory, Greenwich, taken within a couple of years of JSP's visit in 1925. The first is a view looking north towards the Thames: the second is in the Astronomer Royal's garden. Lady Dyson and Sir Frank Dyson are third and fourth from the left; their daughter is on the extreme right. (Copied from F. Henroteau, *JRASC* 21 (1927): 289, where all seven people are identified)

Observatory, Greenwich, as well as the Imperial College and the Royal Institution, and addressed the two astronomical societies – the British Astronomical Association, composed mainly of amateurs, and the largely professional RAS.[77] Then the Plasketts set off for ten days on the continent, stopping at Paris, Brussels, Antwerp, and Amsterdam, and some of their astronomical institutions, returning to England by 14 July for the IAU general assembly in Cambridge. One of JSP's initiatives at this meeting was his proposal that Harry's classification of the O stars be formally adopted. It would, he said, "bring the classification of these stars into conformity and continuity with the rest of the Harvard system … It is suggested that the O-type stars containing broad emission bands … be expressly omitted from the general sequence for the present, as there is no rational basis for their classification."[78] In the absence of Walter Adams, chair of the commission on spectral classification, JSP presided. His proposal was turned down as a result of objections raised by Cecilia Payne, the recent Harvard graduate whom JSP had considered hiring the previous year. She later wondered how she summoned the courage to challenge Plaskett, saying that she felt "like a puppy defying an elephant."[79] All in all, though, the general assembly was a triumph. Schlesinger thought "it was the most important meeting of astronomers that I had ever attended, and possibly the most important that has ever been held. The success of the Union has delighted as well as surprised me, for the preliminary meeting at Brussels gave me a rather unfavorable impression as to its future usefulness."[80]

There was certainly no doubt that the Plasketts' time in Cambridge was socially sensational. They were entertained and lodged by the vice-chancellor of the university, a privilege accorded only a very few of the delegates. On 18 July 1925 they had dinner at the Newalls in the company of ten others including Dyson, and the Schlesingers and Campbells from the United States.[81] And at the final IAU dinner, JSP was seated at the head table with the master of Trinity College and some other special guests. (Reba and the other women had their own dinner at Newnham College.[82]) To top things off, the Plasketts received an invitation to the Royal Greenwich Observatory on 23 July accompanied by a notice that they would be introduced to the king and queen.[83]

The occasion at Greenwich, scheduled to follow immediately after the IAU, was the 250th anniversary of the founding of the Royal Observatory.[84] Their Majesties, King George V and Queen Mary, toured the famed facility. Although all the IAU delegates and their spouses were invited to the event, only nine were presented to the monarch, JSP and

Reba among them. That evening there was a conversazione at the Royal Society, Burlington House, and the next day, the First Lord of the Admiralty presided over a luncheon at the Savoy Hotel. Again, the Plasketts were seated at the head table. A special session of the RAS followed, at which Plaskett and other notables from abroad each spoke about their work.[85]

Before departing for Canada, the Plasketts enjoyed three more weeks of sightseeing and socializing combined with visits to Hilger and Sons, instrument makers, to Sir Charles Parsons, head of the famous firm of telescope makers in Newcastle, and to the Royal Observatory at Edinburgh.[86] They finally sailed for home from Glasgow aboard the Canadian Pacific steamer *Marburn* on 14 August, arriving at Quebec City on 22 August.[87] When they got back to Victoria, eight days later, they had been away almost four months.

JSP lost no time in giving a full report to the deputy minister, emphasizing that he had been representing Canada and that all the honours were convincing evidence of the standing of Canada in the world of science.[88] Despite the whirl of gala gatherings, Plaskett assured Cory that the unequalled opportunity for discussion and exchange of scientific ideas from which he had profited would enhance the value of future research at the DAO. He provided further confirmation by writing a short article, "Canada Prominent at Astronomical Union," for a newsletter of the Department of the Interior. Over the next four years he would write four more of these newsletter articles about accomplishments at the DAO though, as he noted, "it is not easy to dress up an abstract and highly technical scientific research, especially in the small space allowable, into an attractive popular article."[89]

After his long absence Plaskett told Cory that he was looking forward to getting back to work with renewed vigour and increased confidence. The only catch was his celebrity status. Everyone wanted to hear of his experiences firsthand, and he was happy to oblige, with talks illustrated by the "intensely interesting" slides he showed of sights he had seen and people he had met.[90] In an interview with the *Victoria Times* he singled out his chief impression as admiration for the brilliant accomplishments of British astronomers and the recognition accorded to the work of the Canadian observatories (in reality the DAO).[91] The Victoria branch of the Canadian Club to which the Plasketts belonged hosted them at a luncheon in his honour. In speaking about the purpose and results of his trip he hoped to fit in with the club's purpose "to foster an interest in all matters of public concern in order to strengthen Canadian

9.6 JSP (extreme right) on receiving his honorary doctor of laws degree from the University of British Columbia, 16 October 1925. For a complete legend, see http://www.library.ubc.ca/archives/hondegre.html. (University of British Columbia Archives, UBC 1.1/489)

unity and identity, to preserve and promote heritage, and to understand other nations."[92] Before long another honour was bestowed, this time by the University of British Columbia.[93] Though UBC was founded in 1908, it moved to its permanent site at Point Grey, Vancouver, only in 1925. To mark this auspicious new beginning, it conferred its first honorary degrees on 16 October. JSP was among the recipients of an LLD on that occasion.[94]

Back to Work

The excitement about the trip overseas died down, of course, and JSP returned to his research, focusing on B-type stars – those whose spectra were closest to the O stars, which had been his forte. He was continuing

Table 9.1 Five binary stars studied by Plaskett in 1925–6

Number in the Henry Draper Catalog	PDAO reference	Spectral class	Who first recognized its binary nature and when
139892	3 (1925), 179	B8	Wm Christie (4 May,1924)
25833	3 (1925), 179	B3	JSP (10 September,1924)
191201	3 (1926), 247	B0	JSP (18 September, 1924)
112014 (Boss 3354)	3 (1926), 247	A0	JSP (25 March, 1919)
4161 (21 Cas =YZ Cas)	3 (1926), 247	A2	Joel Stebbins (1924)

the program that he and Pearce had begun in 1924 on all the B stars within easy reach of their spectrograph (i.e., brighter than magnitude 7.5 and north of −10° declination). Among these stars, the DAO astronomers discovered many spectroscopic binaries. JSP had calculated orbits for two of them before going to England and, in his absence, had Pearce give a presentation about them in June at the Pacific division of the AAAS held in Portland, Oregon.[95] But now that JSP was back at the observatory, he was able to complete work on three more binaries for a paper published in January 1926.[96] The five stars are listed in table 9.1, which shows that three came from the B-star program. William Christie, then a university student at the observatory for the summer, recognized the first one, HD 139892, as a binary, while JSP made the discovery of the other two. The fourth entry in the table is a star from the original Boss program. JSP had noted its binary nature in 1919 but only recently had followed up with further observations and calculations. The story of the fifth star will be dealt with shortly.

With the exception of HD 4161, these stars were double-lined systems (i.e., the spectra showed lines of both components, permitting the ratio of the stars' masses to be found). As usual with spectroscopic binaries, the inclination of the orbits was indeterminate, so only the minimum stellar masses could be stated. Nonetheless, for HD 139892 and HD 191201, the individual stars were over 13 solar masses, unusually large for B types. The spectrograms for the latter system proved to be the most difficult and uncertain that JSP had ever encountered, with diffuse, weak lines and the spectra of the two components often blended together. Still, by making reasonable assumptions based on Eddington's mass-luminosity relation, JSP was able to estimate that HD 191201 was very bright and at least five thousand light years away. At the time, few individual stars were known to be at such a great distance. The investigation was a nice example of the interplay between observation and theory. Plaskett had contributed vital data to Eddington and now Eddington's conclusions were used by Plaskett to interpret new observations.

Finally, we come to the last entry in the table. We saw in chapter 5 that JSP had studied several eclipsing variables that were also spectroscopic binaries, all suggested by the Princeton astrophysicist H.N. Russell. In more recent investigations, another came to his attention, 21 Cassiopeiae, whose very small light variation was discovered with a photoelectric photometer by Joel Stebbins, the director of the Washburn Observatory in Madison, Wisconsin.[97] In 1924 Stebbins had asked Plaskett to look into the time of the spectroscopic phases of this star to see if they agreed with the times of the photometric phases. (At mid-eclipse the orbital velocities should have no radial component.) Stebbins had found a discrepancy of twenty-four minutes, and he thought that tidal effects between the two stars might account for it. Hints of such distortions, especially for systems where the two stars were close together, were just beginning to surface at this time.[98] However when JSP completed his investigation in January 1926, he found a simpler solution: a perfectly circular orbit rather than an elliptical one would fit the observations just as well and would reduce the phase difference to an acceptable four minutes. An interesting aspect of his research was his realization that a very small distortion in the velocity curve, just as the brighter star was being eclipsed, could be attributed to its axial rotation.[99] It seemed that the period of the star's rotation and revolution was the same, a trait of bodies locked by tidal interaction.

By August 1926 JSP had completed an analysis of another three stars; these had peculiar spectra that did not fit the Harvard classification scheme. Two were definitely not binaries (HD 50820 and 45910). The third, Upsilon Sagitarii, was especially puzzling.[100] The presence of neutral helium lines suggested a B star though the lines were uncharacteristically strong and sharp. The hydrogen lines were somewhat like those in an A star, though less intense, and occasionally showed emission as well as absorption. And then there were elements of an F-type spectrum, with abundant metallic lines, ionized silicon being especially prominent. To prior investigators at Lick and Harvard, the composite spectrum and shifts in the lines had suggested the star's binary nature. Hans Ludendorff of the Potsdam Observatory analysed the data and came up with enormous masses for the two components, one at least 260 and the other 54 times that of the Sun. Unconvinced by these claims, the DAO astronomers acquired more spectrograms between November

1924 and May 1926. These led Plaskett to the conclusion that Ludendorff's very high mass estimate was based on the faulty premise that a 300 km/s velocity of approach in the hydrogen lines was due to orbital motion. In fact the radial velocity of the hydrogen lines was fairly constant, suggesting to JSP that there might be an outflow of gas from the star. The smaller shift in the other lines, however, he found to be consistent with a single star in orbit around a fainter companion whose spectrum did not register.

He wrote about Upsilon Sagitarii and the other two peculiar stars for the RAS, an account that Frederick Stratton presented in his absence to the society on 12 November 1926 in London.[101] The ensuing discussion, duly summarized in the *Observatory*, led Plaskett to investigate further. In particular he answered theorist Arthur Milne's question as to which spectral type most closely matched the intensity distribution of the star's spectrum. By judicious planning JSP photographed the spectra of five stars, each with three different exposures, all on one plate and taken on one brilliantly clear night, ensuring, as far as possible, identical conditions for each of the stars whose known spectral types ranged from B5 to F0.[102] He clearly showed that the intensity distribution of Upsilon Sagitarii matched the F0 spectrum very closely. In other words the temperature of this mysterious star was that of an F0 star. As for the unusual lines in the spectrum, he answered a request from Frank Baxandall for their accurate wavelengths by measuring close to three hundred lines in the spectrum with an accuracy of ± 0.005 nm. With Baxandall's help, he identified most of them with various metals such as iron, titanium, chromium, vanadium, zirconium, and so on.

During Plaskett's directorship, the DAO (though not JSP himself) led the world in discovering spectroscopic binaries and in computing their orbits. In an interesting table compiled in 1930 by Plaskett and Pearce, 319 such systems had been announced by eleven observatories worldwide and 100 orbits had been computed for 76 of these systems.[103] The DAO was responsible for 159 of these announcements – over twice as many as the Lick Observatory, which ranked second with 66. DAO astronomers had computed 27 of the orbits, again well ahead of the next-ranked Allegheny Observatory at 18. Plaskett and the DAO astronomers published several lists of the spectroscopic binaries that they had discovered, but they left the publication of comprehensive catalogues of all these objects to the Lick Observatory, where the tradition of such catalogues had been established in 1905 and continued with subsequent editions in 1910, 1924, 1936, and 1948.[104] As for JSP's personal contribution, during his Victoria years he discovered 94 binaries but calculated only 19 orbits, compared to over a hundred that Harper completed during a similar timespan.[105] If that seems like a modest accomplishment (and it really was not), remember that it was only one aspect of JSP's scientific studies.

Moreover, his administrative duties grew along with his status, even beyond his usual sphere. H.M. Tory, chair of the National Research Council, appointed JSP to head the British Columbia division of its physics committee in December 1925.[106] With UBC still in its infancy, Plaskett was really the only senior leader from that province qualified to fill the role. The committee had been given the responsibility of preparing a list of tasks suitable for study by the NRC, and its members from across Canada were to meet occasionally to discuss their progress and defend their positions. The first gathering took place in Montreal at the end of February 1926, conveniently coinciding with joint meetings of the American Physical Society and the Optical Society of America.

Because there was friction between the NRC and the various government departments where scientific work was done, Tory's move in getting JSP on board was tactful and perhaps tactical. In Tory's view, the Department of Interior had been uncooperative with the NRC. However, Deputy Minister Cory was not a member of the Research Council and for some time had felt he had been left out in the cold. Plaskett, on very cordial terms with the deputy minister, expressed his belief that if Cory were appointed to the council the attitude of suspicion might be altered to one of support. Knowing, as we do, how zealously Plaskett guarded the autonomy of "his" observatory, it seems surprising that he said his wish was "to further the influence of the Council."[107] Perhaps he recognized that if there were any threat from a burgeoning NRC it would be best to have Cory involved in the decision making. As it turned out, Cory was never appointed to the NRC, and the observatory did not come under its authority until 1970.

10
Beyond the Stars, 1927–1930

The year 1927 marked the diamond jubilee of Canadian confederation. The Royal Society of Canada (RSC) celebrated by inviting Plaskett along with a chemist, mathematician, meteorologist, and physicist, to deliver addresses on the progress of their disciplines in the past sixty years.[1] In truth there was no one else in the country well qualified to speak for astronomy. With only a couple of exceptions, the only astrophysical papers presented to the society or published in its transactions until that time were by JSP or by those he had sponsored. Always happy to inform the most renowned scholars in the country of the vital work he and his colleagues were doing, he not only gave his invited paper but also spoke about the latest research at the Dominion Astrophysical Observatory (DAO) on the B-type stars.[2]

As he so often did, JSP combined his trip to Ottawa for the RSC in May with other commitments. On his way there he stopped off to speak to the Royal Astronomical Society of Canada in Winnipeg, repeating a talk, "The Galaxy and Beyond," he had given in April to the RASC in Victoria.[3] The title itself hinted that Plaskett was beginning to embark on another major project, one where he would look for large-scale characteristics rather than focusing on individual stars. He also spent three days in Toronto, where professor C.A. Chant thought there was finally a prospect of getting funding for a large telescope at the university. Discussions between them about telescope design may have influenced Reynold Young, now working with Chant, to personally undertake the actual construction of a 48 cm reflector with many of the features of the DAO instrument.[4] Immediately following the RSC meetings in Ottawa, Plaskett stayed on for deliberations of the Physics Committee of the National Research Council.[5]

Plaskett had not been to the United States for four years and no doubt felt he should reconnect with his American colleagues. So, once his Ottawa business was completed, he joined his co-worker Ed Harper in Reno, Nevada, where he spoke to the Pacific Division of the American Association for the Advancement of Science (AAAS) about a specific double-lined B-type spectroscopic binary, HD 193536.[6] It exemplified a problem that sometimes caused complications for those investigating double stars. If the binary's period of revolution is almost an exact number of days, as happened in this case (2.9849 days), there would be inevitable gaps in the observations during daylight hours – breaks that could be filled in only by extending the observations over at least a year or by enlisting the aid of observers in different time zones. For this particular star, Plaskett obtained thirty-four plates spanning two and a half years before he felt confident enough to calculate its orbit.[7]

Six months after the Reno meeting, the AAAS elected Plaskett to office in absentia. He had joined the association in 1922 and had been elected a fellow the following year.[8] Now, late in 1927, he became one of their vice-presidents as well as chair of Section D (Astronomy) for a two-year term. JSP regarded this position as "the highest honour that may be conferred ... on a man of science with the single exception of the presidency of the Association itself."[9] Coincidentally, in September 1927 he was elected a vice-president of the American Astronomical Society (AAS). Surprisingly, he did not attend any of the society's four meetings during his two-year term, probably because none of them was held further west than Chicago.[10]

Plaskett, now in his sixties, showed few signs of slowing down, but a younger generation of astronomers was coming to the fore, sometimes encroaching on territory where he had previously been unchallenged. One such young astronomer, Otto Struve, came from a long lineage of legendary astronomers.[11] He had recently escaped from repression in the Soviet Union and had earned his PhD from the University of Chicago, where Edwin Frost was able to offer him employment at Yerkes Observatory. Through Struve's connections, in the early 1920s Frost convinced other astronomers, including JSP, to provide financial support to destitute astronomers remaining in the Soviet Union.[12]

By 1925 Struve was an astrophysics instructor at Yerkes, beginning to investigate interstellar matter. He extended a new approach initiated by Donald Menzel – studying the intensity (darkness) of the "detached" K line in O and B stars in different regions of the sky.[13] Using spectrograms taken at Yerkes, as well as at Lick, Mount Wilson, and the DAO (which he visited in 1926), he recognized that the K line intensities were strong in some regions of the Milky Way and fainter in others, suggesting that the calcium clouds where the absorption originated were patchy rather than uniform. He also noted that fainter stars, which tended to be further away, had stronger K lines. His work merged closely with ongoing research by Plaskett and Joseph Pearce at the DAO. Because of the cooperation he had received from Plaskett, Struve naturally shared his findings with him before publishing them. JSP's reaction, backed up by Pearce and Harry Plaskett, was somewhat prickly.

JSP thought, with some justification, that Struve's sweeping conclusions were based on too few spectrograms that he had measured in haste, and he suggested that Struve add a disclaimer stating that, as he did not have enough data, his results were "questionable."[14] On the other hand, astronomers at Mount Wilson and Lick reacted quite positively to Struve's draft, so JSP seemed, in this instance, to be betraying resentment of a much younger colleague who was stealing his thunder. Struve made some tactful modifications, acknowledging JSP's papers of 1924 as "the most important observational investigation in recent years" and praising "the excellence of the Victoria equipment and the special care taken by Dr. Plaskett."[15] JSP was satisfied and Struve's "Interstellar Calcium" was published in the *Astrophysical Journal* in 1927. Time would show that it was a significant step; subsequent analysis of the large Harvard collection of objective prism spectra by Struve and others confirmed his early findings.

The disagreement illustrated the difference in approach between a young, impetuous researcher and a conservative, mature scientist preferring to wait until more complete and homogeneous data could provide more definite quantitative results.[16] But it also was a sign of the times. Struve and his generation were motivated largely by H.N. Russell and A.S. Eddington, each many years younger than Plaskett.[17] Russell, in a key article, wrote in 1920 that "the main object of astronomy, as of all science, is not the collection of facts, but the development, on the basis of collected facts, of satisfactory theories regarding the nature, mutual relations, and probable history and evolution of the objects of study."[18] At almost the same time, Eddington said, "If we are not content with the dull accumulation of experimental facts, if we make any deductions or generalizations, if we seek for any theory to guide us, some degree of speculation cannot be avoided."[19]

JSP would not have found fault with either of these statements but would still have preferred more circumspection. In the end, though, there were no hard feelings between him and Struve; in fact Plaskett and Pearce came to the same conclusion as the young astronomer in their "Motions and Distribution of Interstellar Matter," which was presented to the Royal Astronomical Society (RAS) by Eddington in November 1929 and published the following January.[20] The DAO astronomers were able to use reliable, homogeneous data by basing their results solely on their own plates, captured using a new spectrograph with quartz optics, whose transmission of short wavelengths was much superior to what the Yerkes equipment with its glass lenses and prisms could achieve. These advantages enhanced their paper's authority in confirming Struve's contention that the intensity of the interstellar lines is proportional to distance.

Another example of Struve's interest in Plaskett's work was the O-type spectroscopic binary B.D. 56° 2617 or HD 206267. JSP had published its orbit in 1923 but, thanks to an anomaly that Struve pointed out to him, in 1928 he revised his earlier calculations and increased the mass of the system threefold, bringing it into closer agreement with other O stars.[21]

Administrative Concerns

Plaskett seemed genuinely disposed to get along with his Ottawa counterpart, R.M. Stewart, who, for the most part, ably balanced the conflicting demands of the Dominion Observatory (DO), departmental regulations, and Plaskett's perceived priorities. By 1926 Plaskett was suggesting that the cost of a new microphotometer might be shared between Ottawa and Victoria and that it could be kept in Ottawa but used to measure DAO plates as the need arose. He also suggested that Stewart come to Victoria so they could talk over various staffing options.[22] Though government cost-cutting measures delayed the realization of the visit, the invitation was a sign of improved cooperation between the two observatory directors.

Productivity at the DAO continually increased as improvements in the spectrograph and photographic plates allowed a decrease in exposure time by 25 to 50 per cent. There was even some upgrading of the antiquated calculators when a second hand-operated Monroe machine was purchased for $300 in 1926.[23] By 1929 Harper was able to report that as many as twenty-four spectrograms could be obtained in a single night.[24] Indeed the younger astronomers, especially, liked to compete to see who could secure the most plates in a night.[25] JSP's management style, leading by example and persuasion, no doubt had a lot to do with the success of the DAO's operation. But any impression of the astronomers as mere automatons, observing by night and churning out data each day, should certainly be corrected. They were keenly aware of the context of their work and kept up to date by hearing from guest astronomers or by discussing current topics at seminars held every two weeks.[26] They published research papers, spoke at conventions, gave popular talks, and welcomed hundreds of people each Saturday evening when the observatory was open to the public. The annual visitor count (day and night) exceeded 30 000 for several years.[27] Pearce found it "quite fatiguing to handle the 250–300 visitors who, fairly patient but frequently noisy, awaited their turn for a 10-second peep at the objects we were describing."[28] A local journalist had painted a decidedly rosier picture of a group visit to the observatory in "A Trip to Arcturus." He described Plaskett, sitting on a step ladder, chatting informally and answering questions "with assured answers and a good-humored smile that lights a very pleasant face ... The man is not only interested in his work, he 'loves it.' And he wants others to love it as much as he does, and so the crowd does not bore him and, in turn, the crowd is not bored and takes the knowledge he imparts with a delighted camaraderie."[29]

The DAO continued to benefit from temporary workers who were paid less than the permanent astronomers but who sometimes shared in the routine programs and offered valuable external expertise. In response to an enquiry from Stewart in 1926, JSP clarified the types of temporary staff at the observatory as follows.[30] Research associateships, at $200 per month, were intended for established astronomers who might be persuaded to spend two or three months at the DAO. (He had in mind people of the caliber of Eddington or E.A. Milne but generally ended up with astronomers of somewhat less prominence.) Graduate fellowships worth $150 per month, coupled with the opportunity to use the big telescope, were intended to attract the most promising graduates of Cambridge, say, to come for a year. (There was no doctoral program in astronomy in Canada until the 1950s.[31]) Undergraduate fellowships, at $100 per month, were intended to give summer work to promising Canadian university students. Having help in the summer was essential to keep the telescope in use, as the nights were usually clear while at the same time at least some of the permanent staff were away on vacation. Some of the most notable Canadian students (and their DAO years) were William H. "Harry" Christie (1922–6); Robert M. "Bert" Petrie (1925–8); Peter Millman (1927–9); and Andrew "Andy"

McKellar (1931, a year after getting his BA). All of these men subsequently had distinguished careers as astronomers.[32] Christie must have felt sorely tempted by an offer from Mount Wilson of $2000 per year for a computer's position at a time when he had just two summers at the DAO under his belt. Instead he opted to continue his studies in Canada and his summer work at the DAO.[33]

As well, Canadian physics professors continued to come to the observatory for unpaid experience, usually during the summer months. They included an Oxford graduate and war veteran, Stanley Smith from the University of Alberta in 1924; Alfred E. Johns, who, in 1926 had recently returned from fifteen years as a missionary in China; and two people from McGill in 1932 – Allie Vibert Douglas (generally considered to be the first woman in Canada to be a professional astronomer[34]) and her thesis supervisor, John S. Foster.[35]

Many foreign researchers visited over the years, profiting from the privilege of using the facilities for short periods of time and enriching the intellectual climate for the resident astronomers. Samuel Boothroyd from the University of Washington was the first. Other Americans during JSP's directorship included Clifford Crump (1921) from Ohio Wesleyan; Henry N. Russell (1921)[36] and Raymond Dugan (1929) from Princeton; Otto Struve (1926) from Chicago; and Heber D. Curtis (1931) from the University of Michigan. European astronomers staying at the DAO for at least a couple of weeks included Karoly Lassovsky (1925) from Hungary; Antonie Pannekoek (1929) from Holland; Paul ten Bruggencate (1930) from Germany; and John Harald Petersson from Sweden (1931).[37]

JSP was sometimes able to help foreign astronomers who had no large telescope at their disposal and who could not come in person to the DAO. Before he came to Victoria in person, Antonie Pannekoek in Amsterdam was one of the most persistent, asking for observations as early as 1921 to help him investigate the dependence of spectral line-intensities on the physical state of matter in the stars such as temperature, pressure, gravity, and density.[38] JSP tried to oblige but Pannekoek did not seem to realize the practical difficulties in getting high dispersion spectra covering a broad range of wavelengths. Finally, Plaskett arranged for him to spend six months at the DAO in 1929, during which time Pannekoek himself acquired an extensive series of plates, ultimately authoring an important paper in 1946 for the *Publications of the DAO*.[39] Some other observing requests that JSP fulfilled came from Lord Rayleigh of the Imperial College, London, from Ejnar Hertzsprung on behalf of Adriaan Wesselink in Johannesburg, and from Kurt Walter in Koenigsberg, Germany.[40]

As for the permanent staff, after Young's departure and Pearce's arrival in 1924, there were no changes until 1927–8, when a couple of developments led to an increase in the number of astronomers from four to five. The first occurred as a result of an offer from Harlow Shapley at Harvard inviting Harry Plaskett to lecture there for a year. The offer originated in correspondence four years earlier, when JSP asked Shapley and Russell to send letters endorsing Harry's work on O-type stars to help JSP secure a promotion for his son.[41] Shapley responded right away, not only lauding Harry's qualifications but offering him a position at the Harvard observatory. At the time Harry preferred to stay in Victoria but the seed had apparently been planted. When he finally accepted in 1927, Harry wrote Shapley, "Father thinks there will be no serious difficulty in securing a year's leave of absence for me without pay. In fact the only difficulty about leaving here for that length of time is that neither of us looks forward to the separation very much. We are awfully good friends and have such fun together that it is going to be rather hard."[42] JSP of course felt the same way. "Harry goes to Harvard in February for a year to give a course of lectures on astrophysics," he wrote. "We will all miss him sorely but I am glad he will have the association with the group of astronomers and physicists at Harvard. I consider it an honour for him to be asked to give this course and it will be of great advantage to him and his future work."[43] Cecilia Payne, whom we met as a possible candidate for a DAO position in 1924, had an interesting perspective on the circumstances of Harry's appointment. In her autobiography she explained that when Shapley succeeded Pickering as director of the Harvard College Observatory, part of his mandate was to develop a closer affinity with the university by offering courses and an advanced degree.[44] Shapley himself had no interest in teaching and hired Harry to fulfil those aims. Harry did what was expected, setting up a course of graduate lectures and seminars, though apparently without a great deal of enthusiasm.[45]

The arrival of a new member of the young Plaskett family no doubt made Harry's departure from Victoria even harder to face for JSP and Reba. A new baby, John Stanley Plaskett, was born on 10 March 1927.[46] JSP proudly let Frank Dyson know of his namesake's arrival, and the astronomer royal replied with his congratulations and details about his own six grandchildren.[47] The Dysons, of course, knew both generations of Plasketts well.

Harry's initial one-year appointment at Harvard became permanent, as he advanced to associate professor of astrophysics and, one year later, to a full professorship. Even before these developments, his departure from the DAO freed up a spot that was filled by Carlyle S. Beals.[48] Like the younger Plaskett, Beals had studied spectroscopy with Alfred Fowler in London, though he proceeded to earn his PhD from Imperial College and then taught physics for a year at Acadia University in his native Nova Scotia prior to joining the DAO staff. JSP thought he was "just the kind of man we have been looking for."[49] Indeed, the investigations on which he embarked dovetailed with the work on O-type stars done by both Plasketts. The Wolf-Rayet stars that Beals studied were hot and massive stars like the other O stars but with spectra characterized by emission lines. For this research, Beals designed a Littrow-type grating spectrograph with quartz optics, which, unlike glass, transmitted ultraviolet radiation. JSP contracted Sidney Girling (who later joined the staff as a very skilled instrument maker) to build it. The instrument would be in almost daily use for the next twenty-five years, outlasting even Beals' lengthy career at the DAO.[50]

The other change in staff came about because of an initiative, originally suggested by Harry Plaskett, to encourage young astronomers from overseas.[51] JSP wrote to Frederick Stratton, the well-connected English astronomer:

> Perhaps Eddington has told you we are establishing a yearly Fellowship here worth $1800, payable monthly for the year, to graduate students in astronomy and astronomical physics. This scheme will, I think, be mutually advantageous giving the graduate useful experience in a modern observatory and us the benefit of new viewpoints on our science. Our present idea is to confine the appointees to graduates of English Universities, preferably Cambridge, and we hope to get a regular flow of keen young men from England. The amount with care should easily pay all traveling and living expenses and the arrangement should be of benefit to all concerned.
>
> Eddington has just told me of a young graduate named Redman whom he thinks would be very suitable but he wants to keep him another year for additional training before letting him come. This suits us very well and I hope you also will be on the lookout for bright young men in astrophysics to whom such practical experience as we can give would appeal.[52]

Roderick Redman did come to Victoria, in August 1928, but he was the only person whom Plaskett could ever lure to the DAO in this graduate fellowship category – economic constraints would soon put an end to any hope of further appointments. The projects Redman undertook at JSP's suggestion formed part of his thesis for his PhD, awarded by Cambridge in 1930.[53] Redman was appointed to the DAO permanent staff in 1929 and stayed until 1931 when he returned to England as assistant director of the Solar Physics Observatory in Cambridge. The new post was a fine opportunity for him at his alma mater though JSP was "exceedingly sorry to lose him" as he was "a great favourite" at the observatory.[54]

The Dimensions, Structure, and Rotation of the Galaxy

JSP saw himself in the middle of three phases of astronomical research. "In earlier centuries," he wrote, "the solar system was probably considered the major problem of the day, while the problems of the universe as a whole will probably engross the astronomers of the future."[55] Indeed they were even then beginning to grapple with such questions. But the middle phase, especially prominent during the period between the two world wars, was the determination of the dimensions, structure, and dynamics of the Milky Way.

Astronomers had long understood that the Sun was one of an immense number of stars in the Milky Way. The prevailing idea in the early twentieth century was that the Milky Way, with embedded clusters and nebulae, was the entire universe. But then other galaxies (or "milkies") began to be understood as systems separated by vast distances; they were at first called "island universes."[56] It was also understood that the band-like appearance of the Milky Way implied that our Galaxy was disk-shaped. Its central plane could be thought of as the Galactic equator and the positions of the stars could, in fact, be given as Galactic latitude (the angle above or below the equator that bisected the Milky Way) and longitude extending along the equator from some initial arbitrary point.

The dimensions of the Galaxy were still very uncertain. J.C. Kapteyn's cosmos, in its 1922 form, seemed to be built on irrefutable statistical evidence showing that the number of stars thinned out in all directions as fainter and fainter ones were studied.[57] His model of the stellar system was ellipsoidal in shape, something like two saucers with their rims glued together, about 50 000 light years across and 10 000 light years thick at the hub, with the Sun only slightly off centre. Meanwhile Shapley, using globular clusters, had come to the conclusion that the Sun is in an eccentric position, 65 000 light years from the centre of the system, which lies in the direction of the constellation Sagittarius. He figured the diameter of the Galaxy was 300 000 light years and its thickness was 16 000 light years – such large dimensions that the idea of island universes seemed unnecessary.[58] Ultimately both models had something to commend them – Kapteyn's for its size and Shapley's for the highly eccentric position of the Sun and the direction to the centre. The dimensions found by Shapley would soon be greatly reduced once it was realized that absorption due to interstellar dust dimmed the stars, thus making them seem farther away than they actually are.

JSP's own investigations into the structure of the Galaxy first surfaced in his 1924 paper on the O-type stars, where he included a section on statistical aspects of this class of stars.[59] There he discussed their distribution and found them to lie very close to the plane of the Milky Way (Galactic latitude = 0°) though skewed a bit, suggesting that the great circle adopted as the Galactic equator was not exactly right. By combining these stars' very tiny proper motions (transverse to the line of sight) with their radial velocities, he was able to estimate their distances. He calculated that they were, on average, about 3200 light years away, which in fact explained why they appeared to hug the Galactic plane so closely. Their great distance also implied that the O stars were thousands of times more luminous than solar-type stars (by about 9 magnitudes) – much greater than previously thought. Plaskett also discussed the space velocity of the O stars. After accounting for the well-known motion of the Sun relative to its neighbours in the Galaxy, he was left with sizable, though smaller, puzzling residual velocities for the O stars. A clue to the solution had been provided in 1923 by German astronomers Erwin Freundlich and Emanuel von der Pahlen, who found a dependence of the residual velocities on Galactic longitude.[60] The pieces of the jigsaw puzzle were beginning to fall into place but there was still no clear "picture on the box" to serve as a guide.

No particular insight was needed to imagine that a flattened, disk-shaped system like the Galaxy might be rotating, but proving it did so was another matter. If it rotated like a solid wheel with all parts moving with the same angular velocity, there would be no relative motion to measure. But as early as 1871, a Swedish astronomer, Hugo Gyldén, understood how the motion of stars in the Galactic disk would be perceived if they were revolving around a distant hub much as the planets in our solar system revolve around the Sun.[61] Those closest to the centre, like fleet-footed Mercury, would zip around quickly, while those far from the centre would move at a statelier pace. To appreciate his

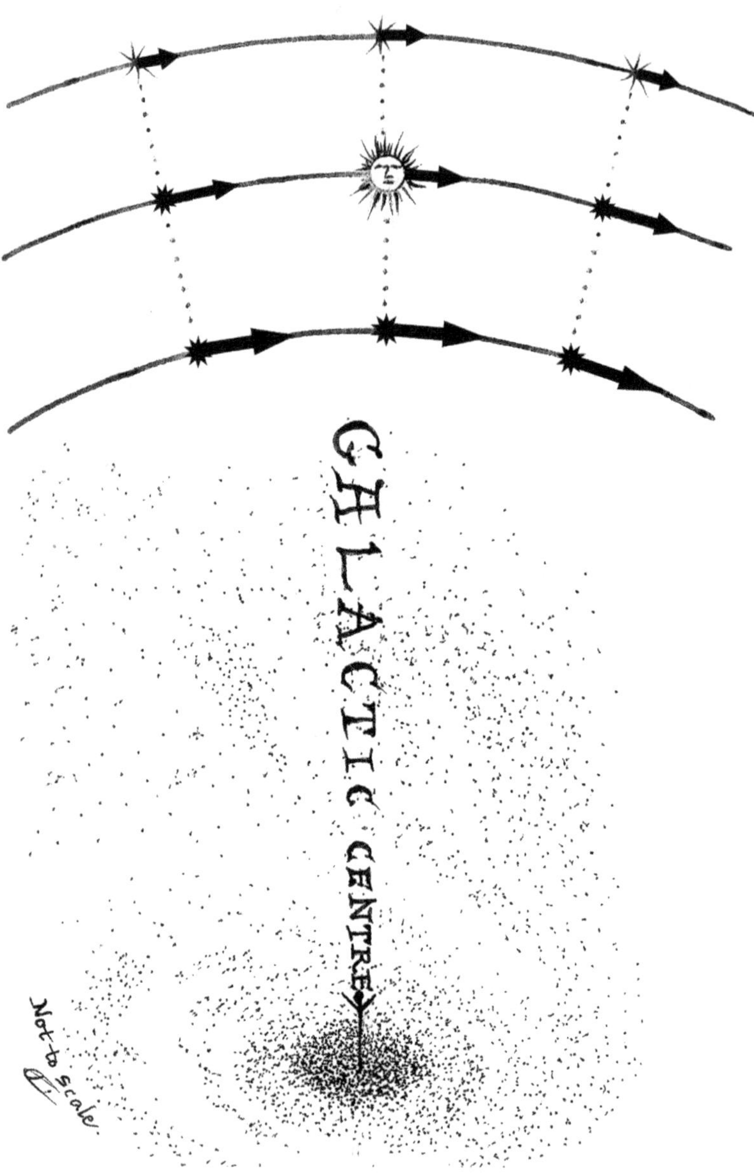

10.1(a, above) The Sun and eight imaginary stars in our Galaxy are shown revolving around a distant centre. As shown by the arrows, those further from the centre are presumed to move more slowly. 10.1(b, opposite) When the effect of the Sun's motion is removed (by vector subtraction), the residual velocities demonstrate the radial velocities that we observe as we travel along with the Sun. Notice that the dotted arrows in (b) are the same length as the corresponding arrows in (a) but point in the opposite direction. (Diagrams by Randall Rosenfeld)

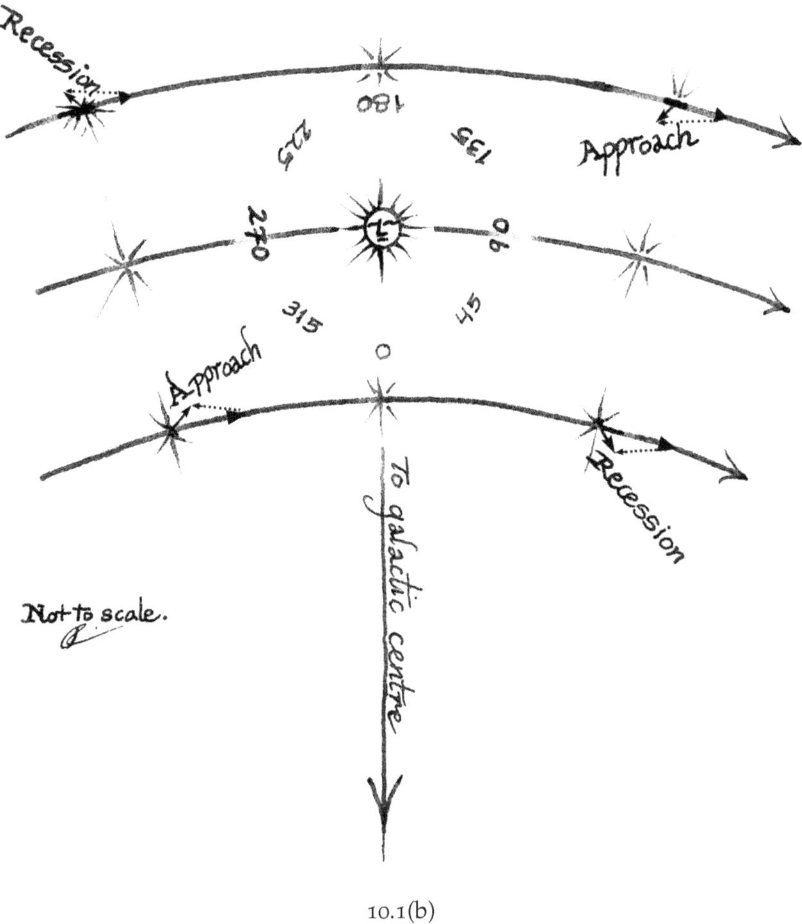

10.1(b)

reasoning, imagine that all the stars, including the Sun, are moving in circular orbits in the plane of the Galactic equator, and think of those not too far from us, at least in comparison to the overall size of the Galaxy. As we look towards the Galactic centre, or in a direction 180° from it, stars would have zero radial velocity, as their observed motion is entirely at right angles to our line of sight. Those observed in the 90° or 270° directions would also have no radial velocity, as they would be on orbits similar to the Sun's. Those at 45° would have positive radial velocities because they are receding from us, and those in the opposite part of the sky, at 225°, would also have positive radial velocities because we are receding from them. In the other quadrants, at 135° the radial velocities would be negative, as we seem to be catching up to the stars further from the centre, and in a direction at 315° from the Galactic centre the radial velocities would also be negative, as those stars seem to be gaining on us.

Another way of expressing Gyldén's theory is that a graph of the expected radial velocities at different longitudes would trace out a sine wave repeating twice in the complete 360° span of galactic longitude. The stars' proper motions would form a somewhat similar curve. Gyldén tested his theory by studying the motions of the asteroids in the solar system without assuming that they circled the Sun; he was able to establish that they were indeed orbiting a central point in a direction only about 6°away from the Sun. With such data, one could come up with a fairly accurate idea of where the Sun was, even if it were invisible. In the same way, Gyldén attempted to find the direction to the centre of the Galaxy by studying the proper motion of the stars. He did find the approximate direction of the line running to and from the centre, but with no way to decide whether the true centre would be in Sagittarius or 180° away in Taurus, he had the misfortune to choose the wrong one – Taurus.

Unfortunately almost no one paid attention to his work at the time, not because they knew any better about where the Galactic hub was, but because he published his findings in Swedish in a journal that was rarely read outside his homeland. Over sixty years after Gyldén's investigation, his compatriot Bertil Lindblad rediscovered his work, but by then Lindblad had already proposed, in 1925, a more complicated model. He proposed a rotating Galaxy composed of a slowly rotating spherical halo populated by globular clusters while faster-moving stars and clusters occupied a rapidly rotating and flattened disk.[62] Of course, from our point of view in the disk, the globular clusters seemed to be the speedy ones while closer objects in the disk appeared to move slowly.

Lindblad's Dutch counterpart Jan Oort provided confirming evidence for a rotating disk.[63] Also unaware of Gyldén's research, Oort reinvented the process of differential rotation and the double sine waves that resulted for radial velocities and for proper motions. From the equations of these two curves, two so-called Oort constants can be found that lead to a value for the distance from the Sun to the centre of the Galaxy.[64] In a series of publications in 1927 Oort examined the radial velocities and proper motions of 700 objects of all spectral types and, after allowing for the Sun's own peculiar motion relative to its immediate neighbours, he concluded that the Sun was orbiting the hub of the

Galaxy and that the direction to the centre was the same as Shapley had determined using globular clusters. The amplitude of the sine wave for radial velocities depended on the distance of the stars being studied but, even for stars 3000 light years away, it did not exceed 20 km/s. That relatively small effect for relative motion should not be confused with the orbital speed at the Sun's distance of about 272 km/s relative to Galaxy's centre. Oort used this latter orbital speed, derived by assuming the system of globular clusters to be stationary, to calculate the Sun's distance from the centre of the Galaxy as 20 500 light years, though with a large uncertainty (mean error) of ± 6500 light years.[65] Furthermore, in the same way that Newton's law of universal gravitation can be used to calculate the mass of the Sun from the velocities and distances of the planets, Oort was able to estimate the mass of the Galaxy (or at least the part closer to the centre than the Sun) as sixty billion solar masses.

While it was natural to assume that stars would orbit the Galactic centre much as planets revolve around the Sun, other scenarios were possible. Even if the stars all had the same linear velocity, Oort could still have derived values for his constants from the observations.[66]

Though Oort relied on DAO data for half of the radial velocities he used in his seminal work, JSP was only vaguely aware of these developments. He skimmed Lindblad's publication but found the mathematics in it rather forbidding, and he read Oort's papers without paying definite attention to them.[67] It took a letter from Frank Schlesinger in July 1927 to alert JSP to the possibility of his confirming Oort's work. By this time, Schlesinger was at Yale, where he had taught young Oort in 1922–4 and where his associate Jan Schilt had recently corroborated Oort's findings using proper motions. Schlesinger thought that Plaskett might do the same using radial velocities of faint and distant stars, providing that they were well distributed in Galactic longitude. JSP knew that he could not cover the complete range of longitudes from the DAO, as the observatory's northern location meant that part of the Galaxy never rose above his horizon, but nonetheless he realized that he already had a wonderful database of O and B stars that he and Pearce had been accumulating. So they were able to begin looking into the rotation of the Galaxy almost immediately. He reported their findings to Oort on 28 November, who replied "I am almost surprised at the accuracy with which your rich and homogeneous material of faint B and O stars confirm the rotation effect. I had not expected that so much important material would so soon be available."[68]

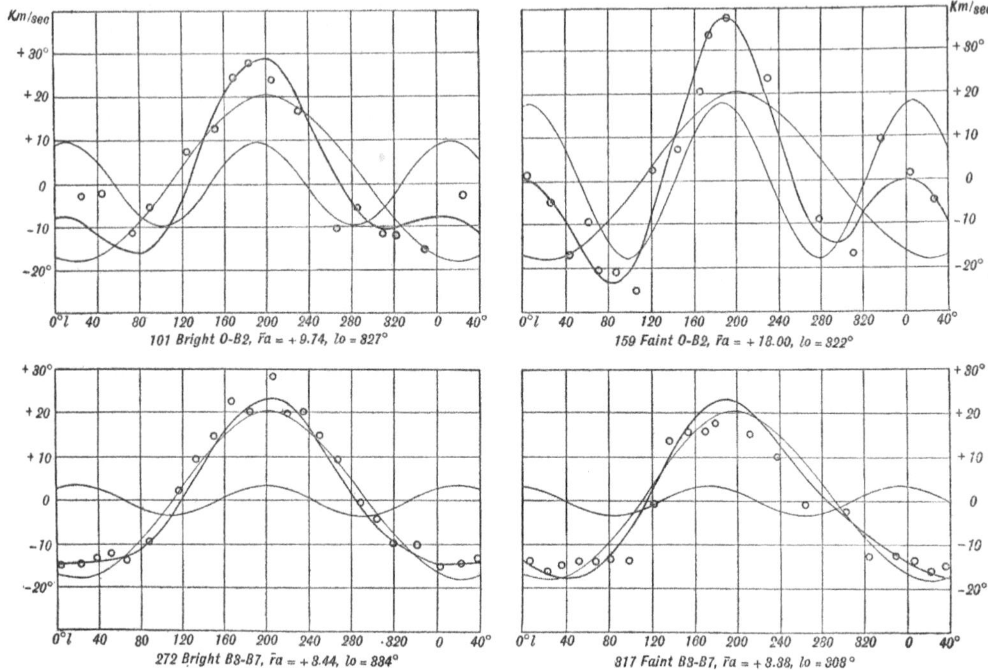

10.2 Radial velocities of various groups of stars observed by Plaskett and Pearce plotted (as circles) against their galactic longitude. In each figure, three curves are shown: the double sine wave due to galactic rotation (with maxima at about 10° and 190°), part of a sine wave (with a single maximum showing at about 200°) representing the projection of the effect of the Sun's motion relative to its neighbours, and a compound wave showing the sum of the two previous ones. Agreement between theory and observation can be judged by comparing the plotted points with the last named irregular wave. (Reproduced from Plaskett and Pearce, *PDAO* 5 (1935): 273)

By January 1928 Plaskett had submitted a paper on the topic to the RAS, and at a meeting of that society on the 13th of that month, Sir Frank Dyson presented a summary of it and led a discussion.[69] The full paper was published in the RAS's *Monthly Notices* in March. JSP's analysis was based on DAO radial velocities of 549 stars of spectral types O to B5 and within 30° of the Galactic plane. About half of the stars were fainter and therefore more distant than the stars that Oort had been able to use in his investigation. Even though the southernmost third of the Milky Way was inaccessible in Victoria, the DAO data did capture two maxima and one minimum in the double sine wave and "left no reasonable doubt that the stars within 3000 light years of the Sun behave exactly as if our Milky Way system was in revolution about a distant centre which lies in the same direction as the centre of the [system of] globular clusters" as determined by Shapley and the centre of the Galaxy as deduced by Oort.[70]

In Canada, members of the RASC Victoria Centre were the first to publicly hear about Plaskett and Pearce's latest investigation. That presentation, given by JSP on 21 February 1928, was a masterful summary in simple language of all the observed motions of stars in the Galaxy. Plaskett concluded that

> the apparently irregular and inextricably confused motions of the stars have been arranged into moving clusters, into stream motions governed by dynamical laws, and the whole may be reasonably explained on the assumption of a rotating galactic system. This is another example of the reign of law in nature, of how the most complicated phenomena may become, as the result of analysis, the most beautifully ordered and relatively simple systems. Generalizations, even if only partial, are what scientific men are always seeking, and even a small measure of success, besides its great value to science, yields great satisfaction and full recompense for any labour.[71]

In light of his interest in the rotation of the Galaxy, which he shared with Jan Oort, Plaskett must have been quite excited that the next assembly of the International Astronomical Union (IAU) was scheduled for Oort's home city of Leiden, Holland, in July 1928. The location was an apt recognition of the remarkable renaissance in Dutch astronomy since the Great War.[72]

Once again there was difficulty in deciding who would represent Canada at the IAU. JSP had not been well for some months. In the autumn of 1927 all his teeth had to be extracted and, as he explained, the shock to his system led to a painful carbuncle and shingles.[73] He had little energy and had lost twenty pounds. His handwriting became shaky and never regained its normal flow. In spite of it all, he found it hard to stay away from work. In another unexpected setback in April 1928, he wrenched the knee that he had previously injured (in the tumble from the ladder in 1922) and synovitis set in. This swelling of the knee confined him to bed for three or four weeks and the doctor told him he needed three or four months of rest.[74]

What was needed, Plaskett thought, was to get away on an extended trip; he was sure that an ocean voyage and change of scene would be as good as a cure. Even though he had to use crutches to get around, he was desperately eager to go overseas. He went through three doctors before finding one who gave him the green light.[75] Stewart, the director of the DO, generously removed himself from consideration as an IAU delegate and strongly supported Plaskett.[76] He wrote the deputy minister, urging him to give the nod to JSP's attendance, subject to the official approval by the IAU National Committee for Canada. This group dragged its feet and made a decision only when its members finally congregated for the RSC meeting in May in Winnipeg. There the committee members made their decision, electing seven delegates to the general assembly in Leiden though in the end only three went – Daniel Buchanan, newly appointed dean of Arts and Science at the University of British Columbia, Chant, and Plaskett.[77] After the Winnipeg gathering, Stewart inspected the department's seismic station in Saskatoon and went on to spend a day at the DAO, 2–3 June 1928. By that time, Plaskett had probably already departed, leaving Harper in charge. It was Stewart's only visit to the DAO during Plaskett's directorship.

Jack and Reba Plaskett sailed from Montreal for Liverpool on 8 June, two weeks ahead of Chant and his wife.[78] While in England, JSP took many photos, especially of the impressive architecture in several cities. In Essex he called on the amateur astronomer Herbert G. Tomkins, whose observatory boasted a new 61 cm reflector. He toured the professional observatories at Cambridge and Oxford with their somewhat smaller, historic instruments, and spoke to the British Astronomical Association on 27 June.[79] He and Reba proceeded to the continent for the IAU general assembly, 5–13 July, where JSP immersed himself in the busy schedule of meetings supplemented by dazzling receptions and pleasant excursions.[80] Leiden, a small city, could not accommodate all the visitors, so many of them, probably including the Plasketts, were lodged twelve kilometres out of town at the summer resort of Noordwijk overlooking the North Sea. The delegates had a good view of the countryside as they were transported to and from the sessions in comfortable railway cars decorated with vases of freshly cut flowers and pulled by electric engines flying orange flags. JSP expressed his amazement at the vast land reclamation scheme being undertaken by the Dutch.[81]

Because of his previous correspondence with Jan Oort about Galactic rotation, he especially treasured his visit to the Oorts' home in Leiden. For their part, the Oorts recalled the pleasant time and JSP's courageous climb on crutches up the long flight of stairs to their apartment.[82]

10.3 About half of the general assembly of the IAU in the concert room of the city auditorium, Leiden, 6 July 1928. Plaskett and Chant may be glimpsed near the centre of the first or second row. (*JRASC* 22 [1928]: Plate XVI)

10.4(a) Astronomers on an excursion hosted by the Astronomisches Gesellschaft to the Kepler monument and birthplace in Weil der Stadt, Germany, July 1928 (*JRASC* 22 (1928), plate XIX)

Right after the IAU, 170 of the 250 delegates went on to Germany for additional international gatherings sponsored by the Astronomische Gesellschaft at Heidelberg. In the postwar period, Germany and eight other central European nations were still not part of the IAU, but whatever feelings of hostility there had been had largely dissipated.[83] Cordial toasts were exchanged at banquets, and meetings at the historic University of Heidelberg and at the nearby observatory on the Königstuhl were held in a spirit of international cooperation. Trains transported the astronomers to Stuttgart and the nearby village of Weil der Stadt, the birthplace of Johannes Kepler in 1571.

In Stuttgart the astronomers were introduced to a new device that simulated the appearance and motion of the heavens by projecting lights representing stars and planets on the interior of a dome. This

10.4(b) Arthur Eddington addressing the group (RAS/Science Photo Library, H405/0209)

planetarium show was a first for most of the delegates.[84] Chant himself commented:

> Of course, I had heard of this wonderful means of illustrating the sky and the motions of the heavenly bodies but had not had the opportunity of seeing it in operation. The representation of the night sky is simply startling. I believe some 15 German cities have Zeiss planetariums, and they are certainly to be complimented. I have seen nothing so realistic and effective for instruction in the motions of the heavens.[85]

Plaskett also was struck by the proliferation of planetariums in Germany but questioned if the money might not have been better spent on observatories – a distinctly different reaction from that of Chant, the educator.

After spending some time in England to further JSP's recuperation, the Plasketts and Chants returned to Canada on the same ship, the recently launched White Star liner *Laurentic*, arriving back in Quebec City on 22 September.[86] Before the final transcontinental leg of his journey home, JSP stopped in at three New England observatories to discuss proper motions of B stars relating to his research on Galactic rotation.[87] After his return to Victoria, Plaskett spoke to the local RASC centre, giving, as usual, an amusing account of his travels.[88] He showed lantern slides of his own photographs, including some of the astronomers gathered at the IAU whose names would have been familiar to many in his audience.

JSP billed his department for $1212 for his European trip, about a quarter of his annual salary, not an unreasonable sum considering the length of his absence.[89] The benefits, as with most scientific conferences, were personal contacts, which sometimes resulted in follow-up correspondence. The Danish astronomer Ejnar Hertzsprung, then working at Leiden, wrote to JSP, wondering if a certain star (HD 115043) was a member of the Ursa Major "stream."[90] Renowned for his eponymous colour-magnitude diagrams of star clusters, Hertzsprung would have hoped to include only stars belonging to the far-flung cluster. Plaskett immediately secured a spectrum, measured the radial velocity of the star, and confirmed that it did belong. Hertzsprung also pointed out that the radial velocities of members of the Pleiades and Hyades star clusters were not well known but JSP could not commit to observing all those stars. JSP also corresponded with Oort: a question from Plaskett about Oort's opinion on the efforts of Freundlich and von Pahlen on galactic dynamics resulted in a lengthy response.[91]

WHITE STAR LINE

GLASGOW TO CANADA

DATE	DAY	STEAMER	FOR	
1928 Sept 15	Sat.	LAURENTIC -	Quebec and Montreal	Passengers will embark at PRINCES PIER GREENOCK

CABIN CLASS PASSENGERS leave ST. ENOCH STATION, GLASGOW, by Train … PASSENGERS from Ayrshire and the West of Scotland … GILMOUR STREET STATION, PAISLEY …

TOURIST THIRD CABIN PASSENGERS leave ST. ENOCH STATION, GLASGOW, by Special Train … PASSENGERS from Ayrshire and the West of Scotland … GILMOUR STREET STATION, PAISLEY, where it is due at 5-35 p.m.

THIRD CLASS PASSENGERS leave ST. ENOCH STATION, GLASGOW, by Special Train … PASSENGERS from Ayrshire and the West of Scotland … GILMOUR STREET STATION, PAISLEY, where it is due at 5-35 p.m.

PASSPORTS CANADA. …

UNITED STATES. … Unless this requirement is complied with, Passengers will not be admitted into the United States.

FRIENDS of passengers wishing to accompany them to Princes Pier Station, Greenock, may travel in the same train provided they obtain railway tickets in the ordinary way.

BAGGAGE. All heavy baggage should be at Princes Pier Station, Greenock, not later than noon two days previous to sailing. … The Company's baggage labels should be affixed to ends of all packages.

Apply to Local Booking Agents or

WHITE STAR LINE,

153, ST. VINCENT STREET (P.O. Box. 72), GLASGOW.

10.5 Embarkation notice for the voyage that brought the Plasketts and Chants back to Canada in September 1928. (Courtesy Gjenvick-Gjønvik Archives, http://www.gjenvick.com/SteamshipLines/#axzz4UbXvH9N5)

JSP continued to apprise fellows of the RSC of developments as they occurred. He did not attend the RSC in May 1928 due to ill health, or in 1929 because of many other commitments. Nonetheless, Plaskett did submit papers on the rotation of the Galaxy to both meetings, the second one jointly authored with Pearce, who because he was not yet a fellow could not attend.[92] JSP's excitement was evident in a letter he wrote to Chant in 1929: "We are now commencing the discussion of our results, computing the rotation of the Galaxy which is becoming more and more absorbing as we proceed. Papers have been presented at the meetings of the Royal Society of Canada and at the A.S.P. [Astronomical Society of the Pacific] (Berkeley meeting). We will also send a more complete paper to the A.A.S. meeting at Ottawa and later one to the Royal Society of England [sic]."[93]

At the joint AAAS/ASP session in Berkeley on 21 June 1929, Plaskett reported on the analysis that he and Pearce had just completed on the rotation of the Galaxy. Their study encompassed 870 stars, an increase of more than 50 per cent over their previously published work.[94] As significant as this paper was, the assembly was probably more excited by the latest news from the Mount Wilson astronomers, who were just beginning to come to grips with the expansion of the universe. Though it is now accepted as fact that the most distant galaxies are receding at speeds that are a significant fraction of the velocity of light, in 1929 any speeds greater than a few hundred kilometres per second seemed inconceivable. JSP commented that, at what was "probably the first scientific meeting after Hubble's announcement of the velocity-distance relation ... no one appeared to have the temerity to suggest that the apparent velocities of some 5000 km/s obtained at that time could possibly represent the recession of the nebulae."[95] Even Hubble himself noted, "It is difficult to believe that the velocities are real – that all matter is actually scattering away from our region of space."[96]

The August 1929 meeting of the AAS was in Ottawa. Plaskett did not go because of his many other obligations but Pearce did attend and he read two of their contributions, "The Galactic Rotation and the K-term for the O and B Type Stars" and "The Motions of the Interstellar Calcium."[97] The first of these two papers was probably similar to the one that JSP gave in Berkeley two months earlier. In it they confirmed the overall rotation of the Galaxy around the accepted centre, but they had to remove from their analysis a group of the brightest stars in the southern Milky Way and Orion – a group that seemed to be part of a moving cluster or stream whose recession masked the rotational component.[98]

The K-term, which they also dealt with, was a small general recession that all investigators had found remained after considering all other known motions. No one knew how to explain it but there were suggestions that it was not a manifestation of velocity at all. Some thought that it might be caused by Einstein's gravitational redshift or convection in the stellar atmospheres or even by errors with adopted wavelengths. In their paper, however, Plaskett and Pearce showed that the effect was absent in the intrinsically fainter B stars and so these elaborate explanations could largely be ruled out. Even so, a small excess redshift in the spectral lines of the most luminous stars persisted and has continued to puzzle some astronomers even in recent times.[99]

In their other paper, Pearce reported to the AAS on the research that he and JSP had done on 261 stars of type O5 to B3 whose spectra showed both interstellar lines and well-determined stellar lines.[100] Following a suggestion from Pannekoek, who had spent six months at the DAO, they calculated the Galactic rotation and the K-term from the velocities of the Ca+ lines.[101] In earlier work they had always assumed that the interstellar matter was at rest but now they entertained the possibility that it, along with the stars, rotated about the Galactic centre. They found that the rotational term derived from the Ca+ lines was exactly one-half the value derived from the stellar lines. Any lingering doubts about the location of the ionized calcium were now quashed: it was distributed more or less uniformly throughout the Galaxy and, as would be expected on average, the absorption occurred midway between the star and the observer.

Plaskett and Pearce published the full story in the January 1930 issue of *Monthly Notices* of the RAS, a paper Eddington referred to as a "remarkable investigation."[102] It posed a very challenging dynamical problem for theorists: to explain the stability of a rotating mass of exceedingly thin gas on the same basis as a rotating system of stars.[103] "To speak frankly," Eddington wrote, "I should have been better pleased to see more discordance, since the closeness must to some extent be put down to rather outrageous luck – as the authors indeed recognize."[104] Their proof that the pervasive interstellar gas participated in the Galactic rotation is another instance where Plaskett and Pearce were anticipated in their findings. In 1927 Oort had found that the calcium clouds tended to participate in the Galaxy's rotation, though he used only forty Ca+ velocities; in 1929 Boris Gerasimovich and Otto Struve, colleagues at Yerkes, used just over a hundred values to discuss the Galactic rotation of the interstellar clouds.[105] In both cases the authors used data that

Plaskett had previously published but not analysed. When Plaskett and Pearce ultimately did study 185 stars whose spectra showed interstellar lines, they were justified in claiming to be dealing with "a much more accurate and homogeneous system than any previously used."[106] An important side issue was their finding that there was no residual K-term for the interstellar calcium.

Though Plaskett did not get to Ottawa in August 1929, he did go east that fall, for a number of reasons. He had to be in Ottawa at the end of October to attend the Associate Committee on Physics and Engineering Physics of the National Research Council (who would pay costs).[107] He then went to Toronto to speak about the rotation of the Galaxy to the Canadian Club on 4 November 1929 – an occasion illustrating Plaskett's adaptability.[108] The person operating the projector put up JSP's slides in reverse order. Plaskett was unable to communicate with the projectionist, who was in a booth, but managed to ad-lib his way through the presentation.[109] After Toronto he travelled to Boston, where he joined Harry and served as guest lecturer at Harvard for a couple of weeks.[110] Then he went on to New York City, Brooklyn, New Haven, Albany, Rochester, Cleveland, Flagstaff, and finally Pasadena, where G.E. Hale had invited him to discuss design features of a proposed 5 metre telescope (later sited on Mount Palomar).[111] During the course of this whirlwind tour, JSP gave five lectures about his research.[112] Once he got back to Victoria he had only a couple of weeks to catch his breath before leaving again for another convention – this one in Des Moines, Iowa – where he gave an address, entitled "The Motions of the B Stars," at the end of his two-year term as chair of AAAS section D (Astronomy).[113] By this time, he was basing his conclusions about Galactic rotation on 815 stars of classes B0 to B5 to see if they gave results consistent with the earlier-type stars. After allowing for the velocity of the Sun relative to its neighbours, he found that, on average, the faintest stars, being the most distant, showed the rotational effect most clearly. He continued to believe that a widespread group of the brightest (and nearest) B stars in the constellations Scorpius and Centaurus showed higher velocities, which he took as an indication that they belonged to a cluster of stars with their own motion. Philip Fox, the secretary of section D, reported that Plaskett's paper "was by far the most important," on account of "the masterful and penetrating treatment" he gave to "an enormous number of observations."[114]

For years Plaskett had been using the accumulating database of radial velocities to study interstellar matter and Galactic rotation. Finally in

Table 10.1 Probable errors in radial velocities

Observatory	DAO	Lick	Yerkes	Mount Wilson
Number of stars	386	306	160	64
Av. probable error of radial velocities (km/s)	± 1.49	± 2.03	± 2.54	± 1.72

1930 he and Pearce felt ready to publish all these radial velocities in the DAO *Publications;* at the same time, they included them in a comprehensive catalogue of 1010 O and B stars from all published sources.[115] Four observatories – the DAO, Lick, Yerkes, and Mount Wilson – provided 90 per cent of the entries. After omitting stars with variable velocities but undetermined orbits, Plaskett and Pearce compared their results (see table 10.1). Clearly Victoria led the world in both quantity and quality of this sort of data.

The two major DAO papers of 1930 not only served as the basis for Plaskett and Pearce's further investigations, but the details they contained about the possible binary nature of nearly a third of the entries were of use to some astronomers for generations to come.[116] Nonetheless, their compilations were largely superseded within a couple of years by *A General Catalogue of the Radial Velocities of Stars, Nebulae and Clusters* by Lick's Joseph Moore, which encompassed 6739 stars of all the spectral classes, including those listed earlier by Plaskett and Pearce.[117] JSP, as president of the IAU Commission on Radial Velocities, concurred in the decision to publish Moore's list, seeing the value and necessity in having a homogeneous and comprehensive catalogue.

The research on the motion of B-type stars that Pearce had been doing at the DAO for the past six years finally earned him credit for his PhD from the University of California at Berkeley in May 1930.[118] While he was in California attending to academic concerns, he was given permission to examine hundreds of spectra of B stars at Lick and Mount Wilson. JSP had written to Walter Adams on 27 March 1930 to say that "Pearce, who has collaborated with me in the work of the B type stars, ... would like, if he may, to look over your spectra of B type stars for the purpose of classifying them uniformly with our own spectra and those of the southern Bs at Mt. Hamilton. He is a very fine fellow, a great enthusiast in astronomy and I am sure you will all like him very much."[119]

Life in Victoria and at the Observatory

Aside from periods when JSP was laid up, and during the summertime when student assistants often took his place, Plaskett seems to have done his fair share of observing. Between 1923 and 1930, he observed and measured about a quarter of all the plates, exceeded only by Pearce, who did about 40 per cent.[120] The director certainly ensured that the telescope never had a holiday. As Pearce recounted, on Christmas morning 1924, "it clouded over at 3:00 a.m. – and I confess that I was glad to close the dome, develop my last plate and walk the nine miles home."[121]

On the lighter side, all was not routine and research at the observatory. Pearce recalled that he and Petrie (a summer student from 1925 to 1928) laid out a nine-hole golf course around the observatory, just for putting:

> We teamed up the A's (Ed Harper and Bert Petrie, who were observing A-type stars) against the B's (Plaskett and Pearce). The A's were usually the victors for Harper was a steady player and Petrie was consistently good, while we were frequently erratic … About once a month we played at the Macaulay Golf Course at Esquimalt; if the sky was cloudy, we left the hill in the early afternoon in time to play nine holes. The A's almost always out-drove us. I vividly recall one occasion when we were one up at the seventh tee – a long up-hill drive across some rocks. We were driving first so Plaskett led off with one of his curious heaving drives and we watched his ball sail in a mighty slice far into the rough where a white goat was browsing. With uncontrolled glee in his voice, Petrie remarked, "That's easy to find, right beside that old goat!"[122] Once he arrived at the DAO, Roderick Redman was another regular who played with great exuberance.[123]

According to Harry, his father took a keen interest in the work of his staff and was always ready to take time to discuss difficulties and to share his expertise in mechanical design.[124] Facts bear out his opinion. JSP consistently spoke highly of the ability, industry, and enthusiasm of his staff and advocated for salary increases whenever he could.[125] He raised their international profile by proposing them for IAU commissions. In 1922 he had advised Schlesinger to put Harper and Young, rather than himself, on the Stellar Parallax Commission, as they were involved in that type of research and he was already on two other committees.[126] Though JSP practically moved heaven and earth to get government financing for his own attendance at conferences, he also strove to get travel grants for the other astronomers, especially after his own status was firmly secured. Harper, for example, was hoping to attend meetings in the east in 1924 and JSP backed him up, noting that Harper had gone to only three scientific gatherings in eighteen years of service. In another instance he wrote to Ottawa, "I do not want in any way to jeopardize Dr. Beals' chances of getting to the meeting."[127] He even encouraged those without much status, suggesting that Robert Petrie, a lowly summer assistant, and Sherwood Hill, a computer, send papers to the AAS meeting in 1926. JSP wrote to the chair of the session, Joel Stebbins, to say that Petrie "is a promising young lad. I would prefer you delete my own [contribution] and accept Hill's and Petrie's if you have too many."[128] In fact JSP had noticed Petrie's reliability even when he was still in high school, paying him a small sum to observe on Saturday nights, thus relieving the regular staff of their least favourite working night.[129]

The Plasketts were kind to all the staff and their families. They would entertain them at tea on Sunday afternoons and include them with other friends for evening slide shows of their travels.[130] Where this entertaining took place is unclear, as the Plasketts had always found their home on the hill inadequate. JSP had tried unsuccessfully for years to have better housing erected at the observatory. By 1926 he and Reba had given up on the idea of a bigger director's residence and decided to buy some property far from the observatory where they would build their own place, designed by JSP. Located on Armit Road, Esquimalt, on land sold by the Hudson's Bay Company only a decade earlier, the situation was ideal – just four kilometres west of downtown Victoria, on a beautiful 0.6 hectare lot with 180 metres of shoreline on the Strait of Juan de Fuca.[131]

The Plasketts' new home brought them closer to their friends who had no connection with astronomy. Some were in the historical society. Apparently JSP had a long-standing interest in Canadian history, judging from his response, years earlier, to a request for endorsement from the publishers of a series of thirty-two popular history books called *The Chronicles of Canada*. He wrote, "I have enjoyed reading the 'Chronicles of Canada' very greatly indeed. They are in my opinion a series which imparts the essence of Canadian History in a most readable and entertaining form, which will tempt even those not inclined to study history to read them from beginning to end."[132] His admiration of these volumes, edited by the Anglophile George Wrong, could

10.6 The Plasketts' house at 318 Armit Road, later 318 Plaskett Place, in Esquimalt. The house was demolished in 1988 to make way for the Birch Buildings. (Courtesy Hallmark Society)

also be taken as evidence of his enthusiasm for the British Empire, a perspective that was often a vital component of Anglo-Canadian nationalism.[133] Indeed, his national pride was always abundantly clear whenever he spoke of the origins and work of the observatory. He and Reba also took an interest in more local history. They took out membership in the British Columbia Historical Association in 1929 and two years later JSP accepted the office of second vice-president, which led to his presidency in 1934–5.[134]

The church played a major role in the Plasketts' lives, as it did for large numbers of Canadians in those days. Evidence of regular family involvement is hard to come by, with the exception of the baptism of the two grandchildren, but when planning for a new Anglican cathedral in Victoria began in 1922, JSP was appointed to the building committee (technically the Cathedral Buildings Limited), which had charge of the

10.7 The first service held at the new Christ Church Cathedral, Victoria, 11 March 1928. (Royal BC Museum and Archives, Image F-08174)

architecture, construction, and financing.[135] He was at the laying of the cornerstone on 25 May 1926 and at the first service in the still uncompleted cathedral on 11 March 1928.

In 1928 and again in 1929, he represented the Parish of Christ Church as a delegate to the Synod of the Diocese of British Columbia and served as one of two churchwardens of the cathedral from 1928 to 1930. It was the wardens who escorted the cross of the cathedral at the head of the colourful procession at the consecration of the new building on 28 September 1929. Thousands of citizens, unable to fit inside the building, lined the streets to see the stream of dignitaries and to hear the service of dedication over loud speakers. Countless other proud Victorians tuned in to a radio broadcast of the proceedings and of a glittering civic banquet that evening attended by the lieutenant governor, the premier, the mayor, and a host of dignitaries, Plaskett among them. The new cathedral was reputed to be the largest in the British Empire outside of

England, but many more years were required for its completion. Some of the remarkable features of this building include a choir screen originally in Westminster Abbey, a pulpit made from a single five-hundred-year-old English oak tree, and a dean's chair from St Paul's Cathedral in London. When JSP stepped down as warden, the dean praised him by saying, "there was no more generous or kindly person than the distinguished scientist."[136]

11
The Big Picture, 1930–1934

On 16 January 1930 the Royal Astronomical Society (RAS) in London, England, announced its prize winners for the year: John Stanley Plaskett was to receive its highest honour – the gold medal. Alfred Fowler had nominated him in November, 1929, and, after a vote, the RAS Council selected JSP above five other outstanding candidates. The society also anticipated that he would deliver its George Darwin Lecture. Named for the late mathematician (second son of the celebrated naturalist Charles Darwin) who applied his talents to the theory of tides and the motion of the moon, this annual lecture had been instituted in 1926 with the wish that "an eminent person from abroad" be invited to speak about his own work.[1] A tradition had emerged that the RAS president would base his address at the society's annual meeting on the accomplishments of the gold medalist, who, if he were also the George Darwin lecturer, would speak at a later date. So it was that, on 14 February, President Andrew Crommelin gave an eloquent summary of Plaskett's career.[2] He noted that Plaskett would receive the medal "not for any single outstanding result, but in recognition of the high merit of a long series of researches, extending over a quarter of a century, and marked throughout by a painstaking striving after the highest accuracy attainable."[3]

The announcement of this latest honour rapidly made its way to Canadian politicians at the highest level, apparently encouraged by JSP or his friends. The Conservative premier of British Columbia, Simon Fraser Tolmie, wrote to the Liberal prime minister, Mackenzie King, in February, "Our Government has been approached with the suggestion that we do something in recognition of Dr. Plaskett's scientific achievements ... We do not wish to take any steps until we know whether you are contemplating any testimony on behalf of the Dominion, since anything he has done must reflect credit on Canada as a whole, rather than on any particular province."[4] King replied:

> Please let me assure you that any recognition that the Government of British Columbia may wish to give, will be cordially welcomed by our Government. It will in no way conflict with anything that we here may find it possible to do.
>
> Now that titles and decorations are barred by resolution of our House of Commons [the so-called Nickel resolution, named after the Conservative member who proposed the motion in 1919], there are distinct limitations upon what is possible in the way of recognition of public servants.[5]

King immediately forwarded this correspondence to his minister of the interior, Charles Stewart, with the note "I believe [Plaskett's] professional services would entitle him to any recognition you might find it possible to give." This elicited a lengthy reply from the minister, explaining that, in spite of the rule in the department that "our technical men are only permitted to undertake overseas engagements once in a five year period," he had come to the conclusion that, as a special case, Plaskett's expenses, estimated at $1000, should be paid for his travels overseas from early April to 13 June 1930. He concluded his letter this way:

> As for Dr. Plaskett himself, I have every appreciation of his scientific achievements, and for a scientist he is the best advertiser I know. The Department has supplied him with first-class equipment, has given him a nice home to live in and a car with a chauffeur who is also man of all work around the Observatory. His son [Stuart], who is apparently unable to qualify under the provisions of the Civil Service Act, is engaged by us as a labourer. In addition, we permit him [JSP] to travel about a great deal of his time conferring with his colleagues. To my mind he is very comfortably situated.[6]

About the same time that JSP heard the good news about the RAS gold medal, he learned he would receive awards from the two most venerable learned societies in the United States – the American Academy of Arts and Sciences in Boston and the American Philosophical Society in Philadelphia.[7] The former invited him to receive its Rumford Premium on 9 April and the latter to participate in a symposium on 26 April and accept foreign membership. JSP got two telegrams from the Harvard College Observatory on 17 February – one from the director, Harlow Shapley, notifying him of the academy's award and the other from Harry saying that he hoped his father would come to Boston in advance so they could have a good visit.[8] JSP fired off a telegram to Deputy Minister Cory informing him of the recognition he was getting and adding (as Harry had told him) that the Rumford Premium "is awarded at infrequent intervals for distinguished work in science, the last recipient being Arthur Compton, recent Nobel Prize winner."[9] He added that he would travel to Boston en route to Philadelphia and London so that no extra leave would be required.

All went according to plan. The Plasketts left Victoria early in April and took the train to Boston, where JSP received his award and delivered a speech, "Diffuse Matter in Interstellar Space," to about fifty members and guests of the academy.[10] The family enjoyed a couple of weeks together while JSP was guest lecturer at the observatory.[11] Then the senior Plasketts went on to Philadelphia, where JSP participated in a symposium on 25–6 April at the American Philosophical Society with three other prominent astronomers: Ernest Brown, Harlow Shapley, and John Q. Stewart.[12] JSP was made a foreign member of the society at the same time as Albert Einstein and Sir Hubert Wilkins, noted Australian explorer. Then JSP and Reba returned to Boston and sailed from there aboard the White Star Line's *Arabic*, arriving in England on 5 May.[13]

Four days later, at the RAS in London, JSP received the gold medal for "valuable researches in stellar radial velocities and important conclusions derived therefrom," and delivered the George Darwin lecture, which he entitled "The High Temperature Stars."[14] In his speech he noted that O stars constituted only 0.1 per cent of the stars in the *Henry Draper Catalog*, and the B0 to B5 stars only 1.3 per cent. He summarized the O- and B-type stars as the hottest, most massive, and intrinsically brightest, and then went on to discuss their motions – the reflex of the solar motion, Galactic rotation, and the mysterious residual motion (the K-term) that showed the tendency of the most luminous of the stars to recede from us in all directions at a velocity of about 5 km/s. He further explained that the sharp calcium lines in the spectra of these stars convincingly showed that interstellar matter was spread throughout the disk of the Galaxy and rotated with it. The research in Victoria, as Plaskett pointed out, led to significant advances in all these areas and "incidentally provid[ed] a new and powerful method of obtaining the average distances of homogeneous groups of these very distant high-temperature stars."[15] While appropriately acknowledging the important contributions of his co-workers Pearce and Beals, and of his son

Harry, JSP said he regarded the gold medal of the RAS as the highest award an astronomer could receive and thanked the society from the bottom of his heart. His lecture was followed by complimentary remarks by some of the big guns in attendance – Crommelin, Dyson, Oort, and Eddington.

After extending their stay overseas, with JSP speaking to the British Astronomical Association on 28 May and taking in another RAS meeting on 13 June, the Plasketts arrived back in Canada on 6 July 1930.[16] They broke their long train trip home in Winnipeg, where the RASC centre had arranged a luncheon in their honour at the Fort Garry Hotel and the city's Scientific Club made JSP its first honorary member.[17] In his historical remarks to the fifty luncheon guests, he concentrated on the determined effort it had taken to make a place for Canada in astronomy.

During his absence from the Dominion Astrophysical Observatory (DAO), a period of almost four months, JSP missed the opportunity to welcome the prime minister to the observatory, a pleasant duty that would have fallen to Ed Harper, as acting director. Plaskett had invited King in September 1920, when he was leader of the opposition, but a decade passed before he actually made the visit.[18] What King sensed when he finally looked through the great telescope illustrates his psychic belief that his deceased parents were somehow still watching over him.[19] In the midst of an election campaign when he made his visit to the observatory, he confided to his diary:

> Chances on [Vancouver] Island are none too good, tho it [is] mostly in Victoria the fault of organization. Still the feeling is better than at any previous time. After the meeting I went out to the Dominion Observatory to see the stars. It was a wonderful experience & worth the trip to the coast. The telescope is the 2nd largest in the world. It is an awe inspiring sight to see this great instrument, moved about like a piece of artillery in the time of war. – It is for constructive work as against the destructive of war. I saw Saturn, a planet in the circle of flame spinning around it. Also a star divided under the telescope by 2 stars – the one of the two & in 2 others. They were at the edge of visibility, thro the telescope they were the two brightest purest bits of light I have ever seen. I have seen nothing like them ever. Light perfection in purity and brilliancy. I thought of dear mother & father, they spoke to me of both of them. – It filled me with awe to see them tho millions of miles away. Then we saw the moon like an alabaster globe. It was deeply impressive and I would not have missed the experience for anything.[20]

11.1 In this evocative photo of Prime Minister Mackenzie King, he seems to be communicating with his deceased mother, whose portrait hangs on the wall. (Credit: Gordon H. Coster, LAC, C-075053)

Even twelve years later King was still writing in his diary of this significant sighting. We will never know if King's unforgettable experience might have made a difference to Plaskett and the DAO in the years of retrenchment ahead, as the Liberals lost the election of 1930 to R.B. Bennett's Conservatives. By then the stock market crash of October 1929 had already led to a drastic drop in investment, production, prices, and employment.

Harry's New Position

During the gloomy years of the Great Depression there were a few bright spots for the Plasketts. In the fall of 1931, Harry Plaskett learned that he had been chosen to be the Savilian Professor of Astronomy at the University of Oxford, a distinguished chair created in the seventeenth century and occupied by such illustrious men as Christopher Wren, James Bradley (later Astronomer Royal), and more recently by the late Herbert Hall Turner, whom Harry had met in 1919 after his war service. As grand as the title was, astronomy at Oxford was in desperate need of revitalization.[21] Harry may have seemed an unexpected candidate, but the brilliant young Oxford theorist Arthur Milne actively solicited and promoted his appointment. Milne and Plaskett had gotten to know one another at a summer school for astronomers in Ann Arbor, Michigan, in 1929 and each recognized how well their research interests harmonized.[22] Early evidence of the close ties between the two men surfaced in 1930 when Bart Bok, a young graduate student on fellowship at Harvard, was highly critical of some work by Milne, and Plaskett sprang to his friend's defence. Bok suspected that Harry's real reason was that he wanted to leave Harvard and thought his chances of joining Milne in Oxford would be strengthened by savaging Bok's criticism.[23]

While the Oxford position initiated new collaborations for Harry, it did mark the end of the close professional association with his father. As we saw earlier, JSP had enjoyed two or three weeks as a guest lecturer at the Harvard Observatory in November 1929 and again in April 1930. Harry had reciprocated by returning to Victoria in the summers of 1930 and 1931 to visit his parents and to use the DAO telescope in an attempt to obtain spectra of small regions of the Sun's surface known as granulations. This research required some ingenious modifications to the instrument so that the optics would be subjected to the heat and light of the Sun only for very short periods of time.[24]

11.2 The 1.75 m reflector in the Perkins Observatory in Ohio. Note its overall resemblance to the DAO instrument, the first of many close copies. (Courtesy of Ohio State University Archives)

With Harry's imminent departure, JSP was glad when a further opportunity arose in November 1931 to reunite once more before the young family left for England.[25] The Perkins Observatory, part of Ohio Wesleyan University in Delaware, Ohio, was about to get a new 1.75 m mirror for its telescope, the third largest in the world.[26] Its director, Harlan Stetson, asked Plaskett if he would be willing to go to Brashear's in Pittsburgh, at the observatory's expense, to test the new mirror. He told JSP that he knew of "no one in this country whose judgment of optical excellence would be more universally accepted here and abroad."[27] The need for independent assessment was especially vital in this case, as J. Walter Fecker, who was figuring the mirror, was new to optical work on a large scale; Plaskett knew him for his mechanical expertise on the DAO mounting at Warner & Swasey. After getting approval to use three weeks of his annual leave, JSP left Victoria on 6 November 1931. He travelled by train to Pittsburgh to conduct tests on the mirror and then went on to Boston, where he reunited with Harry.[28] The optical testing went well – so well that Plaskett, on his way home, reported to Stetson and his colleagues that the mirror was superior to the one at the DAO.[29] Although the mirror was successfully installed in the telescope on 14 December, a poor secondary mirror meant that the reflector did not live up to its initial promise.[30] The overall design was fine and closely resembled the DAO instrument, though "hardly as well proportioned," as JSP proudly pointed out.[31]

Depressing Constraints

In the early 1930s, with Canada's national debt standing at about four billion dollars, the Conservative government of R.B. Bennett looked for savings in the public service.[32] Ottawa's Dominion Observatory was hit hard, though Director R.M. Stewart tried valiantly to make rational decisions about staff reductions.[33] The DAO was, of course, not immune to these financial measures; consequently, JSP had to juggle priorities and defer expenses, sometimes with almost no notice.[34] Indeed, on 29 December 1931 he received a telegram that, effective 1 January, the observatory car could no longer be used and must be disposed of as soon as possible. This change was very inconvenient, with the observatory a sixteen-kilometre drive from the city, and so those with their own cars had to be pressed into service. Barely six weeks later Stewart wrote JSP to say he had learned that there had been drastic cuts by the Treasury Board. Present salaries were to be cut by 10 per cent, summer assistants were to be dropped, and staffing might have to be reduced. JSP's

11.3 DAO staff in 1931. Left to right, front row: J.A. Pearce, W.E. Harper, JSP, L.M. Blake, C.S. Beals. Second row: S.N. Hill, T.T. Hutchison, and F.S. Hogg. (*JRASC* 62 (1968): 291; scan courtesy of Mike Peddle, DAO)

objections were to no avail. Stewart told him frankly, "It appears that you have entirely failed to realize the government's view of the seriousness of the economic situation, and the fact that the many favours that you have received in the past, both for yourself and your family, cannot be expected to continue under such conditions."[35]

The salary cuts were instituted in 1932 and other costs were stripped to the bare essentials, with no travelling expenses to be paid. At least JSP managed to ensure that no one was let go. (In the country as a whole, one in five workers was unemployed.) The following year there were further cuts. No summer assistants were hired until 1936, DAO publications were suspended, and housing allowances (affecting Harper,

Pearce, and Beals) were cut in half. JSP's hopes of proceeding with postponed projects were thwarted. He wrote to Stewart "I do not see how it will be possible to carry on with a reduction of over $2000 when our controllable expenditure, beyond salaries, is only some $5000."[36]

Austerity certainly meant that the director could not contemplate hiring extra staff. However, after R.O. Redman returned to England in 1931, Plaskett was able to hire a replacement in the person of Frank Hogg, another of C.A. Chant's protégés, like Harper and Pearce. A recent PhD graduate from Harvard, he came with his New England bride, Helen Sawyer Hogg, who also had a new doctorate in astronomy.[37] In due course she carried out important research at the observatory, though she could not join the staff because of an ironclad government rule that forbade spouses from both being on the payroll. JSP tried mightily to get some funding to enable her to do astronomy. He tried the National Research Council in Ottawa with no success; he wrote to Frank Schlesinger at Yale, who asked others. Helen recognized that chances were very slim but told her family, "It is awfully nice of him to try, and I do appreciate it. He is the dearest man." In the end it was the National Academy of Sciences in the United States that came through with $200. Helen explained:

> I used the $200 to pay a full time maid for a year. $200, full time, one year! That's what the dollar was in those days. To start with, I stayed home, when our first child [Sally] was born in '32. Then I quickly discovered that the child slept much of the time, and while she was sleeping I might as well be doing something else …
>
> I went to the observatory every day. And I also went at night when my husband observed, for the first year. But after the baby came, I didn't go with him when he was doing his observing program, but I continued [my own]. Dr. Plaskett gave me use of the big telescope, the 72-inch, for as many nights as I wanted. I was the first woman to use it, you see, and that was a very wonderful thing to have him trust me with.
>
> My husband helped me prepare the observational program. There was always a night assistant too at the telescope. Working at the Newtonian focus with the big telescope required three people so one wouldn't have to be going up and down stairs …
>
> I observed the first month. We got in the end of August, and I began with a program in September that year and continued in October, and then continued when the globular clusters came around again next year. They're summer objects.[38]

Helen gave an idea of life on the west coast in a letter to her family when she and Frank took the ferry to the mainland for a short holiday in Vancouver: "Well, here I am on the great North American continent once more. And after nearly two years absence, Frank and I have seen traffic lights, a big train, skyscrapers, and hordes of people."[39] Apparently Victoria was tame by comparison but its natural beauty more than compensated for any lack of sophistication. Helen loved the fact that "our office window here commands the most glorious view … I tell Frank that I would come out here every day even if I didn't work, just for the view."[40] The night-time prospect of the Milky Way was equally glorious and Helen had a special perspective when she was observing from her perch at the Newtonian focus high above the observing floor.

Raising Spirits

Plaskett did what he could to advance the Hoggs' careers. In 1932 he got them appointed to commissions of the International Astronomical Union – Frank on one dealing with nebulae and Helen on variable stars. His personal kindness was clear after the Hoggs' baby, Sally, arrived on 20 June 1932. The Plasketts were attentive, sending Helen two dozen red and pink carnations, visiting her in the hospital, inviting the family for dinner, and so on.[41] During their four years together at the DAO, the Hoggs and Plasketts often entertained one another, and Helen felt she had to try hard to keep up with Reba's "grand" dinners. The Plasketts brought home a dress for baby Sally from one of their trips and passed on their granddaughter's wicker chair for the wee one who unfortunately mistook it for a potty.

Given the closeness between the two families, the Hoggs were, of course, acquainted with Stuart, the Plaskett's younger son. Stuart apparently never married and always lived at home. He loved boats and fishing and he eventually may have earned his living as a fisherman.[42] He had a seven-metre launch, and friends, family, and observatory staff sometimes went out on it with him. When Stuart caught a twenty-six-pound salmon in Brentwood Bay, both family and staff found they were eating it for days to come. As Helen Hogg said about the four-pound piece they were given, "It is awfully good, but of course there is a limit to how much of even the very best salmon you can consume."[43]

JSP's generosity towards his staff was not directed only at the Hoggs. A further indication of his consideration was recalled by his brother Frank decades later.[44] One of the astronomers (Hill perhaps) had been ill for some time and was worried his absence would cost him his pay. He was going to struggle back to his duties even though he was really too ill. JSP solved the problem by taking a calculating machine to his home. There he could compute a binary star orbit without taxing himself unduly and yet feel he was continuing to contribute.

Yet another example of JSP's concern is the help he gave to Arthur Beer, a Czech astronomer who had been dismissed in 1933 from the Naval Observatory in Hamburg on the grounds that he was non-Aryan. In replying to letters from Beer, Plaskett wrote,

> I was very sorry to hear that you are leaving Germany and seeking employment in England … You cannot do better than to apply to Sir F.W. Dyson … He will know well all possible openings in astronomy in England and will I am sure do all he can for you. I am sending off in the same mail as this letter to him, asking him to do what he can for you [which in fact he did] … I will tell my son, Prof. H.H. Plaskett, who succeeded Turner at Oxford to look out for you and do anything he can to help you.[45]

Beer appreciated Plaskett's efforts and wrote to him again in July 1934, saying, "I am very thankful for all your kind advice and help during the past time, and also for all the so very kind endeavours which, on your kind suggestion, have been taken by Prof. H.H. Plaskett in Oxford in my case, and which helped me so very much, especially in my negotiations with the Academic Assistance Council in London." Beer's one-year appointment at the Solar Physics Observatory in Cambridge was the start of a new life in England.[46] He pursued his astrophysical research, partly during visits to the DAO, and became the founding editor of a unique and important periodical *Vistas in Astronomy*, "the first modern platform for history of astronomy."[47]

Plaskett's thoughtfulness seemed to extend to anyone asking a favour. He answered enquiries of a general nature (Can you explain relativity to me? Can you recommend a good book on comets?); he dealt with unsolicited articles from amateur theorists; he supplied photographs for articles and books, and lantern slides for illustrated lectures, and he arranged special daytime visits to the observatory for those who were unable to attend on Saturday evenings, when it was open to the public. Reading Plaskett's voluminous correspondence leaves the impression that he concerned himself with all aspects of the observatory's administration no matter how small or petty. Given that he had limited assistance, he may have taken this hands-on approach out of necessity. Still, it is hard to imagine him happily delegating authority, and surely the staff was quite pleased with the arrangement.

Personal Faith

JSP's cheerful readiness to help others was a reflection of his strong Christian faith. In a public lecture sponsored by the Royal Society of Canada in 1931, he described the structure and rotation of the Galaxy as "a wonderful example of the reign of law in the physical world, of the beneficent rule of a Supreme Power."[48] On another occasion, when asked to explain the vastness and movements of the universe, he said, "There must have been a mastermind behind it all. Most astronomers believe there is a power behind it all – the heavenly father."[49] As he wrote to a friend, "I am very glad indeed to be classed as a scientific man who is a Christian and good Churchman, and I never lose an opportunity in any address of emphasizing the fact that the study of science should tend to strengthen the belief in a Supreme Being."[50] For a closer look at his beliefs, we can turn to a sermon he delivered at Christ Church cathedral entitled "Religion and Science." He saw no conflict between the two, a coexistence that may be more common than general opinion might suggest.[51] Taking for his text the first verse of Psalm 19, "The heavens declare the glory of God and the firmament showeth His handiwork," JSP began by describing the wonders of the world, from the atom to the cosmos, and the sense that all these marvels could not have happened by chance. Continuing, he declared:

> Another aspect of creation is that spark of the Divine – life; that mysterious essence which, united with certain forms of matter, produces something that is more than matter, that endows matter with new and miraculous properties …
>
> The evolution of life, should give to thoughtful people a much higher and nobler conception of the wisdom and power of the Supreme Being than one which imagines the whole created, ready made, in a moment …
>
> God … has given to man … the faculty of reason, to enable him to penetrate into the interior of the atom on the one hand or into the inconceivable depths of space on the other. But in addition to this gift of the intellectual faculties, man has been endowed by a loving God with the sense of moral and spiritual values, with the urge towards the higher ideal of life and with a conscience always fighting against the forces of evil. Man has been

given, if in a very imperfect degree, some of the attributes of his Creator; man has been made in the image of God …

It need not seem presumptuous then to compare the qualities required of a good scientist with those necessary in a true follower of Christ. Singleness of purpose, patience, unselfishness, honesty, all essential in a scientist are also some of the qualities required of a Christian.[52]

JSP's sermon was published in *Canadian Churchman*, the national newspaper of the Anglican Church in Canada. One reader probably spoke for many when he wrote to the editor, "The reprint of a few more sermons such as Dr. Plaskett's will do much more good than the 'wishy-washy' sentimentalism of so many published sermons. It is a manly, straightforward, scholarly presentation of an urge to faith, convincing in its obvious truth."[53]

Plaskett was never reticent about speaking publicly about his beliefs, which were clearly shaped by his scientific understanding (rather than the other way around). Religious faith was not unusual among his contemporaries, including Eddington, Milne, and Russell, for example.[54] The University of Toronto physicist J.C. McLennan, speaking in 1926, said "I know most of the leading physicists in the British Empire … and I do not know any more devout and religious men."[55] C.A. Chant stated that he was "almost sure that the distinct majority of outstanding astronomers are men of Christian faith – but, and mark this – they don't talk about it."[56] Nonetheless, McLennan and Chant were evidently willing to talk about *their* beliefs, at least when asked. It is highly doubtful that anyone nowadays would make the same claims about the faith of scientists, or even that interviewers would think to ask about such matters.

Team Effort

Throughout the Great Depression, Plaskett and the DAO staff managed to continue their work. The telescope was in nightly use. That said, JSP personally did no observing from 1930 on, and his papers for the duration of his career required no new data.[57] Instead he was able to make ever-broader interpretations of information that he and Pearce had compiled and published in October 1930, as "A Catalogue of the Radial Velocities of O and B Type Stars."[58] In the last five years of Plaskett's directorship it is difficult to separate his scientific accomplishments from Pearce's as they collaborated so closely. Their final joint paper, published in 1934, provides a couple of clues as to the distribution of labour. In it Pearce is given sole credit for using class B stars to investigate solar motions and for one of their several determinations of the Galactic rotation constants and direction to the Galactic centre.[59] Pearce specifically studied 680 stars of type O to B5 whose radial velocities had small statistical errors (probable errors less than 3 km/s). However, the initial conception of the program and interpretation of the results does seem to belong to Plaskett.

Almost everything they published in the five years from 1930 to 1934 related to the collective motions of stars and the material between them. The culmination of their investigation of interstellar matter was reached at the end of 1931 in "The Problems of the Diffuse Matter in the Galaxy." They began their paper with an extensive summary of historical observations of interstellar lines, including a nice (though belated) tribute to the pioneering work by Vesto Slipher, who, they acknowledged, was far ahead of his time when he suggested, in 1909, that absorption occurred "in an interposing cloud covering at least certain extensive regions of the sky."[60] They then discussed the recent efforts of Struve, still insisting that his results were dubious because they came from spectra of very different dispersions from several observatories. In their own analysis, they used 314 stars from their own catalogue, all chosen because their spectra showed clear interstellar lines. For each one selected, they gave the stellar velocity as well as the velocity derived from the interstellar K line and an estimate of its intensity. From this, they confirmed a direct linear relation between the intensity of the K lines and the distances of the stars, giving unequivocal evidence that interstellar matter was widespread and probably distributed uniformly. Moreover the relationship could be turned around to provide a reasonable method of obtaining individual distances of very remote stars – novas, for example – located close to the Galactic plane.[61] But primarily their data were central to their studies of the rotation of the Galaxy.

Plaskett and Pearce's work, as well as Struve's, seemed to be "remarkable and important developments" to the secretary of the Smithsonian Institution in Washington, Charles G. Abbot. Using the title "The Contents of Interstellar Space," he summarized the DAO paper along with Struve's "Matter in Interstellar Space"[62] in an appendix to the institution's *Annual Report … for the Year Ending June 30, 1933*. This may seem like an odd place to do so until one reads at the beginning of the appendix that it had been "a prominent object of the Board of Regents … from a very early date to enrich the annual report" with papers such as this.

In the preface to the volume, Abbot expressed the hope that such miscellaneous memoirs would interest "collaborators and correspondents of the Institution, teachers, and others engaged in the promotion of knowledge."[63]

The Royal Society of Canada Celebrates

In 1931 the RSC met in Toronto. The Society had first assembled in 1882, so this was its fiftieth annual meeting, even though the RSC would not officially turn fifty until the following year. JSP and Reba decided to use the opportunity to spend some time in the city where they began their married life together. After leaving Victoria on 11 May, they were entertained in Winnipeg, where JSP spoke to the local RASC centre on the structure and rotation of the galaxy," the same topic he would use in Toronto.[64] For the RSC meetings, he brought four papers by members of the DAO staff – one each by Pearce, Harper, and Beals, and a joint one by himself and Pearce entitled "The Motions of the Diffuse Gaseous Matter in the Galaxy" that he presented at a combined session of the astronomers in the RSC section III and members of the RASC.[65] In addition, he was chosen to deliver the RSC popular lecture, an annual event since 1891. Associated Press covered this lecture and, as a result, a summary appeared in several newspapers, including the *Sentinel Review* of Woodstock, his old hometown, which reported his comments about O and B stars:

> The motions of these stars agree so exactly with those that would be given by a rotation of the galaxy, that there can be no reasonable doubt of its presence. This rotation causes the sun and neighbouring stars to move about the distant centre of the galaxy at a speed of nearly 200 miles per second … So vast is the galaxy, that it will take the sun some 250 million years to make one revolution … Our conviction of the reality of this rotation is much increased when we learn that the direction to and the distance from the centre of the galaxy which can be calculated from the motions of the stars observed at Victoria, are almost exactly the same as those earlier obtained from the distances of the stars and the dimensions of the galaxy.[66]

At the RSC meetings, JSP also acted as the delegate from the BC Historical Association, reporting to the Historical Section of the RSC on the activities and publications of the provincial association during the past year. Perhaps because of all his official duties, on this occasion the RSC paid over half his travel expenditures, which was fortunate because the rule of the Department of the Interior was that "people who are invited to give lectures by societies must look to those societies for payment of their expenses."[67]

After the RSC meetings, Plaskett spent some time in Toronto. There, Chant took him to the site of the observatory that he had been trying to establish for decades and introduced him to the benefactor, Jessie Donalda Dunlap, widow of the late lawyer and mining magnate David Dunlap. Plaskett was very pleased with the hilltop site surrounded by farmland, sixteen kilometres north of the city limits, and wrote an appreciative letter to Dunlap for her magnificent gift to the university and to astronomy generally.[68]

After the convention in Toronto and some departmental business in Ottawa, the Plasketts had a couple of week's holiday, including a visit to JSP's brother Fred in Rochester. They then headed west for brief stopovers with Philip Fox in Chicago and the Sliphers in Flagstaff, Arizona, before reaching Pasadena for more scientific conferences in the middle of June.[69] These were the usual summertime joint assemblies of the Astronomical Society of the Pacific (ASP) and the American Association for the Advancement of Science (AAAS). JSP had obtained prior authorization from the Department of the Interior for Beals to make a presentation in Padadena, and now he joined him, presenting a paper by himself and Pearce, "On the Motions and Distribution of Interstellar Calcium."[70]

More Accolades

No sooner did JSP get back to Victoria than he learned that the ASP had decided to award him its highest honour, the Catherine Wolfe Bruce gold medal, in 1932.[71] The announcement was made on 28 November 1931. As was the custom, the society's president, Alfred Joy, gave a lengthy account of the medalist's accomplishments at the annual meeting the following January.[72] Speaking frankly, he said that Plaskett's services had "seldom been rendered in new and untried fields, but each of his many achievements constitutes a notable advance over previous attainment ... [His] success was attained by painstaking mastery of detail as the work progressed, rather than by the more brilliant but sometimes less thorough and less scientific process by which the goal is envisaged at the outset." Further, Joy referred to Plaskett's organization, improvement, and direction of astrophysics in Ottawa and in Victoria as a type of service "rarely found in such generous measure among men of scientific attainments."[73] The actual presentation of the Bruce medal and the recipient's speech were scheduled for a later date. Plaskett went at his own expense to San Francisco, where, on 20 April 1932, he received the medal and gave a lecture on the rotation and structure of the Galaxy.[74]

His speech was widely appreciated. This time the Smithsonian's secretary, Abbot, reprinted the entire paper in the same appendix, immediately preceding his article on interstellar material referred to earlier in this chapter.[75] Many of the items in the Smithsonian report were originally lectures given either to the institution or other bodies, and many had already been published elsewhere. So for Plaskett to have his ASP lecture reprinted in this way was not strange but in fact was an independent recognition of its worthiness. The only problem was that, by the time this *Annual Report* was published, two years had elapsed and very significant revisions were required. Abbot gave JSP the opportunity to update the story, which he did in a supplement, dated April 1934. In it, JSP pointed out that there was now "general acceptance by astronomers of the presence of some kind of an absorbing medium in the galaxy, principally in the neighbourhood of the central plane, which both dims and reddens the light of the distant stars." He referred to the photometric work of Joel Stebbins at Mount Wilson that showed the globular clusters (forming a halo around the Galaxy) were only about half as far away as previously thought. This meant that "the actual diameter of the Galaxy is only about 100,000 light years instead of the 200,000 obtained without consideration of the presence of absorbing material." The revised size of the Milky Way was much closer to the dimensions of our neighbouring galaxy in Andromeda.[76]

When the Smithsonian report for 1933 was finally issued on 30 June 1935, many newspapers picked up on Plaskett's paper (though none of the others in the report). The *New York Times* highlighted the picture that it gave of our "tremendous Galaxy" as a "circular disk about 200,000 light-years in diameter and 10,000 light-years in thickness, containing about 200,000,000,000 stars."[77] Clearly they (and the *Washington Post*) had completely missed JSP's update.[78] Other papers, like the *Chicago Tribune*, did better, using a headline "Star Distances Overestimated, Study Indicates."[79] The Toronto *Globe* also highlighted Plaskett's revision, but used the unfortunate headline "Astronomers Err."[80]

Plaskett's 1932 ASP paper contained something, quite apart from its main subject, that gave it lasting appeal – a table and accompanying figures that were part of a reprint in the ASP *Leaflets* in 1939.[81] It was

PLASKETT'S SCALE OF THE UNIVERSE

—9	.000 000 000 000 06 inch	1.6×10^{-13} cm.	Electron
—8	.000 000 000 006 inch	1.6×10^{-11}	
—7	.000 000 000 6 inch	1.6×10^{-9}	
—6	.000 000 06 inch	1.6×10^{-7}	Atom
—5	.000 006 inch	1.6×10^{-5}	Soap bubble
—4	.000 6 inch	1.6×10^{-3}	Tissue paper
—3	.06 inch	1.6×10^{-1}	
—2	6.3 inches	1.6×10	
—1	.01 mile	1.6×10^{3}	
0	1 mile	1.6×10^{5}	
1	100 miles	1.6×10^{7}	
2	10 000 miles	1.6×10^{9}	Earth
3	1 000 000 miles	1.6×10^{11}	Earth-Moon
4	100 000 000 miles	1.6×10^{13}	Earth-Sun
5	10 000 000 000 miles	1.6×10^{15}	Solar system
6	1 000 000 000 000 miles	1.6×10^{17}	
7	17 light-years	1.6×10^{19}	Nearer stars
8	1 700 light-years	1.6×10^{21}	Local cluster
9	170 000 light years	1.6×10^{23}	Galaxy
10	17 000 000 light-years	1.6×10^{25}	Nearer spiral nebulae
11	1 700 000 000 light-years	1.6×10^{27}	
12	170 000 000 000 light-years	1.6×10^{29} cm.	Einstein Universe

11.4 Plaskett's Scale of the Universe (from *Leaflets of the ASP* 3 (1939): 172)

a sort of a static version of the "Powers of Ten" videos so well-known today even on *YouTube* and *The Simpsons* – graphic reminders that the knowledge of the very smallest entities in the universe are vital to understanding some of the largest and most distant objects. Plaskett had something similar in mind as early as 1927 when he showed a series of specially prepared slides depicting the remarkable range of astronomical research from sub-atomic dimensions to the largest distances in the universe.[82]

Within a month of his San Francisco trip, JSP was off to RSC meetings in Ottawa, once again paying out of his own pocket.[83] Assembled at the new National Research Council laboratory on Sussex Drive, over one hundred fellows presented papers to section III, necessitating a split into three subdivisions to accommodate them all. To celebrate the society's golden jubilee, delegates were invited from Great Britain, the United States, and France. Prime Minister Bennett was in attendance and received an honorary fellowship. To mark the occasion, the RSC published a volume entitled *Fifty Years Retrospect*, whose twenty-four chapters, each written by a distinguished fellow, covering all facets of intellectual life in Canada (JSP's contribution was entitled "Fifty Years of Canadian Astronomy"). The jubilee year was memorable too for the first RSC annual fellowships, each worth $1500, which were awarded to ten promising young scientists, and for the presentation of the society's Flavelle Medal for distinguished service to science to JSP.[84] The same day that he was awarded the medal, he went to McGill University in Montreal to receive an honorary LLD along with the famed philosopher Alfred North Whitehead.[85]

Visitors from Abroad

The IAU, if it had followed its triennial schedule, would have met in 1931 but the assembly was delayed by a year to fit in with the solar eclipse of 31 August 1932. The path of totality included parts of the province of Quebec and the New England states, so the location and time of the IAU general assembly at Cambridge, Massachusetts, on 2–9 September was ideal. Because this was the first international gathering of astronomers to be held in North America since 1910, opportunities were provided for overseas guests to look in on many observatories. Twenty astronomers, mainly from England, visited Victoria prior to the general assembly in the east.[86] Though most of them stayed only a short time, the Dysons came early and stayed with their old friends, the Plasketts, for two weeks. Indeed, JSP had a room added to the house on the hill to accommodate them.[87] Sir Frank Dyson, about to retire as Britain's Astronomer Royal and as president of the IAU, was determined to have Lady Dyson by his side, "even if he had to earn her fares himself by lecturing his way across Canada and America," which in fact he did.[88]

On one short part of their trip, naval regulations forbade Lady Dyson from travelling with her husband. JSP had arranged for the Astronomer Royal to accompany him to the small settlement of Bamfield on the west coast of Vancouver Island, the site of the eastern terminus of the trans-Pacific cable.[89] The occasion was a ceremony, organized by the BC Historical Association, at which Lieutenant Governor Fordham Johnson would unveil a plaque marking the thirtieth anniversary of the globe-girdling cable connection. The plaque, supplied by the Historic Sites and Monuments Board of Canada, paid tribute to Sir Sandford

11.5 Members of the IAU and RASC Victoria Centre members on the steps of the DAO, 10 August 1932. Sir Frank and Lady Dyson are on either side of JSP, and Reba Plaskett is on the extreme right of the front row. Joseph Pearce is the highest figure on the steps; his wife is in front of him in a short-sleeved dress. To the right of Pearce are Frank Hogg, Sherwood Hill, Marshall (first name unknown), Carlyle Beals, Guido Horn d'Arturo, Ed Harper, and Theodor Niethammer. Miriam Beals is in the second row, wearing a long string of pearls, and Helen Hogg is next to her, with a pin on her lapel. (Published with complete key in *JRASC* 26 (1932): 360)

Fleming, who, as early as 1879, had envisioned the trans-Canada and trans-Pacific telegraph links as fundamental to the security and cohesion of the British Empire. Yet, while wires connected Bamfield to the world, there were no roads between it and Victoria in 1932, so the Royal Canadian Navy accepted the responsibility of transporting a party of twelve in the destroyer *Vancouver*, from the base at Esquimalt to the remote location 200 km away. Regulations forbade any women to travel overnight on a naval ship, so Lady Dyson and Reba Plaskett remained in Victoria.

In addition to time at the DAO, all the astronomers who came to Victoria were treated by the lieutenant governor to tea at his grand residence, to an elegant dinner given by the RASC Victoria Centre at the Empress Hotel, and to a tour of the Butchart family's famous garden.[90] After a three-day stay, the visitors were off to the east. The Plasketts left on 21 August. They joined other astronomers at Conway, New Hampshire, to view the eclipse but were thwarted by cloudy skies. However, the Dysons and the Schlesingers, located seventy-five kilometres away in South Poland, Maine, had a perfect view in a clear sky, giving the Astronomer Royal a perfect record on six out of six solar eclipses. At the IAU, there were nine Canadians in attendance, including JSP.[91] He was chair of the Radial Velocity Commission – in fact the only Canadian to chair a commission – while Harper was on two commissions. JSP tried to get funding of $300 for each of them, though that was, as we have seen, a difficult prospect in a period of economic retrenchment.

With all these events, it is hardly surprising that Plaskett had to miss a meeting he would have liked very much to attend – the "triple A-S" gathering in June in Pullman, Washington, in conjunction with the ASP. He had been asked to give a general review of science, a topic that he found involved too much preparation, deciding in the end to restrict himself to a hot topic that had grabbed his attention in Berkeley three years earlier – the expansion of the universe.[92] In his absence, his speech was read by Ferdinand Neubauer, who had been doing similar work to Plaskett and Pearce using O and B stars, but in the southern hemisphere.[93] Though this talk (and a modified version for the RASC in Victoria the following March[94]) was Plaskett's first and only foray into cosmology, he gave the topic more than cursory consideration. After it was published in the RASC *Journal*, Shapley wrote a lengthy letter to JSP praising his excellent and entertaining article, his recognition of Slipher's pioneering work, and generally sharing in Plaskett's healthy scepticism.[95] However, he took issue, in a somewhat defensive

11.6 This photo shows approximately one-quarter of the astronomers assembled for the IAU at Harvard in September 1932. Several Canadian astronomers can be seen: Beals (in the back row, centrally placed against the white column), Harper (tall, bald, clean-shaven man three rows in front of Beals), R.M. Stewart (on the left-hand edge with the stand-up collar), and C.A. Chant third in the same row. Prominent in front of the tree is JSP with W.S. Adams on his right and A. Eddington on his left. Next to Eddington is E. Hertzsprung, and behind them are P.M. Millman and his wife Peggy. (Full group photo and key at https://www.strw.leidenuniv.nl/oortfotos/)

way, with Plaskett's characterization of what has come to be known as the Great Debate between Shapley and Heber Curtis in 1920.[96] JSP had said (as was generally accepted by 1933) that Curtis had prevailed, but Shapley in his letter insisted that, because the debate was on the scale of the universe rather than on the nature of the spiral galaxies, he was right in arguing for a much larger Galaxy than Curtis.

The earlier doubt that the red shifts in the spectra of galaxies were measures of their enormous velocities was dissipating, thanks in part to the widely publicized theory of an expanding universe by Belgian cosmologist Abbé Georges Lemaître.[97] Plaskett, as we have seen, was never happy with what he saw as unfounded speculation. In his lecture, he agreed that the observational evidence supported what we now call Hubble's law, namely that the redshift is proportional to the distance, but he saw several challenges in assuming that this was evidence for an expanding universe: (1) Edwin Hubble had found that the redshifts, interpreted as radial velocities, increased by 560 km/s for every million parsecs (or 172 km/s per million light years), implying that the universe was less than two billion years old, much too short a timeframe for the stars to have evolved according to theories of stellar evolution. (The presently accepted value of the Hubble constant is only 68 km/s per million parsecs, and the age of the universe about 13.8 billion years, so JSP was right to be critical.) (2) There were alternative physical theories, proposed by Fritz Zwicky and William Macmillan, for instance, that could explain the loss of energy of photons in their journey through vast distances. (3) The various mathematical models that cosmologists had devised to explain an expanding universe were conflicting and based on unrealistic assumptions. "Models of the universe," he said, "should not be taken too seriously."[98] Though the reasons for his scepticism have since been largely resolved, the conclusion to his talk, though taken out of context, seems startlingly prophetic: "The astronomer may feel reasonably assured that the space, to the limits to which he can observe, is Euclidean and that he need not worry about the curvature of space-time." For most of the rest of twentieth century that point of view would have seemed out of date, but results in 2003 from the Wilkinson Microwave Anisotropy Probe confirmed its truth on a vast scale, far beyond what Plaskett could have imagined.

Plaskett very nearly had a further major commitment during the busy summer of 1932, as the program organizer of another group of particular interest to west coast scientists: the Pacific Science Congress, encompassing the countries bordering on the Pacific Ocean. There was already a realization that the United States was replacing Europe as the world's economic powerhouse and that westward lines of communication were becoming increasingly vital.[99] The objectives of the congress were "to initiate and promote co-operation in the study of scientific problems relating to the Pacific region, more particularly those affecting the prosperity and well-being of the Pacific peoples," and "to strengthen the bonds of peace among the Pacific peoples by promoting a feeling of brotherhood among the scientists of all Pacific countries."[100] The first congress was held in Hawaii in 1920 and subsequent triennial meetings were held in Australia, Japan, and Java. JSP first became involved with the organization in 1930, when he chaired the astronomy committee and was charged with organizing a program for 1932 in British Columbia. His invitation to one of the speakers shows that he was keen to promote research in new directions. He wrote to Harlan Stetson, director of the Perkins Observatory and a pioneer in studying solar effects on radio reception, "I wonder if I might put you down for a paper on the correlation between radio activity and the sunspot cycle ... I well remember your beautiful curves at Des Moines a year ago."[101]

As things turned out, the congress was postponed for a year because of economic constraints imposed by the Great Depression. When it was ultimately held in June 1933 in Victoria and Vancouver, total attendance was 409 including 259 Canadians.[102] As chair of the Victoria reception committee, JSP pulled out all the stops. The lieutenant governor hosted a garden party on the king's birthday, and the premier and the mayor each hosted a lunch or dinner for the delegates, as did the Kiwanis and the Canadian Clubs. There were tours of the city, the museum, and the Butchart Gardens, excursions for botanists, geologists, and entomologists, and of course, visits to the DAO both in the day and at night. Most of the scientific sessions were in Vancouver and they included talks about astronomical research in Australia, Java, Japan, and California (at Lick and Mount Wilson). The media were especially taken with Walter Adams' finding that oxygen constitutes less than one-tenth of one per cent of the Martian atmosphere and that life as we know it could not be sustained there. They quoted JSP as saying the news would be "a shock to amateur astronomers, but the truth must be told."[103] JSP would have been happier if the press had interviewed him about the work of the DAO or a paper contributed by Pearce and himself under the unwieldy title "Determination of the K-term, Solar Motion, and Galactic Rotation from the Radial Velocities of 849 O to B7 Class Stars"[104] but in the end the recognition that their rather esoteric work received in professional journals was what really counted.[105]

Plaskett's Tour de Force

At almost the same time as Plaskett and Pearce had been preparing their catalogue of O and B stars for publication in 1930, Neubauer had been doing the same for radial velocities from observations he had made over the previous half-dozen years at the southern hemisphere station of the Lick Observatory at Santiago, Chile.[106] The telescope there was a 94 cm Brashear reflector whose relatively small size restricted Neubauer's research to the brighter stars. Of the 351 stars of all spectral classes in his catalogue, the Canadian astronomers used radial velocities of 64 of his B0 to B5 stars, otherwise unavailable to them, thus allowing them to extend their analysis of Galactic motions to the southern part of the Milky Way. The combined data provided the foundation for their final investigation into the rotation of the Galaxy, an effort that took them four more years to complete.[107] Their compilation was, as Eddington put it, "a *tour de force*."[108] The sheer size of the database – nearly a thousand stars – was one reason the work ate up so much time. Such a large number of stars was necessary because they had to be subdivided. They were first divided into four groups to ensure that the brighter and nearer stars could be studied separately from the dimmer and farther ones – an important distinction, as the differential rotation effect depended on the distance of the stars under consideration. Then the stars had to be further split up into subgroups according to intervals of Galactic longitude. Because, for statistical purposes, a reasonable number of stars needed to be in each interval, and, at the same time, each interval had to be as narrow as possible, the need for a large number of stars was obvious.

Another time-consuming task was reclassification. The astronomers could hardly say they were studying the motions of the stars of classes O to B5 if in fact some of the stars were late B or A type. So Pearce undertook to examine or re-examine the spectrum of each star before definitely including it in the selection. This meant he had to spend a considerable amount of time at other observatories inspecting their plates. In addition, other factors led to the exclusion of certain stars – some had high Galactic latitudes, some had radial velocities with large errors, some were suspected binaries. In the end they were left with 849 stars. Plaskett and Pearce's final analysis was more complicated than before because they realized that the three types of motion – the Sun's peculiar motion relative to its local standard of rest, the K-term, and the rotation terms – all had to be calculated simultaneously, as they were to some extent interdependent.

Plaskett and Pearce each talked about their work on many occasions as they engaged in this mammoth project on the rotation of the Galaxy. They prepared presentations for the AAS meetings in Chicago in 1930 and 1933 (attended by Pearce) and for joint gatherings of the AAAS and ASP in Pasadena in June 1931 and in Salt Lake City in June 1933 (attended by Plaskett) and gave more general talks for the public.[109] Closer to home, JSP continued to keep the RASC in Victoria up to date and, during a quick trip to the mainland in February 1933, spoke to the RASC Vancouver Centre as well as the University of British Columbia Physics Club and the Vancouver Institute. A student reporter began his write-up of the latter presentation to "the largest audience of the year" this way: "'Iolanthe' met competition on Saturday night from a stoop-shouldered, genial little man ... He had no fairy wand, no sweet instruments, no colourful costumes; on the contrary, he asked his audience to follow him into the farthest deserts of mathematical hypothesis ... Dr. Plaskett was probably as untechnical as his subject allowed and made his main points clear to everyone."[110]

True to form, JSP was proactive in publicizing his achievements. As this final stage of his work was nearing completion, he wrote an article, "The Observational Confirmation of the Rotation of the Galaxy," for the *Observatory* in November 1933. In it he summarized the development and present status of the problem. After praising Bertil Lindblad's "useful and interesting" theories and Jan Oort's analysis, which "provided strong evidence, at least among the more distant stars, of a differential rotation," he went on to write about his own accomplishments. He started with his 1928 paper (published in the *Monthly Notices* of the RAS) that "seemed to show unmistakably the presence of a systematic trend in the residual velocities most simply and satisfactorily explained by a rotation of the whole system about a massive centre, coinciding in direction with its geometrical centre"; he continued by describing how the work on O to B7 stars at Victoria "added considerable observational material," and that the interpretation of the interstellar lines in their spectra "provided what seems to be the best published evidence of the presence of the rotation of the Galaxy." Finally, he spoke about the analysis he and Pearce had just completed "with satisfactorily low probable errors and showing the undoubted presence of the differential rotation effect."[111] In a review paper of this sort, a more modest person might have let his publications speak for themselves but Plaskett was never shy about promoting his own accomplishments. To be fair it must be noted that, in the same article, JSP included the relevant research of other contemporaries, Redman and Nordstrom, Stebbins and van de Kamp, and praised Joy's very recent and as-yet unpublished study of Cepheid variable stars as "the most remarkable confirmation of the galactic rotation yet advanced."[112]

As we have seen, Plaskett, despite budgetary constraints, was able to attend many meetings during the Depression years, but he managed to attend only one AAS session – the joint meeting with the AAAS in Boston just after Christmas 1933.[113] Thousands of scientists from all disciplines attended the "triple A-S" and consequently there was keen media attention. Plaskett's speech, "The Distance and Direction to the Galactic Centre from the Rotational Constants of the Class B Stars," was featured as the opener for the astronomy sessions and elicited coverage on three consecutive days in the *New York Times*. In addition, *Time* magazine excitedly announced that "an estimate 600% higher than any previous census of stars in the Milky Way (the galaxy to which Earth belongs) was given by Drs. J.S. Plaskett and J.A. Pearce of the Dominion Astrophysical Observatory at Victoria, B.C. Their total was 170 billion stars. The Milky Way is apparently rotating round a centre once every 220,000,000 years. From this centre they find the solar system 30,000 light years away."[114] The claim in the opening sentence, the attention grabber, was false but the other figures were accurate.[115] The *Science News-Letter*, in its review of progress in 1934, also highlighted Plaskett and Pearce's research and featured their more realistic conclusion that the Milky Way was no larger than other spiral galaxies.[116]

In England, too, the work of the Canadian team was lauded by the president of the RAS, Frederick Stratton, as "a fundamental contribution of the highest importance." The occasion was a meeting of the society on 8 June 1934, when Harry Plaskett introduced and outlined a paper by his father and Pearce, "On the Distance and Direction of the Gravitational Centre of the Galaxy." Harry gave unequivocal credit to them for their "full solutions … including the K-term, the solar motion (the projection on the galactic plane only) and the galactic rotation," noting that "the K-term has now dropped down to 1.1 km/s, a quantity which may be completely accounted for by the Einstein gravitational displacement in stars of these masses and densities."[117] The thirty-three-page paper was subsequently published in the society's *Monthly Notices*. An even longer, definitive, and final version, though complete at this time, did not appear in the *Publications* of the DAO until the following year due to financial cutbacks.

Though Plaskett valued international recognition most highly, he certainly appreciated his Canadian honours too. On 9 May 1934, Queen's University awarded him an honorary LLD – his fifth and final honorary degree.[118] He combined the trip east to receive this recognition in Kingston with a number of other duties, travelling first to Cleveland at Warner & Swasey's request to confer with them on the design of a new 2.0 m telescope (see chapter 12) and to Toronto to discuss the university's nearly completed 1.88 m reflector with Chant and R.K. Young.[119] At the same time he tried to sound out Chant about the details of the offer that the university made to Frank Hogg to join the staff in Toronto.[120] (Within a couple of months and with Ottawa's agreement, JSP was tempting Frank with a generous counter-offer, but in the end the Hoggs decided to leave Victoria in December. Frank began his new position as lecturer at the University of Toronto on 1 January 1935 but there was no position for Helen. The university policy during the Depression, similar to that of the federal government, was against hiring married women.)

Leaving Toronto, Plaskett went on to the RSC meeting at the Chateau Frontenac in Quebec, addressing the fellows with a talk entitled "The Dimensions of the Galaxy."[121] It was one of very few presentations that year to the whole of section III. By this time, the several subdisciplines in section III were holding separate sessions, though with only thirty-four fellows in attendance, four of whom were astronomers, one might wonder why. The few astronomers who were FRSCs included Beals, Chant, Harper, Pearce, Stewart, and Young. So, although the gatherings were intimate, Plaskett no longer had to carry the ball almost alone.

As Plaskett and Pearce progressed towards a fuller description of the Galaxy, they had to incorporate data from many other astronomers, whose work they naturally took care to cite.[122] Their extensive examination of the stars' proper motions (at right angles to their radial velocities), combined with the radial velocities and certain assumptions, led them to new estimates of the size and mass of the Galaxy. To determine these parameters they needed to know the actual velocity of revolution of the Sun about the Galactic centre relative to external systems, for which they adopted a value of 275 km/s. Plaskett and Pearce concluded that the distance from the Sun to the centre was 32 500 light years and that the Sun would take 224 million years to complete one circuit. Weighing the evidence from others' results, they reasoned that the Sun's distance

11.7 C.A. Chant on the steps of the administration building of the David Dunlap Observatory (DDO) a year before its official opening. (From UTA, DDO scrapbook; the entire scrapbook, designed and produced by Rous and Mann, is available online at https://archive.org/details/ddobookooastr)

from the centre was about two-thirds of the overall radius of the Galaxy and that its total mass was about 165 billion solar masses.[123] Partly because they had now adopted a smaller orbital velocity for the Sun, their dimensions and mass of the Galaxy were reduced by about 20 per cent compared to what JSP had supposed in his address to the ASP in 1932. (Currently accepted values for the Sun's distance from the centre and the time required for one circuit are respectively 27 200 light years and 200 million years – not far from the values Plaskett and Pearce found.)

In a sense, Plaskett and Pearce merely solidified others' earlier results on the structure of the Milky Way, but their analysis was so thorough that there was little room for doubt about the reality of the Galaxy's rotation or its approximate size and mass or about the pervasiveness of the interstellar medium and its participation in the Galactic rotation. Other astronomers had taken bold and pioneering steps but Plaskett and Pearce tied everything together in a neat package, completing it in December 1934, just as JSP was about to retire. In the closing twenty-four-page section of their final joint publication, they put their work into the context of studies over the previous 200 years. They also compared the Milky Way Galaxy to what was by then understood to be its neighbour, the spiral galaxy known as M31 in the constellation of Andromeda. This well-known object, dimly visible as a mere smudge to naked-eye observers in the northern hemisphere, is seen in photographs to be very much like our own – a flattened, rotating system of billions of stars accompanied by clouds of obscuring matter. In 1934, M31's diameter was thought to be about 65 000 light years, or two-thirds of JSP's value for our Galaxy. (Now, it is recognized that both galaxies extend much further and that M31 is the larger of the two.)

A significant testament to the long-lasting value of the final DAO paper of Plaskett and Pearce was written seventy years later. One of the outstanding astronomers of the time, Allan Sandage, wrote that Plaskett and Pearce's paper "remains a treasure of astronomical literature."[124] In another remarkable end-of-the-twentieth-century review, Plaskett's diagram of the Milky Way was cited as "essentially the one we all still sketch in introductory courses."[125] And astronomer-historian Owen Gingerich selected this drawing to be among those illustrating twentieth-century physical science.[126]

Though Plaskett and Pearce were clearly at the forefront of astronomical research and had made a major contribution in understanding the

11.8 Plaskett's illustration of the Galaxy, which he used in his Halley lecture in 1935. In the diagram and in Plaskett's own description that follows, the numbers are in parsecs; to convert to light years, multiply by 3.26. "The effective diameter is 30,000 parsecs with a thickness 1,000 to 2,000 parsecs, but with a spheroidal enlargement at the centre 5,000 parsecs or more in thickness. There are however scattered high velocity stars and the globular clusters extending beyond these boundaries and a thin stratum of absorbing material some 500 parsecs in thickness is interspersed among the stars near the central plane." (From the adapted version in *JRASC* 30 (1936): 153)

Milky Way and its relationship to external galaxies, they rightly recognized that there was much more to be done. They noted that

> it seems certain that the system has by no means reached a condition of dynamical equilibrium, a "steady state," the presence of clustering and of moving groups of stars being sufficient to contradict this assumption. No really satisfactory explanation of the spiral form, so common a feature of extra-galactic systems and probably present in the galaxy, has yet been advanced, and it is obvious that we have yet much to learn about the organization of the system of stars in which we are situated.[127]

Even during JSP's lifetime, some astronomers recognized difficulties with some of the methods he and Pearce used. When Harry introduced their paper to the RAS in 1934, he may have unintentionally planted some seeds of doubt about his father's finding that "the southern B stars showed a big peculiar effect, which was attributed to the motion of the Scorpio-Centaurus group."[128] The RAS secretary for the meeting, William Smart, may have picked up on this point at the time, but his objections will be considered in the next chapter, as they developed only after JSP's retirement.

Reviewing JSP's Directorship

In closing this chapter, we come to the end of JSP's career as director of the DAO. It is an appropriate point to look back over the previous eighteen years of his stellar career. Historian Richard Jarrell pointed out that the observatory's tremendous output owed much to the staff that JSP chose and to the amicable relationships among them that he fostered:

> When one looks over the bulky bound volumes of the DAO's *Publications* ... and reflects upon the extent of first-rate research they hold, it seems difficult to believe that it was achieved by a handful of astronomers and one telescope. Unlike the great astrophysical centres further south along the Pacific coast, Lick and Mt Wilson, the DAO did not possess a battery of large instruments, the backing of a Carnegie Institution, or the relationship with renowned scientific schools like the University of California or the California Institute of Technology.[129]

Jarrell noted the challenges that Plaskett faced in working in an era clouded by war and economic depression. Throughout JSP's years of leadership, the DAO operated on a shoestring. The first year's operating expenses, including salaries, was no more than $13 000.[130] This was about a quarter of the Harvard College Observatory's budget, or 40 per cent of that of Lick or Yerkes.[131] The amount did increase over the years as the staff increased and the temporary cutbacks during the early years of the Depression were rescinded. (In June 1934 the Bennett government decided to try to alleviate unemployment by spending on public works, including $36 000 to improve the observatory grounds.[132]) By the time of JSP's retirement in 1935, the total annual DAO expense had just about doubled to $24 500 – $20 000 for salaries, $3500 for research and scientific maintenance (including printing the publications), and $1000 for office expenses.[133] So, he did manage to achieve some modest growth in the operation in spite of very challenging times.

Moreover, JSP's own scientific output, averaging over seven papers per year during his directorship, was remarkable, considering the massive amount of data that many of his publications contained and the fact that he simultaneously gave five or six talks per year and had the full responsibility for administering the observatory. His example and the very high standards he set ensured that the DAO set the course of Canadian astronomy for decades to come. The telescope design at the University of Toronto's David Dunlap Observatory was strongly influenced by the blueprints that JSP lent to C.A. Chant in 1927. The research pursued there was very much like that at the DAO – hardly surprising, as its first directors (after Chant retired at age seventy on the observatory's opening day in 1935), namely Reynold Young and Frank Hogg, had been among JSP's handful at the DAO. Thus Plaskett's influence lived on not only in Victoria but in Toronto, where the university continued to educate the majority of astronomers in Canada until the 1960s. As those graduating from that university spread out across the country and the globe to train their own students, Plaskett's indirect influence went with them.

PART FOUR

The Fruits of His Labour

12
Retirement, 1934–1941

The summer of 1934 in Victoria was marked by drought, and the forests were tinder dry. About 2:45 p.m. on 23 August, a municipal truck driver spotted fire and smoke on the west side of West Saanich Road opposite the observatory.[1] All available municipal fire trucks were rushed to the scene, but by the time they arrived the flames had jumped the road and were licking up the side of Observatory Hill. The parched brush was soon ablaze, and before long the flames reached close to the summit. Helen Hogg, who was there, described the tense situation: "About three o'clock, some of the staff at the Ob[servatory] looked out the windows and saw flames and burning trees on top of the hill a couple of hundred feet from the director's house. There were only about five people on top of the hill at the time, and believe me, they did some scampering, getting out fire hoses and turning in the alarm. The first hour was the perilous time, before the necessary equipment and number of fighters had reached the scene."[2] Huge clouds of smoke were clearly visible from the city, but the thousands of curious spectators who tried to get close were not allowed through police lines. Throughout the night two hundred men toiled in shifts to bring the fire under control, their efforts severely hampered by a limited water supply. But with a change in wind, the firefighters prevailed and by morning the danger was over. About five square kilometers had been burned over, including the electric light poles. The observatory gates were ruined but no buildings were damaged and no one was injured. It had been a very close call.

Plaskett was apparently not at the observatory on the day of the fire – perhaps he was on holiday. If the government had had its way, the sixty-eight-year-old director would have been retired. A new regulation mandating retirement at age sixty-five had come into force the previous year when the government was trying to find every conceivable cost-saving measure in the depths of the Great Depression. JSP was not about to go quietly from the observatory that had been such a huge part of his life and he latched onto a couple of loopholes. The rules provided an exception if

> Treasury Board, on the recommendation of the Minister of the department concerned, approves of the continuance in the public service of any such employee on the ground:
>
> (a) that it is impossible to arrange that the duties of a position be carried on by rearrangement of staff or other means, or
> (b) that the qualifications and experience of a person employed in an administrative position, are such that the public interest will suffer if his services are terminated.[3]

By appealing to these provisos, and with the support of Dominion Observatory director R.M. Stewart, Plaskett managed to get his formal retirement date put off until 31 January 1935, when he would be in his seventieth year. He was entitled to six months retiring leave, which kept him on full pay until the end of July, during which period Ed Harper would serve as the acting director.[4] Plaskett's plans already included the International Astronomical Union (IAU) meeting in Paris in July 1935, but in the meanwhile he received yet another illustrious honour affecting his retirement arrangements – he was invited by the University of Oxford to be the Halley Lecturer for 1935.

Ultimate Honours

The Halley Lecture was created in 1910, a year when Comet Halley made one of its rare apparitions. The comet had a special place in the history of the university, as Edmond Halley had been Savilian Professor of Geometry there in 1705 when he first announced that the comet was periodic and would return approximately every seventy-five years. So a wealthy benefactor, Henry Wilde, decided to celebrate the appearance of the comet in 1910 by establishing an eponymous lecture to be given annually on a subject connected with astronomy or terrestrial magnetism. Each year a group of five Oxford academics met to decide who should be invited to deliver the prestigious address the subsequent year. When the five electors met in June 1934, one of their number was naturally the Savilian Professor of Astronomy, Harry Plaskett. No details survive regarding how proactive Harry was in proposing his father, but historian Roger Hutchins suggests that Harry and his colleague Arthur Milne used the lectureship "to invite the most distinguished international investigators in astronomy ... to Oxford."[5] In any event, the group did make JSP their first choice for the following year.

JSP was, of course, thrilled to be selected, but at the same time he was a bit apprehensive. Harry had told him the previous Halley Lecture, delivered by Edwin Hubble, was excellent but that the one before it, by Henry Norris Russell, was not.[6] With such precedents, JSP wished to avoid any gaffes and sought advice from some of his peers. He asked to borrow slides from Mount Wilson astronomers Frederick Seares and Alfred Joy[7] and, in an interesting letter to Edwin Hubble, asked for clarification on the presence of globular clusters in low Galactic latitudes where interstellar dust ought to obscure them. Hubble replied,

> Personally, in view of the uncertainties regarding obscuration, I have felt that the dynamical evidence for distance to the center which you have presented, outweighs the evidence from the clusters and that the clusters must eventually be forced into the picture you sketch ...
>
> I am delighted to learn that you have chosen the galactic system for the subject of the lecture. It is as important a subject as it is difficult. Your discussion in connection with your last big paper is by far the best summary available and we will look forward with the greatest interest to a restatement of the situation by one who speaks with such authority.
>
> We had the most delightful visit in Oxford with Harry. He has a very lovable character in addition to ability and charm. Everyone I talked with had the same tale – the right man for the right place – a happy combination.[8]

Given that JSP was at the pinnacle of prestige, it is worth noting that about this time he joined with ninety other fellows of the Royal Society in signing a petition designed to reform some of the restrictive policies and autocratic management of that august body.[9] Perhaps, on the verge of retirement, he knew he would face no professional repercussions if those on the powerful council took offence, or perhaps he just wished to draw attention to the limited rights that overseas fellows could exercise. He was, however, not so rebellious as to join with a core group of thirty-three, who demanded more far-reaching changes. Harry, presumably for opposite reasons, declined to sign even the original memorandum, and before a year had passed the Royal Society Council decisively rejected all of the reformers' ideas as too revolutionary.

As JSP laid his plans to travel to Europe, first to England to deliver the Halley Lecture in June, and then to attend the IAU meetings in Paris in July, he once again asked the government to pay his expenses. His pleas for $900 to cover his trip were refused through the normal channels on the grounds that no one on retiring leave had ever been sent on a similar mission. From the government's perspective it seemed a very logical way to restrain spending from which they could see no lasting benefit. But Plaskett, as usual, stressed that he was being recognized, and that he was president of an IAU commission (on radial velocities), because of his outstanding work as a civil servant. Therefore, he argued, the government should accept its obligation to provide the means for him to attend. Furthermore, he pointed out, his scientific record had brought worldwide attention to the progress of science in Canada; the international community would look on his absence as a snub by himself and the country.

No one in his department seemed to be listening, so (as he had on previous occasions) Plaskett went to the very top. In spite of his assurances to his former deputy minister, back in 1927, that he would not "make any appeals to others than yourself," he wrote Conservative prime minister R.B. Bennett a long letter on 5 December 1934.[10] He introduced himself as having received the Flavelle Medal of the Royal Society of Canada at the same time as Bennett was made an honorary fellow. He outlined all the recognitions he had received in the past and the reasons why he should make this trip at government expense. He concluded he "would be much happier on leaving the service of Canada if this final request was granted and I could thus feel that extra-ordinary services had really been appreciated by the Govt."[11]

The letter worked. Bennett replied within ten days that he thought Plaskett had made his case. Not only did Bennett get Plaskett's expenses approved but he recommended the astronomer for one of the highest honours the government could propose, second only to knighthood. JSP was named a Companion of the British Empire (a civil as opposed to a military CBE) in the king's honours list published on New Year's Day 1935. It happened in a narrow window of opportunity; Bennett took the strictly legal view that the Nickle Resolution of 1919, which prohibited Canadians from accepting awards from foreign governments, was not an act of Parliament and so felt justified in proposing names for royal recognition between 1933 and 1935. After his defeat the Liberal government reverted to the former policy and Canadians were no longer formally recognized in this way.

To mark JSP's retirement and his CBE the Mens' Canadian Club of Victoria held a gala dinner in the Crystal Ballroom of the Empress Hotel on 11 January. Over two hundred attended, including the lieutenant governor, cabinet ministers, clergy, and many other prominent citizens in public and private life. The Women's Canadian Club simultaneously entertained Mrs Plaskett "in the alcove of the main dining room" and later joined the men for the formalities. When JSP rose to speak after an impressive introduction and tumultuous cheers, he paid tribute to the observatory staff, past and present. His colleagues – Beals, Harper, and Pearce – were at the head table but the Hoggs had already moved to Toronto. JSP declared that he had been graciously guided by the "hand of Divine Providence ...Without His aid and guidance I never could have accomplished what I have." He then insisted that Reba rise and receive her share of the praise, a gesture that was warmly applauded. Never missing an opportunity to promote scientific understanding, in his speech, "The Purpose of Modern Astronomy," he included a "sad commentary" that "so many people believe in such rubbish" as horoscopes.[12]

On the eve of JSP's departure for Europe, having received word that his expenses had been approved, he wrote Prime Minister Bennett expressing his gratitude in the most obsequious terms imaginable:

> I am deeply grateful to you Sir for interceding in my behalf, for without it I feel sure this would not have been granted by the Department. I will always recall, with the keenest pleasure and gratitude, the personal interest and kindness of the Prime Minister of Canada, first of all for recommending me to His Majesty the King, for the distinction of the C.B.E., and now for this recent grant. I cannot tell you how much it means to me, on leaving the Government Service, to feel that my work for Astronomy in Canada is thus appreciated by the Head of the Government. I assure you that I can never forget your thoughtfulness and kindness and will always prize your letters.
>
> With best wishes for the fullest measure of success in your endeavours for the well being of our country.[13]

The circumstances were reminiscent of 1930, when Prime Minister King had interceded with officials on Plaskett's behalf. And history was repeated in another sense as well, as the Americans once again seemed to follow the lead of Britain in identifying JSP for one of their most distinguished awards. The National Academy of Sciences (NAS) awarded its Henry Draper gold medal to Plaskett "for his able and consistent labors in stellar radial velocities, and related studies energetically pursued for nearly 30 years."[14] It was especially fitting for JSP to receive this recognition named for Henry Draper, who had obtained the first photograph of a star's spectrum in 1872. The presentation was to take place in Washington, DC, on 25 April. W.W. Campbell was president of the NAS at the time and Vesto Slipher, as a member of the academy and a previous Draper medallist, prepared JSP's citation.[15] These men had each helped Plaskett launch his career nearly thirty years earlier, so it is easy to appreciate how much the honour meant to Plaskett. Regrettably, he could not fit a trip to Washington into his itinerary and the medal was accepted on his behalf by McGill geologist Frank Dawson Adams, a foreign associate of the NAS.[16]

JSP, Reba, and Stuart left home at the end of January. The steamship *California Express* carried them down the coast, through the Panama Canal, and across the Atlantic to England, on an ocean voyage lasting five weeks.[17] JSP wrote to Helen Hogg from the Regent Palace Hotel at Piccadilly Circus in London:

> The weather was splendid until the last week when we ran into heavy gales around the Azores – ships in distress all around. But we had a fine boat and good captain and the Plasketts never missed a meal. Indeed I enjoyed watching the ship buffeting the waves.
>
> We spent nearly three weeks with Harry and his family at Oxford, ten days with a friend in Sussex and have been here a week seeing the sights with Stuart. After the R.A.S. meeting on Friday, we spend the weekend at Oxford where we are picking up a car and touring about the country.[18]

One of the joys when the family reunited was the opportunity for JSP and Reba to see their growing grandchildren. They gave young John Stanley, for his eighth birthday, a telescope and microscope, which he proudly kept for seventy-five years before passing them on to his daughter and two teenage grandsons.[19]

While the Plasketts were in England, the 1.9 m mirror for the University of Toronto's David Dunlap Observatory was nearing completion at the Grubb-Parsons works in Newcastle-upon-Tyne. The Hartmann tests showed there were some small aberrations, and C.A. Chant was undecided as to whether to accept the mirror as it was or to ask the firm to continue trying to improve it. Neither he nor anyone from Toronto was on hand to make personal inspections, though apparently Harold Knox-Shaw, an experienced astronomer at the Radcliffe Observatory in Oxford since 1924, acted as an intermediary. Probably at Chant's suggestion, Knox-Shaw showed JSP the results of the mirror tests. Plaskett wrote to Chant saying that under normal seeing conditions the defects would not matter. He even went so far as to say that he faced a somewhat similar decision with the DAO mirror nearly twenty years earlier and in hindsight thought it would have been better not to have insisted that Brashear continue their polishing – work that reduced the aberrations but introduced secondary irregularities and delayed delivery by six months.[20]

JSP attended three monthly meetings of the Royal Astronomical Society (RAS) when he was in England. In addition to the one on 12 April referred to above in his letter to Helen Hogg, he was at an earlier one on 8 March and a later one on 10 May. The March meeting featured the ongoing joust between the two great knights of theoretical astrophysics, Sir James Jeans and Sir Arthur Eddington. Each protagonist, with typical British understatement and cutting wit, defended his own point of view and challenged the other's, in this case on vastly different ages of the universe. Rounding out the event were extensive comments by their younger rival Arthur Milne, Harry Plaskett's associate at Oxford.[21] At the May meeting, JSP gave a brief account of Beals' paper on the peculiar variable star, P Cygni, and listened intently to a presentation by Milne on "Stellar Kinematics and the K-effect." Milne thought that JSP's results supported his contention that the K-term showed there was Galactic expansion, but Plaskett maintained that the analysis he had carried out with Pearce showed there was not much observational evidence for it.[22]

On 24 April the British Astronomical Association invited JSP to speak and afterwards treated the Plasketts as honoured guests at its annual dinner held at Frascati's, a renowned restaurant on Oxford Street.[23]

These engagements were all secondary to the Halley Lecture. On 5 June, when JSP made his presentation, entitled "The Dimensions and Structure of the Galaxy," he must have felt a certain sense of incredulity. He had come such a long way from his roots in colonial Oxford County to the podium of the oldest and probably most famous university in the English-speaking world to deliver one of its most lofty lectures. He later told Chant, "I think I spent more time and thought over this than any lecture or paper I ever gave, and I was very glad when it was delivered."[24] At Harry's instigation, his father's address was published in a small book along with four other recent Halley Lectures on astronomy.[25] The stellar list of authors (and topics) shows what exalted company JSP had: Sir Arthur Eddington ("The Rotation of the Galaxy"), E. Arthur Milne ("The White Dwarf Stars"), Henry N. Russell ("The Composition of the Stars"), and Edwin Hubble ("Red-Shifts in the Spectra of the Nebulae").

The visit to Oxford was special for reasons other than the lecture and family reunion. Harry himself had something to celebrate. When he had assumed the Savilian Professorship in 1932, he had shrewdly stipulated that the university provide a new solar observatory where he could pursue his investigations. On 11 June 1935, this promise came to fruition in an official opening ceremony.[26] The year also marked the closure of the old Radcliffe Observatory to allow expansion of a neighbouring infirmary. This was a step in the long process towards the opening of a new Radcliffe Observatory near Pretoria, South Africa – a move strongly advocated by JSP, as president of the commission on radial velocities at the 1932 IAU.[27] After long delays caused by court battles and the Second World War, this important southern facility would get a 1.88 m telescope closely modelled after the one at the DAO and a virtual twin of the David Dunlap instrument. It would eventually begin spectrographic work in 1951. Its spectrographs were deliberately designed like the DAO instruments in order that Plaskett and Pearce's research on Galactic structure and rotation could be extended to the southern hemisphere.[28]

The Plasketts proceeded to Paris, where JSP found the IAU general assembly very successful and pleasant. In a letter to Helen Hogg, he noted, "Good progress was made in the Commissions' work and the business was harmoniously conducted. You were named on the Commission on clusters as well as on variables … There were three excursions, three receptions and two banquets, which with the heat, made it remarkable any business was completed."[29]

R.M. Stewart and the Chants were in Paris too. Chant was celebrating after his retirement and the successful opening of the David Dunlap Observatory on his seventieth birthday on 31 May. At a spectacular IAU dinner at the Eiffel Tower on 14 July (Bastille Day), he asked some of the assembled astronomers to provide their autographs.[30] The American magazine *Time* covered the assembly, giving a general idea of the IAU and what was accomplished at the meetings.[31] In amusingly vivid prose, the article portrayed the president, Frank Schlesinger, as "square, heavy-jowled" and vice-president Walter Adams as "wry, hollow-eyed, pucker-mouthed, who still talks with a New England accent." Heading the commission on stellar spectra was "tall, spare, eager Henry Norris Russell, 57." Although the article called Mount Wilson "the astronomical capital of the world," as the home of the largest reflector, it did recognize that Canada was home to the second- and third-ranking telescopes. The article went on to report on Plaskett's work, repeating an earlier error that his estimates had increased the population of the Milky Way by 600 per cent.

A number of American astronomers went to England after the Paris meetings had ended, Frank Schlesinger among them. His diary depicted a happy time in Oxford among old friends, including the five Plasketts (JSP, Reba, Stuart, Harry, and Harry's wife, Edith), Mabel Smith (Edith Plaskett's sister), two Turners (probably the widow and daughter of the late Herbert Hall Turner, Harry's predecessor), and Annie Cannon. After a lazy day in Harry and Edith's garden at 11 Belbroughton Road, just a few blocks north of central Oxford, they set out for tea at the historic Red Lion Inn at Long Compton and a visit to the nearby Rollright Stones, a neolithic site. The next day, Schlesinger saw Harry's solar observatory and enjoyed a dinner at his home with the Turners, Annie Cannon, Herman Zanstra (who had spent the summer of 1927 at the DAO), and a brilliant young fellow from Cambridge, Subrahmanyan Chandrasekhar.[32]

JSP would have been happy to stay longer, but Stuart and Reba were eager to get home. The three of them left England on 3 August and, after arriving in Montreal, took the transcontinental train west. As usual, Plaskett was happy to give a talk about his experiences to the RASC's Victoria Centre, illustrated with slides he had taken.[33] The group ate up his humour and his personal account of the silver jubilee celebrations of the coronation of George V held in May.

"Retirement"

Although Plaskett's retirement marked the end of his active research, he continued to attend astronomical conferences and, as we shall see, once again took up his long-standing interest and expertise in telescope design and testing. He also dropped in at the observatory quite frequently – too often to suit his successor, as we may judge from a letter that the new director, W.E. Harper, wrote to R.M. Stewart in Ottawa in 1937: "Everything goes smoothly apart from the visits of JSP. Thank goodness the blocked condition of the roads [because of snow] has kept him at home the past two or three weeks."[34] Under Harper's directorship, relations between Ottawa and Victoria definitely took a turn for the better, perhaps helped by a government reorganization in 1936 that saw the dissolution of the Department of the Interior and the creation of a new Department of Mines and Resources, but certainly by Harper's docile acceptance of the rules.[35] Other members of staff seemed more pleased with JSP's frequent calls at the observatory and "enjoyed many keen and helpful discussions" regarding their research.[36] Joseph Pearce regarded 1935, the year of Plaskett's retirement, as the start of a new era for the DAO for in that year Bert Petrie and Andy McKellar, former summer students, returned as full-fledged astronomers. In due course, Petrie and Kenneth O. Wright, who arrived at the observatory the next year, would become DAO directors, overseeing the construction of new photometers, measuring machines, spectrographs, and telescopes.[37] A later perspective, provided by DAO astronomer Alan Batten, is that the Plaskett era really ended when Wright retired.[38] He was the last director to know JSP, and he tried to preserve the Plaskett tradition of careful and thorough work, getting the maximum from the available instrumentation, pursuing fundamental programs, such as the one on Galactic rotation and of course, the study of spectroscopic binaries, which remained one of the principal areas of research until 1976.

Even in retirement, Plaskett continued to do some work on the Galaxy. He may have felt forced into it by criticisms raised by William Smart, already alluded to in chapter 11. In a series of papers between 1936 and 1940, Smart seemed to show that the southern B stars did not form a moving cluster (or "stream") and that the stars' apparently anomalous velocities were actually a reflection of the Sun's motion.[39] If he were correct, Plaskett and Pearce should not have removed the supposed cluster motion from these stars before considering the Galactic rotation and the residual K-term. (It is worth noting that recent research has vindicated their approach – while the aggregation of stars is presently known as an association rather than a cluster, its stars do indeed have systematic motions.[40]) Smart also objected to the way they handled the equations. One aspect of his criticism was that all their results were affected by using only a projection of the solar motion on the Galactic plane. In responding to the first of Smart's papers, Plaskett acknowledged the theoretical correctness of Smart's analysis but believed the end justified the means. He wrote an implausible, weak defence: "Although [my] procedure … is theoretically wrong, there must be some significance in the fact that this incorrect method gives improved values of the solar motion and galactic rotation, and reduces the K-term to the value of the gravitational shift [predicted by relativity]."[41] Perhaps his argument was a sign that theoretical considerations, never JSP's strong point, were overtaking him. And, in the end, Smart's analysis demolished Plaskett's claim that he and Pearce had successfully explained the very small K-term, though it did not greatly affect their conclusions about the Galactic rotation. Plaskett wrote two more papers in scientific journals in which he gave overviews of "modern conceptions of the stellar system." He did not mention Smart in either of them, though he did conclude both papers by acknowledging that his proposed model "must be considered only as a preliminary attempt at the solution of this difficult but very important astronomical problem." [42] Pearce's final contribution to Galactic rotation was an invited address in Seattle to the Astronomical Society of the Pacific (ASP) in June 1936, a meeting that was jointly hosted by the DAO. Plaskett attended but left all the presentations to those actually on staff.[43]

Though Plaskett and Pearce's model of the visible Galaxy has stood the test of time, the overall size and mass of the Galaxy are now thought to be several times greater than the values they derived, mainly because of unseen mass that makes its presence known only through its gravitational pull. Curiously, it was in 1933, just as Plaskett and Pearce were neatly tying everything up, that the iconoclast astronomer Fritz Zwicky made the first foray into "dark matter," as he called it, a concept that ultimately revolutionized galactic dynamics, though it became a viable topic after only Vera Rubin's trail-blazing work in the 1970s.[44]

Reflections

Consultation on telescope manufacture was an important aspect of JSP's later career and extended into his retirement years. We saw in the previous chapter that he had done the testing on the mirror for the Perkins Observatory at the end of 1931. He subsequently became involved in another large project whose story began several years earlier. In 1926 the University of Texas had inherited a million dollars from a wealthy banker, W.J. McDonald, who stipulated that the money be used for an astronomical observatory, an idea his family contested on the grounds that he was not of sound mind.[45] The university's dean (and later president), Harry Y. Benedict, asked astronomers to help bolster the university's case. JSP replied that the foundation of a great observatory would be a most effective way to perpetuate McDonald's name and that there were other precedents for such memorials.[46]

Once the dispute was settled in the university's favour, a wonderful site was chosen atop the two-kilometre-high Mount Locke in the southwestern part of the state. Then trouble arose over the design and manufacture of the telescope. An influential member of the university's board thought the revolutionary Ritchey-Chrétien telescope design would be the best system and recommended that George Ritchey be engaged to manufacture it.[47] But Otto Struve, the University of Chicago astronomer who would direct the new observatory, knew of Ritchey's reputation for slow work and his controversial background at Mount Wilson.[48] Struve favoured a conventional design and solicited letters from JSP and Walter Adams to show to the board in support of his view.

Struve had no proficiency in optical design himself and had to rely heavily on others, especially Plaskett, for outside advice. Being thoroughly familiar with the excellent mechanical work done by Warner & Swasey on the Ottawa and Victoria telescopes, JSP was in favour of them doing that aspect of the work on the new observatory. However the firm had recently set up an optical component of the business with the hiring of Robert Lundin. The 2.1 m mirror for the McDonald telescope – second in size only to the Mount Wilson telescope and far larger than anything he had ever undertaken – would be Lundin's first project with the firm, and Plaskett was sceptical at first. He wrote to Struve, who had asked for his opinion: "So far as I know Lundin is capable enough with small surfaces but I do not believe he has figured anything over two feet. The problem of an 80-inch surface is of quite a different order and I would be personally loath to take a chance on him being able to handle it."[49] However, by June 1933 Plaskett fully supported the decision of Struve and the others overseeing the Texas project to go with Warner & Swasey.[50] Indeed, he soon served as consultant on the project, spending a week at the company in Cleveland going over the design of the McDonald telescope in May 1934.[51]

The rough grinding of the Pyrex disk, which began in October, went well, but when it came to figuring the mirror to final paraboloidal shape, Struve found Lundin's performance to be a disaster.[52] Months turned into years and JSP had to keep assuring Struve that delays were to be expected with large mirrors. Yet, try as he might, Lundin could not get the mirror into shape.[53] A major problem was his insistence that the focal length had to be as specified, even though Plaskett and Struve told him that a few centimetres here or there were not important.[54]

According to JSP, two years were wasted because of Lundin's desire for perfection but Struve thought there were other problems. He told Plaskett, "I hope that you will continue to act as scientific consultant for the Warner and Swasey Company, and your participation in this project is the only bright spot in the picture. I wish you could be more permanently in Cleveland to advise Mr. Lundin from day to day. I fear that the effect of your advice lasts only for a short time and must be renewed in order to produce the desired effect."[55] JSP balanced the occasional trip to Cleveland with more domestic matters. After he and Reba returned home from a trip to Pasadena in March 1937, they were busy refinishing some of their "furnishings in the house to say nothing of work in the garden and grounds."[56]

On one of his periodic trips east to check on the mirror's progress, he also attended a memorable ceremony in Philadelphia, on 30 April 1937. The Westinghouse Electric Company had invited him and other notables, including Nobel laureates Albert Einstein and Robert Millikan, to formalities marking the completion of the tube for the giant Palomar telescope.[57] The company, with experience in manufacturing huge turbines for power generation, had the facilities and know-how to make the mounting and tube for this new leviathan.[58] At 5.0 metres, the telescope would become the world's largest, doubling in one leap the diameter of the previous record holder on Mount Wilson.[59] Almost a decade had passed since the Rockefeller Foundation had granted six million dollars for the Palomar project, and another decade would elapse before the optics would be complete. G.E. Hale, having launched and guided the endeavour just as he had for the Mount Wilson Observatory, was regrettably too ill to go to Philadelphia. Within a year he would be dead but, when Palomar finally opened, the telescope would be dedicated to his memory and named after him.

While heading back to Victoria in June 1937, Plaskett stopped off in Denver for the joint meeting of the ASP and the AAAS, where he was elected president of their two-thousand-member-strong Pacific Division. He must also have felt very proud to chair a session at which Carlyle Beals presented an astonishing seven papers by DAO astronomers. Though JSP did not make a formal presentation, he and Warner Seely of Warner & Swasey did talk publicly about the progress of the McDonald telescope. They mentioned a couple of its unique features – the coudé focal arrangement, which was simplified by not having the declination axis pierce the polar axis, and the drive mechanism, which allowed for differential atmospheric refraction.[60]

12.1 Ceremonies marking completion of the tube for the Palomar 5 m telescope, Westinghouse works, Philadelphia, 30 April 1937. (Westinghouse Electric Corporation, Steam Division photograph collection (1969.170), Hagley Museum and Library, Wilmington, DE, 19807)

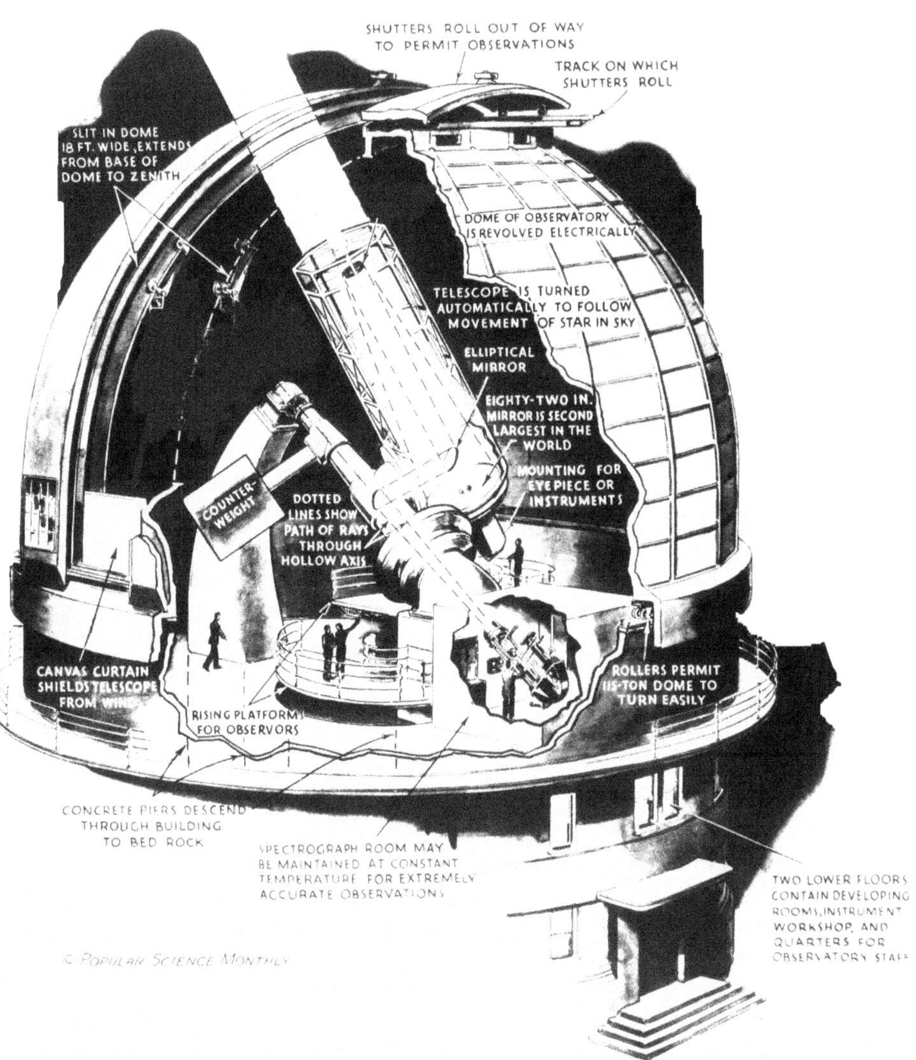

12.2 Cut-away drawing of the McDonald Observatory. The telescope is nearly 2 m shorter than the DAO instrument and can be used at the coudé focus (shown here as "Spectrograph Room"). (University of Chicago Library, Special Collections, apf2-05058, modified)

Back in Cleveland at the Warner & Swasey factory, the mechanical work on the 2.1 m McDonald telescope was going well. Plaskett thought it was the masterpiece of the company, with several original features executed by Edward Burrell, the same engineer whose ingenuity had impressed him twenty years earlier.[61] It would be the last project for Burrell and for Swasey, who both died in the spring of 1937. JSP marked the end of an era by writing obituaries for both men.[62] Progress on the optical work was another matter entirely, and Plaskett was again summoned in November 1937, when his suggestion to use a smaller tool seemed promising, and again in March 1938, when there were further disappointments.

Around the end of that month, he reported, Lundin was finally about to accept his advice to do a knife-edge test that would clearly show where further figuring was required when Struve and his colleague George van Biesbroeck visited the shop with a new graphical method to guide the process.[63] Although it turned out to be unhelpful, Struve accepted none of the responsibility for the diversion his intervention had caused and confided to a couple of colleagues his "utmost disgust" with Lundin and "the entire Cleveland crowd" and his fear that Plaskett, "this grand old man whom I respect perhaps more than any other spectroscopist in the world," was "losing his former keen insight."[64] Exasperated, Struve insisted that the optician from Yerkes Observatory, E. Lloyd McCarthy, supervise the work; the Warner & Swasey management agreed. While McCarthy was in Cleveland, from May until September, Plaskett made only sporadic appearances at Warner & Swasey. It is uncertain if JSP felt superseded during this period or whether he simply had too many other things to do. During his absence from the woes at Warner & Swasey's, Plaskett seemed to be constantly on the move. He was in Toronto in early May 1938. The Chants drove him and Helen Hogg down to Hudson, Ohio, where the little Loomis Observatory was celebrating its centennial.[65] In June, as retiring president of the AAAS Pacific Division, he gave a public address in San Diego, "Modern Conceptions of the Stellar System," and chaired a session of the concurrent meeting of the Astronomical Society of the Pacific.[66] He and Reba along with about forty others were able to visit Palomar Mountain and have a tour of the immense new dome that eventually would house the 5 m telescope.[67] They then travelled across the continent by train, perhaps stopping in Cleveland, before boarding the MV *Britannic* in New York bound for Southampton.[68] The six-week trip enabled JSP to attend the IAU general assembly in Stockholm on

3–10 August and provided the opportunity for the Plaskett family to reunite in England once again. At this point, Harry was firmly part of the scientific establishment – a secretary of the RAS and a fellow of the Royal Society. The Plasketts (both generations), Chants, Hoggs, and Harpers (with their two daughters) all went to Stockholm for the IAU.[69] The total attendance, 282 delegates and 139 guests, exceeded all previous assemblies. The Swedish parliament buildings (Riksdagshuset) served as headquarters for the business sessions, while, for the first time in IAU history, colloquia were included. As always there were tours and gala receptions – one at the Golden Room at city hall, the venue of Nobel banquets, and another at the Royal Palace. Plaskett, as president of the British Columbia Historical Association, took a special interest in Skanson, a district of Stockholm that recreated the old days in the city.[70] And of course he was delighted to see two of Sweden's observatories, at Saltsjobaden and at Uppsala, the latter itself a historic university city about two hours north of the capital.

By the summer of 1938, Germany had invaded Austria and was eyeing Czechoslovakia, yet the astronomers, reflecting the optimistic attitude of some, betrayed no premonition of war. A year later the world would be thrust once again into dreadful hostilities. The Plasketts could not know it at the time, but JSP had attended his last IAU, as the war meant that the Union would not reconvene until 1948, and he would never again see Harry and his family. In the meantime, a more immediate plight was that Harper was stricken with pneumonia while overseas and had to be hospitalized in Germany for several weeks.[71]

After Plaskett returned from Europe at the end of August, he checked on the situation at Warner & Swasey's and then went to AAS meetings in Ann Arbor, Michigan, before returning to Cleveland to do the final optical testing.[72] According to Plaskett, by the time of Struve's next appearance at the plant on 6 October, "the surface looked fairly good by the knife edge and Ronchi test – so good that [Struve] appeared very satisfied with it, much more satisfied than Lundin and I who felt that it could be made much better."[73] Struve formally accepted the mirror on 15 October 1938.[74] It was then aluminized (a new process that replaced the old silvering technique) and Plaskett carried out the traditional Hartmann test. He was impressed with the results: in his article, "The 82-inch Mirror of the McDonald Observatory," he declared the mirror to be unequalled by any previously made and tested.[75]

This paper, though not Plaskett's last, in a sense brought his career full circle. He had published some of his earliest papers on instrumentation in the *Astrophysical Journal*. Then for twenty years, he submitted nothing to this respected American periodical, preferring to turn to England and the equally renowned *Monthly Notices of the Royal Astronomical Society* when he wished to publish outside Canada.

The published details of how the mirror was tested and reshaped caught the attention of Albert Ingalls, widely admired by hobbyists for his monthly magazine columns and books on how to make telescopes. Ingalls was sure that amateur telescope-makers would be very interested in Plaskett's article, as they performed similar work on a smaller scale as they ground and polished their own mirrors. So, he arranged to republish JSP's paper in *Scientific American*, where his column regularly appeared.[76]

When JSP returned home on 21 November 1938, he was elated by the successful conclusion to four years of work.[77] He was looking forward to going to Texas in three months' time for the arrival of the mirror at the McDonald Observatory and "first light," when the telescope would start its useful life. Following his stay in Texas, 23 February to 3 March 1939, Plaskett gave a full report on the installation and testing of the optical parts of the McDonald telescope, reiterating the essential details in a letter to Edward Burrell's widow. He wrote to her that, on 2 March, "Dr. Struve used the knife edge test himself [at the prime focus] and expressed himself as thoroughly satisfied … On Friday March 3, the Cassegrain mirror was attached and on Friday night the two mirrors were tested with the knife edge showing a practically perfect figure. Even Struve, who is rarely given to enthusiasm, said it was perfect. And all the mechanical parts, wholly due to the genius of your late husband worked to perfection."[78] At the prime focus of the telescope, George Van Biesbroeck took one of the first photographs – a five-minute exposure of the region near S Monocerotis.[79] In the published photo, the stars appear perfectly round, showing no sign of coma or astigmatism.

The formal dedication of the McDonald Observatory was held on 5 May. A large number of astronomers attended and remained for a three-day symposium sponsored by Warner & Swasey. A stellar cast of speakers presented papers on galactic and extra-galactic structure.[80] JSP and Reba were not the only Canadians present for the occasion: Joseph Pearce and Helen Hogg were there too. Hogg described the vast open spaces surrounding the dome and some of the events – a Mexican-style chuck wagon dinner, a rodeo, and even an encounter between Jan Oort and a rattlesnake.[81] One of the American astronomers, Peter van de Kamp, took movies of the festivities, including what is probably the

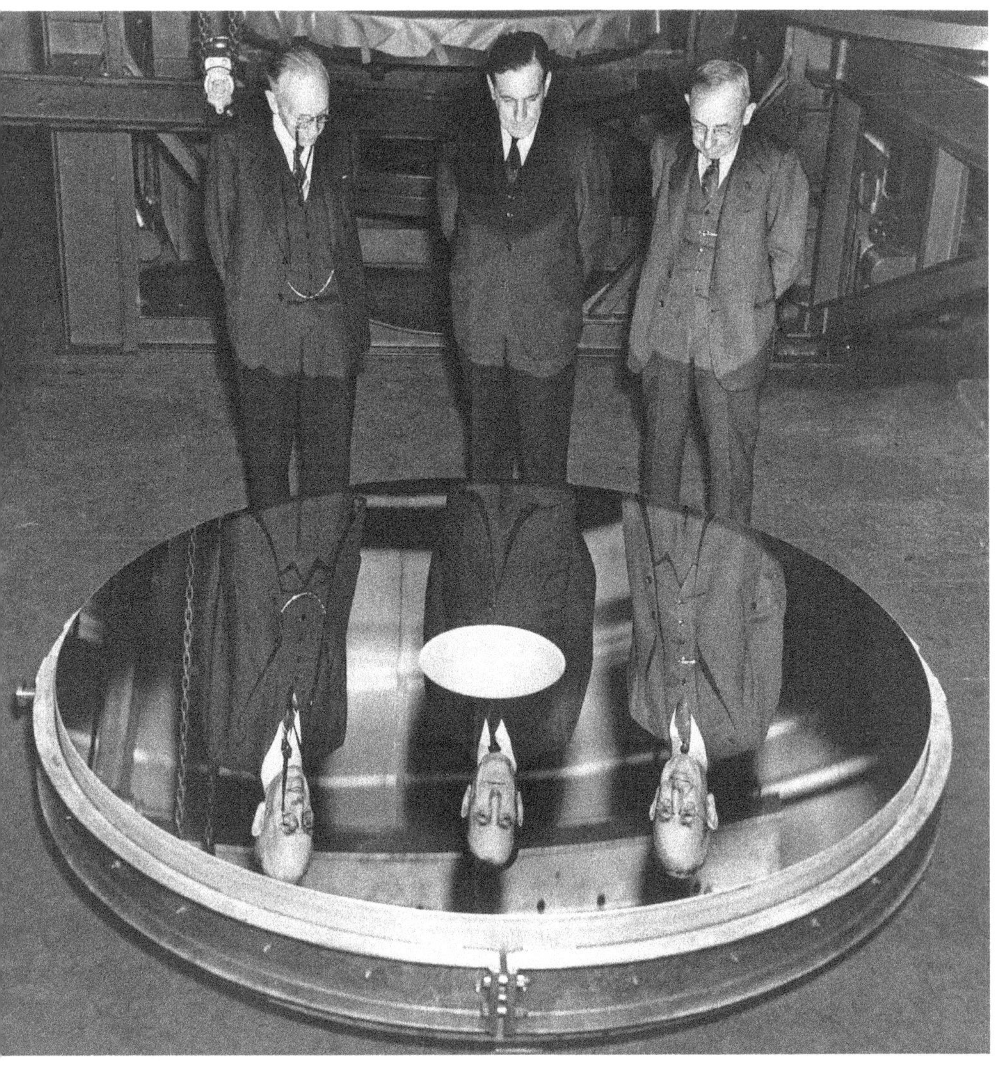

12.3 JSP, Robert Lundin (optician), and George Decker (engineer) admire the 2.0 m mirror at the Warner & Swasey works in Cleveland before it is shipped to Texas (CWRU, Warner & Swasey collection, box 43, folder 1)

12.4 Gathered on the steps of the McDonald Observatory at its official opening are (left to right) Henry N. Russell, JSP, Otto Struve, Harlow Shapley, and Homer P. Rainey, president of the University of Texas. (CWRU, Warner & Swasey collection, box 42, folder 7)

only extant motion picture of JSP.[82] He surely would have chuckled to have the tables turned, as it was he who always loved to catch his colleagues in informal shots.

At the public opening, 400 guests gathered inside the dome to hear the distinguished speakers address them from the observing bridge. Struve, as the new director, made the interesting observation that the year 1939 marked the hundredth anniversary of two of the world's great research facilities: the Harvard College Observatory and the Pulkovo Observatory near St Petersburg, Russia, whose founding director had been his great-grandfather, F.G.W. Struve. The present head of the

Harvard observatory, Harlow Shapley, was present at the ceremony, but the late director of Pulkovo was not: Boris Gerasimovich, a collaborator of Struve's, had been executed by Stalin's firing squad, "a recent victim of the most barbarian dictatorship of all times."

As Struve continued, he spoke about the optics of the new telescope in glowing terms:

> We wanted to get a mirror which would be perfectly free of all distortion when used visually or photographically. Corning supplied us with a Pyrex disk which changes little with temperature, and Mr. C.A.R. Lundin has put on it a figure so close to the true mathematical ideal that our experts assure us we have the finest astronomical mirror ever made.
>
> The spectrographic requirements have been constantly in the foreground during the design of the telescope. That we have succeeded in our aim is amply demonstrated by the work of my colleague, Professor Kuiper, during the past six weeks. He has secured some 600 spectrograms of stars never before examined by any astronomer, and has added several brilliant discoveries to the list of his former achievements.[83]

JSP followed Struve with a talk entitled "Some Features of the New Mirror."[84] He also had high praise for Lundin whose presence at the ceremony may have necessitated these tactful remarks. Lauding the optical quality of the main mirror, Plaskett compared it to others of similar size.[85] The Hartmann tests, he said, had shown that it was 2.5 times better than the DAO mirror, three times better than the one at Perkins, and four times better than the University of Toronto's new telescope at the David Dunlap Observatory.

The McDonald optics, like the Perkins mirror seven years earlier, did not behave as well in actual use as the tests had initially indicated. JSP would not have been surprised; in fact he had told Struve in 1932 that "the figure [of a mirror] when in use will practically never approach that obtained in the optical shop,"[86] although how this happens is debatable. Somehow, between the McDonald telescope's first light in March and the dedication in May, a "problem of deformed images" emerged. The observatory's historians, Evans and Mulholland, merely say that "even after he had pronounced the telescope performance to be perfect, Struve was not satisfied – and with cause. Warner & Swasey's expansion into the field of large optics had not been fully successful."[87]

A year after the telescope's inauguration, Struve wrote about its beginnings in the observatory's first *Contributions* and sent an inscribed copy to Plaskett.[88] JSP thought that Struve did not give adequate credit to the vital role of Warner & Swasey. He wrote to the president of the company, Charles J. Stilwell, to say that "the success is ultimately due to the ability and experience of the Company and their determination, regardless of cost, to make [the telescope] the best yet."[89] After Stilwell replied with more information about the difficulties the firm had experienced with Struve, Plaskett, in a subsequent letter, went further. He referred to Struve's ingratitude towards Edwin Frost, who had done so much for him in his younger days at Yerkes Observatory but, after blindness and retirement, was shunned by his younger successor. Plaskett thought that Stilwell's experience was "probably not out of character."[90]

Two decades after Plaskett's death, Struve would write that the secondary mirrors had been "tested quite superficially. The 82-inch mirror has an excellent shape, but its original supporting mechanism frequently allowed it to become astigmatic. We knew how to correct this but during World War II there was rarely enough time and manpower available to do the job. A new supporting mechanism was built after I left the observatory in 1950, and my successor, G.P. Kuiper, has told me the performance of the mirror is now excellent."[91] Two years after he wrote that, the French optician Jean Texereau performed tests showing errors in the figure of the primary, which, he alleged, would have been there right from the start – errors of nearly a wavelength, or twenty times Plaskett's claim.[92] He felt that Plaskett may have performed his tests at certain favourable temperatures and failed to double check the results from the Hartmann test by also performing an additional knife-edge or Foucault test. His criticisms seem to belie the hope that temperature changes would have had little effect on the figure of the Pyrex mirror. Furthermore, Plaskett's correspondence and his paper on the subject show that he employed many different tests. Fortunately the faults, whatever their origin, did not prevent excellent work being done in spectroscopy, where the sharpness of the images was not as important as it would be for direct photography.

Winding Down

After retirement JSP had more opportunity to pursue his interest in history. He completed his term as president of the BC Historical Association in 1935, when it encompassed only Victoria and vicinity, and then embarked on a round of first vice-presidencies of the province-wide association, ultimately becoming its president in 1938–9.[93] As he stepped down from office at the group's annual meeting on 13 October 1939, he gave an address, "The Astronomy of the Explorers," that was probably an expanded version of a speech he had given the association in 1932. It was subsequently published in its *Quarterly* and, forty years later, in an unwitting infringement of copyright, republished in the RASC *Journal* under the title "The History of Astronomy in British Columbia," when Plaskett's manuscript copy came to light.[94] In it he recalled the vital part astronomy had played in navigation and how essential it was to explorers of the Pacific coast such as James Cook and George Vancouver. He spoke of the inland surveyors too and concluded with a substantial section on the establishment of the observatory that he knew so well. In his final public speech, given to the RASC on 13 December 1939 at the Victoria Centre's annual dinner, he took the history of the Dominion Astrophysical Observatory as his topic.[95] At age seventy-four, he was able to look back on his life's work in a broader context.

During his retirement, many of JSP's astronomer friends passed on, Edwin Frost, G.E. Hale, W.W. Campbell, Frank Dyson, and Annie Cannon among them. The death of his brother Tom at Christmastime 1939, and his own serious heart condition, which became apparent in 1940, reminded him of his mortality but did not deter him from making plans.[96] He wrote to Struve in April 1940 that he had been under observation in hospital for over two weeks and had taken a while to recover from the drugs and hospital fare.[97] Nonetheless, he continued, "My wife and I had planned for some years to attend one more meeting of the Royal Society of Canada, and this seemed to be the year." So they arranged to leave Victoria about 23 April for Pasadena and then travel across the continent via Cleveland, where he had been asked to spend a couple of weeks providing advice on a proposed Schmidt telescope at the Warner & Swasey Observatory.[98] He would, as he put it, bone up in Pasadena on his "pretty sketchy" knowledge of this new design. As he and Reba crossed the United States, he sounded out opinion about the war, finding those in the west were strong "pro-allies but in the east indifference and isolationism [were] marked."[99] Still, when he got to the Warner & Swasey factory he was struck by the "tremendous increase in the production of turret lathes ... vital to the mass production of shells."

After the Plasketts arrived in Toronto, Chant drove JSP to London for the RSC meetings on 20–2 May. As they passed through Woodstock, JSP must have been filled with memories of his family and childhood. Perhaps the two astronomers spoke of the horrors of the war against totalitarianism gripping the world. No doubt they agreed with the pronouncement of RSC president H.M. Tory that knowledge could grow only in an atmosphere of freedom and supported Tory's pledge of the services of the RSC to the government war effort. And without question they would have taken profound pleasure at the election of W.E. Harper as vice-president of section III.[100] Each man, in his own way, had shaped Harper's career. Sadly, illness kept Harper from the meeting and he died only two weeks later at age sixty-two. He had never fully recovered from the pneumonia he had suffered while in Europe in 1938.

When the University of Western Ontario opened its Hume Cronyn Observatory in October 1940, Chant was in attendance but Plaskett was not. Nor was he among those sending congratulatory greetings. Official word came from Pearce, by this time the director of the DAO. JSP's friends in Victoria realized his health was rapidly failing. However, as Beals remarked, "this growing physical weakness made no difference to his happy disposition … and it was apparent to all that he was facing the end with the same cheerful courage with which he had always faced the tasks of life. His faculties were undimmed to the last."[101] Indeed, he was still able to attend some local meetings. He and Reba were at the head table at an RASC banquet in December 1940, and he is on record as attending Christ Church Cathedral Building Limited meetings up to March 1941.[102]

The End

JSP died of heart disease and fibrosis of the lungs on 17 October 1941. Officially, the diagnosis was arterial degeneration complicated by hypertrophy of the prostate and subacute cystitis.[103] The private family funeral was conducted by the Reverend George Biddle at the Church of St John the Divine.[104] In accordance with Anglican tradition at the time, there were no eulogies. Some might have thought a grand service celebrating the life of one who so loved the limelight would have been appropriate. A world at war may have militated against it, and perhaps at the end JSP and Reba recognized that fame is fleeting and modesty more seemly. It may have been that Stuart served as a cogent reminder that one's intellectual capacity is a godsend rather than a source of personal pride. The service was followed by cremation, and no cemetery memorials were erected to his memory. The observatory itself and its *Publications* that came out during his directorate, along with the hundreds of papers and talks he prepared (listed in the appendices to this book) are undoubtedly a more telling tribute to his influential career than any tombstone could be. Besides, as we shall see in the concluding chapter, in the years to come Canadian astronomers would ensure that John Stanley Plaskett's name lived on in a variety of ways.

Colleagues and astronomers around the world mourned their loss. Shapley was one of the first to write to Reba, referring to the "high respect at Harvard for [his] personality and scientific attainments." When he wrote to Harry the next day, he corrected himself with his characteristic touch of humour, "I see that vainly I again confused Harvard and the cosmos, because certainly that respect is universal." When Harry replied, he wrote about his father, "I am afraid I am going to miss him terribly, more and more as time goes on. We were very good friends and even when we couldn't see each other, exchanged gossipy letters every week."[105]

How wonderful for us if we could read those private letters too, but they appear not to have survived. We do learn from Harry's letter, just quoted, of the trials of wartime for both himself and Shapley. Harry would not be able to get to his father's funeral but his daughter Barbara did cross the Atlantic safely at the beginning of September 1941, and was married in Ottawa on 25 October to Arthur Leslie Pidgeon.[106] Arthur had graduated from McGill in 1937 and had spent the next year as a Rhodes scholar in Oxford, where he met Barbara. With the wedding only eight days after JSP's death, it seems almost certain that Reba did not attend. Perhaps she later was able to visit her granddaughter and great-grandson in Toronto, where Arthur worked for the Canadian Broadcasting Corporation.[107]

Bart Bok, who first met the Plasketts in Leiden in 1928, and who got to know the family well after he moved to Harvard, wrote, "It is impossible to appreciate JSP without considering his happy family relations. Mrs. Plaskett and his son Harry played a very important part in his life. Plaskett was a strong man who could on occasion be sharp and blunt, and he needed the sweetness and kindness of Mrs. Plaskett. I doubt whether any astronomer of the present generation will ever think of him without picturing his wife somewhere in the background, not very far away."[108]

Reba was the sole beneficiary of JSP's estate, amounting to $27 522, including insurance policies, annuities, and a 1940 Dodge sedan.[109] Though the house was not included in the valuation, the amount would barely have allowed her and Stuart to live in comfort. Perhaps the modesty of her means figured in the decision she must have made, or at least acquiesced in, to have her husband's medals melted down and the proceeds used to design and make stained glass windows in his memory.[110] This may seem to be a shabby fate for these tangible tokens of Plaskett's celebrated career but the family evidently believed his immortality was rooted in his faith in the Light of the World not in the lustre of worldly gold. Reba did spare the CBE medal from the refiner's fire and had replicas made of the others, all of which she presented to the British Columbia Archives. She and the boys covered the cost of installing the three windows midway along the south aisle of St John the Divine Anglican Church in Victoria.

Members of the observatory staff and of the RASC and Canadian Club attended the service of dedication on Sunday morning, 4 April 1943.[111] Light, which had been at the focus of Plaskett's achievement in photography, optical design, and spectroscopy, cast its rays through the images of St Paul to illuminate the lives of the worshippers. The dedicatory prayer embraced the theme: "Almighty Father, Who has called us out of darkness into Thy marvellous light, mercifully accept our service, and graciously receive at our hands these windows for the adornment of Thy Sanctuary, in memory of Thy servant, John Stanley Plaskett, and in honour of Him, the Brightness of Thy Glory, Whom Thou hast given to be light to lighten the Gentiles."[112] Sadly the Plaskett windows were destroyed along with the church building itself in a frightful fire in 1960, but when the church was rebuilt a new memorial window was installed. Though not of the original design, the new windows continue to revere the saints, and the one to St Paul is in memory of John Stanley Plaskett.

Reba and Stuart stayed on at 318 Armit Road in Esquimalt. A newspaper article provides a glimpse of Reba at her ninetieth birthday party, which was hosted by a neighbour. The newspaper described her as "the bright-eyed star of the party" and noted she was "an active member of the Overseas Club, the B.C. Indian Arts and Welfare Society, and the Victoria Natural History Society."[113] There are still those who recall Reba going down the street (renamed Plaskett Place) to catch the bus, hosting fancy tea parties, and managing Stuart, who in his childlike way seemed to alternate between surliness and sweetness. The two of them courageously carried on in the big house on the windswept point until Reba's death at age ninety-nine, on 16 June 1967.[114]

In her will, Reba stipulated that Stuart would receive a monthly amount of $125 from her estate, or more if the trustees she appointed deemed it desirable or necessary.[115] A year after his mother's death, Stuart moved to the Salvation Army lodge and subsequently to a room in the Hotel Yates, "a sanctuary for Victoria's aged citizens," whom one

12.5 The present Plaskett memorial window in the Anglican church of St John the Divine, Victoria, showing scenes from the life of St Paul. It is one of twelve windows along the sunny south aisle portraying early disciples of the Christian church. (Photo by author)

source described as sitting across from the reception desk "watching their surroundings with distant, unblinking faces and slightly watery eyes."[116] Stuart, suffering "gross mental deterioration," died on 6 July 1976, leaving an estate of under $13 000 to his far-away niece and nephew.[117] Apparently no one felt close enough even to write a notice of his passing in the local paper.

Peer Reviews

After JSP's death, tributes appeared in many of the leading scientific and astronomical journals, in the *Canadian Churchman*, and in many newspapers, including the *Times* of London and the *New York Times*, which referred to Plaskett as "one of the world's outstanding astronomers."[118] His professional peers eulogized him as being among those "responsible for the rise of modern astronomy," as "a leader who secured for [Canada] a prominent position in the astronomical world," as "one of the foremost designers of astronomical instruments," and as opening "the way to a revolution in our knowledge of the universe."[119]

It is best, in summing up Plaskett's character and tremendous astronomical contributions in Canada and abroad, to let some of those who were his colleagues at the DAO and who knew him well speak for themselves. R.M. Petrie said, "His name will always be associated with the pioneering era of stellar spectroscopy and with development of great reflecting telescopes for astrophysical research ... The founding of a major observatory placed Canada in an enviable position in the astronomical world and awakened pride and interest in her national contribution to science."[120]

Beals, in his obituary, referred to JSP's "lovable personality" and his "leading part in the international organization of astronomy," noting that, "by his efforts at founding a great modern observatory, he was able to place unusual opportunities for research in the hands of others."[121] In a less formal situation, Beals recalled that Plaskett "was fundamentally a modest man with a very happy disposition and a wonderful sense of humour. He had a sort of high cackling laugh which was extremely infectious and he used it to great effect."[122] (Characterizing JSP as "modest" may seem to contradict the frequent impression given in the pages of this book. However, it is an appropriate description of his public persona, especially in the years after he had become famous.) Frank Hogg wrote that, "apart from Plaskett's great instrumental achievements as illustrated in several large modern telescopes and spectrographs, and the great mass of major research papers he has published, Plaskett made two other notable contributions – his cultivation of the interest of the general public in astronomy, and the guidance he furnished to many other astronomers, especially youthful Canadians."[123] R.O. Redman spoke of JSP's "kindliness, his happy disposition, his youthfulness of mind, his keen boyish sense of fun, and that delightful chuckle which so readily changed into the heartiest of high-pitched laughs. He had a host of friends in all astronomical circles."[124]

Redman's homage, of course, came from England, where he had returned after leaving the DAO in 1931. Other British astronomers writing obituaries of JSP included Frederick Stratton, an apt spokesman for the international astronomical community as a past general secretary and president of the IAU, and Harold Spencer Jones, the Astronomer Royal. The latter praised Plaskett's telescope, which "profoundly influenced the design of all large reflecting telescopes since that time," and the former recalled Plaskett as "a strong supporter of the scheme to place a large companion telescope at the Radcliffe Observatory in Pretoria to complete the study of the southern regions of the Milky Way."[125]

Several American journals included formal tributes to JSP's accomplishments. Roscoe Sanford at the Mount Wilson Observatory pointed out that Plaskett's adult life "spanned the development of astronomy in Canada from the most meager beginnings to world-wide eminence. This remarkable achievement was due in no small measure to the inspiration of his enthusiasm and untiring industry."[126] Robert Aitken, director of the Lick Observatory in 1930–5, echoed Bart Bok's thoughts, quoted earlier, when he wrote, "Though [JSP] was a strong man, who could on occasion 'speak his mind' very bluntly, he was also a very kindly, friendly man, welcomed by his associates everywhere, on the golf course, at his clubs, and at scientific meetings. He was, withal, a wise man who exercised great influence in astronomical councils and in civic affairs as well. And his home life with Mrs. Plaskett and their two sons was ideal."[127] Bok noted that Plaskett's temperament was well suited to the responsibilities and duties that came with heading a large observatory (in the days before modern management techniques): "I shall always remember his saying, 'As a director it is your privilege to ask advice from many people, consider all suggestions carefully and then do as you please' … Above all, however, we shall remember Plaskett as the man who placed Canada among the leading nations in astronomical research. We expect of Canadian astronomers, present and future, that they carry on the tradition established by Plaskett."[128]

As fair as these evaluations seem to be, it is a truism that obituaries of famous people are not known for unbiased assessments of their accomplishments. It may be more telling to look at their staying power after decades have passed. As we shall see in the concluding chapter, the Plaskett research tradition has carried on, as many Canadian astronomers have continued research in the areas he initiated, albeit with the benefit of technological improvements in equipment, detectors, and computing power that could not have been imagined in Plaskett's day. Their studies, still based on stellar spectra, have formed an important element of Canadian astronomy. And, in a broad sense, the ambition, belief, collegiality, and dogged determination that were Plaskett's traits have served his followers well.

13
Regeneration, 1942–

The astronomical community has commemorated John Stanley Plaskett in many ways since his death. The first posthumous tribute took place on the afternoon of 26 June 1952, when members of two American astronomical societies met at the Dominion Astrophysical Observatory. Along with members of the Victoria Centre of the RASC and the British Columbia Historical Association, they gathered beside the telescope, as eighty-four-year-old Reba Plaskett unveiled a plaque honouring her husband and spoke a few words of thanks. Walter Sage, head of the History Department at the University of British Columbia and provincial representative on the Historic Sites and Monuments Board of Canada, also played a prominent part: it was at his suggestion that the board had officially made the recommendation to the federal government for the plaque. At the unveiling, Sage explained the role of the board and Joseph Pearce paid tribute to JSP's career and attainments. An open house with exhibits and a supper rounded out the festivities.[1]

Perhaps, it might be said, this ceremony reflected a traditional Canadian reserve when it comes to honouring its heroes. Although, for example, the giant telescope on Mount Palomar had been named after its visionary, George Ellery Hale, a similar recognition for Plaskett and his telescope would not occur for another forty-one years. Indeed, it was not until the celebration of the DAO's seventy-fifth anniversary in 1993 that the venerable instrument was officially named the John Stanley Plaskett Telescope.[2] When he had received his first honorary degree in 1912, Plaskett had been called the Keeler of Canada for his pioneering work in spectroscopy. With the 1993 commemoration, he might be thought of as Canada's Hale for establishing the country's leading astrophysical research institute. Fittingly, the ceremony was hosted by the

13.1 Some of the guests gathered at the DAO on 26 June 1952 for the unveiling of a plaque honouring JSP include (from L to R): American astronomers Otto Struve and Joel Stebbins, DAO director Joseph Pearce, and Reba Plaskett. (Photo by S. Horace Draper, UCSC, G1-41)

country's own national professional astronomical society, the Canadian Astronomical Society / Société Canadienne d'Astronomie (CASCA), which held its meetings in Victoria to coincide with the dedication. The fact that this organization, which had not even existed twenty-five years earlier, registered 221 delegates was clear evidence of the expansion of astronomy in Canada during the previous quarter century. One of the speakers in 1993 was CASCA president Lloyd Higgs, director of the Dominion Radio Astrophysical Observatory, a facility allied with the DAO as part of NRC's Astronomy and Astrophysics Program. As a Rhodes scholar at Oxford, Higgs had been Harry Plaskett's last graduate student.[3]

JSP was also celebrated by creating academic awards.[4] Respecting his interest in young astronomers, CASCA partnered with the RASC in 1986 to establish the Plaskett Medal for the most outstanding doctoral thesis in astronomy or astrophysics of the two previous years by a graduate of a Canadian university. Fellowships were also established. Upon appointment as DAO director in late 1977, Sidney van den Bergh established regular appointments of postdoctoral scholars for two-year terms under the NRC's Research Associate Program. By the late 1980s these became three-year appointments. Over the next twenty years, the program established an enviable record of attracting outstanding talent and was renamed the Plaskett Fellowships, a term applied retroactively to the original appointees under van den Bergh.[5]

Plaskett's name is to be found far outside the walls of the observatory or the halls of academe, thanks to the inspiration of Canadian astronomer Donald Morton. He got the idea during an expedition in the Canadian Rockies with some astronomer friends in the summer of 1967. They had pitched camp at Lake Dimsdale, in eastern British Columbia, at an altitude of 1370 m on 21 July and the next morning set off for an unnamed peak at altitude 2880 m. Within eight hours they had arrived at the summit. Discussing what to call it, they agreed it would be appropriate to name it for Dr Robert M. Petrie, a former director of the DAO who had died unexpectedly in April 1966. They knew him well and admired his leadership and the advances he had made in astronomy in Canada.

Climbing Mount Petrie was easy, quite the opposite of what the expedition faced in the next few days. Sometimes struggling in pouring rain through thickets, sometimes across loose gravel or climbing through snow, they reached a second pinnacle on 7 August. Morton recounted how they built a cairn on the summit: "As we were building [it] a hail

13.2 The two faces of the Plaskett medal. The dark areas are polished surfaces where the recipient's name is engraved. (RASCA)

storm broke upon us so that we hurriedly began our retreat. Just then the clouds cleared briefly and we saw the rest of the ridge. There was one high point back towards camp and another to the north beyond a deep cleft. It was not certain which of the three points was highest, though the map indicates it is the north summit. We decided we should like to name the peak we climbed after Dr. John S. Plaskett, the first director of the Dominion Astrophysical Observatory." [6] Morton later contacted the Canadian Permanent Committee on Geographical Names, which, within a year, accepted the designations Mount Petrie (54°08′42″N, 120°33′06″W) and Mount Plaskett (54°08′15″N, 120°34′30″W) for the two summits.[7] As it turned out, Mount Plaskett was the higher of the two, at 2938 m. The close proximity of the two peaks seems all the more apt when we recall how JSP encouraged the eager high-school student, Bert Petrie, in the 1920s.

Beyond Earth's highest mountains, in the celestial realm, Plaskett is honoured too. Of course there is Plaskett's Star, whose great mass was discovered by the man himself in 1922. Even though astronomers still refer to this spectacular system as Plaskett's Star, the International Astronomical Union does not officially recognize any personal names attached to stars. It does, however, approve designations of lunar features, and Plaskett Crater is one of these. The opportunity to christen many new lunar landmarks came about after Soviet spacecraft began to secure images of the far side of the Moon, first in 1959 and into the 1960s. Once they had obtained photographs with reasonable resolution of far-side features, the Soviet Academy of Sciences produced a list of 153 objects with proposed names, which they submitted to the IAU for formal approval in 1966.[8] The Soviets were remarkably fair in the names they selected. With the exception of a dozen astronauts (six Soviet and six American), all were famous scientific figures of the past. Only one Canadian made the cut, and that was Plaskett, whose name they attached to a crater 109 km in diameter close to the Moon's north pole.[9] Because the IAU meets only every three years, it took until August 1970 for formal approval to be given. An IAU working group was kept busy in the meanwhile making extensive additions to the list, and by the time of the general assembly in Brighton, England, in 1970, the number of Canadians had grown to six, including Plaskett and C.A. Chant.[10] Five more were added later, for a total of eleven Canadians whose names are perpetuated on the Moon.[11] Of all these craters, Plaskett was by far the largest. In fact the other ten could fit comfortably inside it. Because of Plaskett Crater's situation with its northward-facing rim always in

13.3 Earthrise over Plaskett Crater. The view from NASA's Clementine spacecraft shows Earth over the north pole of the Moon. The crater with a central peak in the foreground is Plaskett Crater. (Image credit: US Geological Survey Credit: Lunar and Planetary Institute)

13.4 Mount Plaskett in the Canadian Rockies (Photo courtesy of Donald Morton)

darkness, it has more recently been suggested as a site of buried water ice – a vital concern if our Moon is ever to sustain habitation.[12] In 2006, the IAU officially attached the Plaskett label to four other smaller "satellite" craters in the vicinity of the main feature.[13]

In 1984 J.S. Plaskett and his son Harry were jointly accorded another celestial honour when minor planet number 2905, discovered in 1982, was named for them both.[14] Canadian amateur astronomer Chris Spratt came up with the suggestion, and the American discoverer, Ted Bowell, agreed, subject to IAU approval (which was granted). Though the

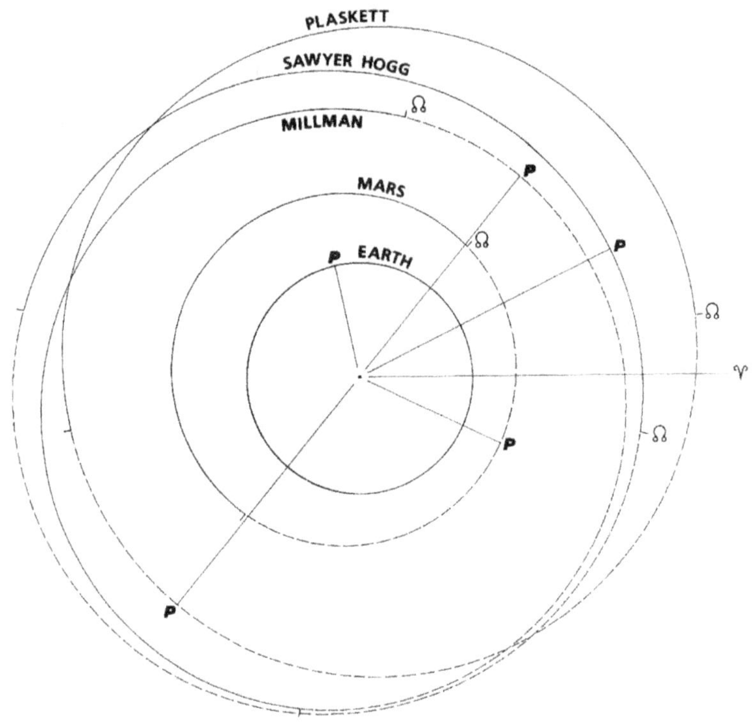

13.5 Orbits of minor planets Millman (number 2904), Plaskett (2905), and Sawyer Hogg (2917) shown relative to the orbits of Earth and Mars. All the orbits are inclined slightly to the plane of Earth's orbit; the portions below are shown as dotted lines with the points, marked Ω, where the minor planets would pierce the plane as they traverse their orbits. Perihelion points are marked P. (Copied from I. Halliday, *JRASC* 79 (1985): 27)

ever-increasing number of minor planets (or asteroids) dilutes the strength of having one's name immortalized in this way, the Plasketts were among the first few Canadians to be so recognized.

Plaskett's Legacy

The emergence of the history of science as an academic field of study in the late 1960s has brought about a clearer contextual appreciation of Plaskett in the development of an international role for Canadian astronomy. York University's Richard Jarrell initiated and developed this understanding almost singlehandedly.[15] He noted that the links to the United States, which began in 1906 when Chant and Plaskett toured American observatories and spent time at Lick, remained strong for decades. Those stateside always seemed to think of their northern colleagues as "honorary American astronomers," to use Jarrell's phrase. There was never any hesitation by the AAS to appoint Canadians to committees, to elect them to office, and even to meet north of the border on occasion. Canadians have benefited greatly from this friendly neighbouring giant, with its much larger population and pool of resources; it has always been generous in providing advice when asked and in offering a large stage for Canadian astronomers to showcase their work and interact with colleagues.

Close ties to the American Astronomical Society, especially evident in the first two decades of JSP's career, flourished through the 1960s. Jarrell found strong Canadian participation in AAS meetings, with the DAO astronomers accounting for over 60 per cent of the national total.[16] A subsequent decline, he suggests, was partly a result of the formation of CASCA in 1971, a development made possible by the tremendous growth in science generally in the wake of the space race. The first artificial satellite, launched by the Soviets in 1957, propelled a great desire in the western world to catch up. One Canadian manifestation of this was the rapid growth in university astronomy departments and the founding of CASCA reflected this growth. As of 2017, its membership stands at 503, a healthy number though far less, of course, than the AAS's 7000.[17] Jarrell made the interesting observation that "CASCA's annual meetings have a flavour quite distinct from the semi-annual conferences of the AAS … While the AAS meetings remain important venues for Canadians to announce research results, CASCA meetings, despite their outward appearance, are really devoted more to politics and networking."[18] Without doubt these enduring characteristics were firmly established by JSP in his relentless efforts to get the federal government to commit more resources to astronomy; Jarrell credited their continuance to relatively smaller budgets in Canada and to the closer liaison between scientists, civil servants, and legislators on this side of the border.

Former DAO director and CASCA president James Hesser sees further features of CASCA meetings as outgrowths of Plaskett's vision and ideals. He notes that they provide an especially good venue "for younger astronomers to give their scientific results and receive constructive feedback. For instance, for coveted oral contributed presentations, students and post docs are given priority. Student papers are judged for quality by the CASCA Board and prizes awarded to best oral and poster presentations." The meetings "create an annual opportunity for its committees to meet with the membership to hone national positions and understanding, the better to negotiate with international colleagues in the joint research and instrumentation projects now common."[19]

In considering JSP's legacy, it is also important to consider instrumentation. Jarrell considered the Victoria telescope as "the ancestor to an evolutionary line."[20] Plaskett himself spoke of it as marking "a new era in telescope design and construction … [setting] a new standard of accuracy and efficiency," and Worcester Warner, co-founder of the firm that built the telescope, believed it would be "the standard for a hundred years."[21] The main features of the DAO 1.83 m Cassegrain reflector, with its f/5 focal ratio and its asymmetrical mounting, were imitated in a dozen large telescopes up to 1960. These instruments, on every continent except Antarctica, were the workhorses of astrophysics for half a century. Most are still in use, though as a result of huge technological advances, no one builds them like that anymore.

As a result of JSP's vision and drive, Canada became known internationally for its first-class work in astrophysics. The constraints he experienced from very tight budgets seem to be a continuing trend that can be both a frustration and a source of pride. Recent articles in the popular media show that, among astronomers, Canadians deliver the most scientific bang for the buck.[22] Estimates show that Canada's expenditure on astronomy in 2010 was $87 million, the lowest in relation to GDP of any G7 nation. Yet, "Canadian astronomy and space science has the highest impact of any of the sciences. Canadian astronomy ranks highest in impact of any country in the G7. Canadian astronomical facilities have had an enormous impact on the development of the field over the past few decades. And, Canadian astronomers do disproportionately well in garnering prizes and distinguished chairs, compared to other disciplines."[23]

When it comes to lobbying for financial support, JSP's subversion of the normal chain of command has no lessons for today's astronomers, but Plaskett can teach some valuable lessons about the importance of cultivating friendships, collegiality, and public interest. The community atmosphere on which JSP thrived is an essential ingredient to researchers everywhere, and international collaboration is now the norm. Regular colloquia, forums, and group meetings tracing their origin to seminars instituted by Plaskett are regular features on the DAO's Observatory Hill. As a master publicist, JSP was always willing and able to tell colleagues and the public about his chosen field and its exciting developments. In fact, slightly over half of the 188 talks he gave (listed in Appendix B) were for public audiences. He would surely be pleased to see that the general public, from children to seniors, still enjoy tours of the observatory. For twelve years, their visits were enriched through an interpretive facility with the catchy name "The Centre of the Universe." Sadly, it had to be closed in August 2013 due to NRC's financial constraints and operating priorities but RASC members are doing their best to keep an element of JSP's vision alive by offering, during the summer, free Saturday-night public viewing centred on the Plaskett Telescope. The University of Victoria's Science Venture program has stepped in to offer summer camps for children in the Centre of the Universe building, and a not-for-profit organization, Friends of the Dominion Astrophysical Observatory, has formed and is actively raising funds to restore outreach to schools.[24]

For Plaskett, the interplay between astronomy and instrumentation was always important. It was an essential aspect of work at the DAO during his directorate and has remained so ever since, with the construction of new photometers, measuring machines, spectrographs, control systems, and telescopes. To carry out this work, a building housing a machine shop was erected in 1956 and an optical shop was built in 1973.[25] There a new mirror was ground and figured to replace the original one, then about sixty years old. The new material, called Cer-Vit, had a much lower coefficient of thermal expansion than glass, resulting in a much more stable shape for the mirror when temperature changes occur. The same material and expertise were used to prepare the 3.6 m mirror for the jointly operated Canada-France-Hawaii Telescope (CFHT), one of the most successful and productive telescopes in the world.

Changes on Observatory Hill

Up on the hill
There's a big plan still
In the house that Jack built.

Those lyrics to Aretha Franklin's memorable song can serve to remind us that the institution established by JSP still inspires advanced research. If Plaskett were able to see Observatory Hill today, he would still recognize the domed building as the centerpiece of the DAO. Inside, the telescope would still seem familiar in its overall appearance. Scientists in other disciplines might be amazed that a century-old instrument could still be doing current research – Parks Canada even designated the observatory as a national historic site.[26] But it is far from a museum piece. Transferral of government observatories to NRC in the early 1970s formalized their operations as national scientific facilities open to qualified observers (primarily university researchers and their students). This broadened usage has produced, on average, about seven refereed scientific papers per year since the Plaskett Telescope was "christened" in 1993.[27]

Other than the mounting and frame, many parts of the telescope have been replaced – the drive, the mirrors, and especially the spectroscopes are completely different. Upgrades have made the instrument about 10 000 times more sensitive than it was originally, one of the main reasons being very efficient CCD detectors that record spectra digitally rather than on photographic plates. Plaskett would also be amazed that today's astronomers no longer need to stand in the cold, dark dome, carefully keeping watch through an eyepiece, ensuring the telescope keeps on target. Now they (or often a proxy operator) can sit comfortably in a warm room watching data accumulate on a computer screen. In fact plans are now afoot to enable the telescope's operation remotely from anywhere in the world.

Since 1962 there has been another dome on Observatory Hill, housing a 1.2 m telescope used to obtain high-dispersion spectra through its coudé focus. Since about 2010 it has been operable in fully robotic mode. Both telescopes are in use every clear night (about 200 per year, as in Plaskett's day) and the overall purpose of the DAO remains the gathering and analysis of data. That said, however, the scope of projects (which includes positions of comets and asteroids, studies of variable stars, spectroscopy and polarimetry of peculiar stars, supernovae, globular clusters and active galactic nuclei) and the way the analysis is done would be unrecognizable to Plaskett. James Hesser explains:

> In fulfilment of the parliamentary mandate to NRC, staff responsibilities today now strongly focus on research with, and the operations and enhancements of, the international partnership facilities to which Canada is committed. As a result only a small number of staff regularly use the DAO telescopes for their astronomical or instrumentation research. Nonetheless, competition from staff astronomers and qualified observers in other parts of Canada and the world for observing time exceeds the number of nights available. Accordingly, an allocation committee apportions the use of the DAO facilities."[28]

The modest office building that Plaskett got after years of insistence is now incorporated in a large, multi-story, quadrangular building. Additions in the 1950s, 1980s, and the first decade of the new century provided specialized spaces for instrumentation development, data archiving and distribution, unique library holdings, offices, and meeting facilities.

Reaching Out

The office building not only provides space for the DAO staff, it houses the Canadian Astronomy Data Centre (CADC) and the National Research Council's Herzberg Astronomy and Astrophysics Programs (NRC Herzberg), the umbrella organization under which the DAO operates.[29] That organization's responsibilities include Canada's participation in some of the world's largest optical observatories, owned and operated by international consortia at the best astronomical sites in the world: the 3.6 m CFHT in Hawaii, the two 8 m Gemini telescopes in Hawaii and Chile, and the Thirty-Metre Telescope funded in 2015 and still under development. NRC Herzberg also actively supports research at various universities. In particular there are strong interactions with the Department of Physics and Astronomy and the Faculty of Engineering at the University of Victoria as well as with universities and industry throughout Canada. There are also close relations with institutes in partner countries involved in the operations and development of Canada's international facilities.

The CADC, established in 1986, creates specialized software for astronomical data archiving, processing, calibrating, and interpretation. It makes available online digitizations of some of the more than one hundred thousand DAO photographic spectrograms in an archive carefully maintained by the volunteer efforts of Elizabeth Griffin. The CADC also provides complete access to the archived images from the Hubble Space Telescope and from the large Earth-based telescopes in which Canada has a share.

A very important part of NRC Herzberg's mandate is the development of advanced scientific instruments for other astronomical observatories and industrial partners – clearly an outgrowth of Plaskett's goal of high-quality, high-efficiency instruments. Its employees have helped to build (under contract) some of the most sophisticated equipment presently in use. Through collaboration with engineers, these projects have been very rewarding, financially and intellectually, to Canadian companies interested in research and development. NRC Herzberg staff continue to play a leading role in the development of next-generation optical and radio observatories. Plaskett could never have imagined developments such as space telescopes and the huge role played by electronic computers. With modern techniques, astronomers at the DAO can measure radial velocities of stars 10 000 times fainter than those Plaskett studied and can detect changes in radial velocity (for some stars) of less than a metre per second, a thousand times smaller than anything Plaskett could achieve.

Human Heritage

Though JSP was never a professor, he is the progenitor of a lasting line of intellectual descendants, starting with his son. Harry always gave a great deal of credit to his father's influence. Harry's lifelong achievement in stellar and solar spectroscopy, and his advocacy of large telescopes, stemmed directly from experience he gained from his father. His genetic inheritance in addition to the environment in which Harry matured gave him "a degree of personal assurance and savoir faire that unquestionably eased his passage through life" and helped to make him an inspiring leader.[30] For his part, JSP was immensely proud that Harry followed in his footsteps, never more so than at his election to fellowship in the Royal Society in 1936.[31] They had the unique father-and-son distinction of receiving both the gold medal of the Royal Astronomical Society and the Halley Lectureship at Oxford, though JSP was long gone by the time his son received those awards in the 1960s. Their attainments seem all the more remarkable as neither formally studied for any advanced degree beyond their University of Toronto BAs. Harry was able to capitalize on new physical theories emerging as he began his career. As for personality, one trait definitely distinguishing Harry from his father was his "utter modesty."[32] It may have been a conscious aversion to JSP's love of the limelight; if so, it was a singular departure from filial admiration. After Harry, the scientific tradition of the family carried on through his son, another John Stanley Plaskett, a mathematical physicist with a productive career at the University of Sussex.

Many other astronomers could be considered JSP's intellectual heirs. His young colleagues at the DAO – Carlyle Beals, Frank and Helen Hogg, Joseph Pearce, Bert Petrie, Reynold Young – all had distinguished careers. None of them would have got their start in Canada without Plaskett's success. They very likely would have chosen to locate in the United States, and Canada would probably have been deprived of their talents. Naturally, each of these astronomers in turn mentored others – such is the legacy of a pioneer. And, of course, the Plaskett Telescope, the DAO, and NRC Herzberg continue to be a valuable training ground for future scientists, engineers, and technicians.

Some researchers at the DAO and elsewhere continue in fields familiar to Plaskett – analysing stellar spectra, understanding spectroscopic binaries, and studying the interstellar medium – though the details have become far more complex.[33] After Plaskett's death and the Second World War, Milky Way research got a tremendous boost with the advent of radio astronomy. By 1951, this new tool was beginning to trace the Galaxy's spiral arms. In the years since, it has not only defined the Milky Way's structure in ever greater detail but recently has established that the Galaxy is about ten times more massive than Plaskett had thought. Although the extra stuff is elusive dark matter, the Galaxy seen in optical wavelengths is still pretty much as he pictured it at the end of his career.[34]

What Have We Learned?

JSP maintained that his greatest contribution to science was "the capturing of Dominion Government funds for the erection of the 72-inch telescope."[35] Indeed, all his other major accomplishments depended on that – his catalogues of radial velocities and spectroscopic binaries; his studies of the most massive stars, which provided data for the mass-luminosity relationship; and his reliable confirmation of the rotation of the Galaxy, including the pervasive interstellar matter within it. He would have been the first to acknowledge that all great successes are supported, sometimes consciously and sometimes not, by predecessors or contemporaries.

The complexities of astronomical research and the means of funding it are now incomparably greater than those Plaskett faced. Still, it is relevant to recall that he saw a need for more and better data as an opportunity for Canada to make an important international scientific contribution if it could be equipped with a major telescope, and he pursued this goal relentlessly. But just as we, as a nation, have outgrown the reputation of being mere suppliers of raw materials to foreign manufacturers, we now also have our own university departments and institutes that analyse collected data to produce stimulating theories of the behaviour of the universe and its constituent parts.

In the preface to this book, I invited you, the reader, "to question the role of nature and nurture, luck and pluck, challenges and opportunities in life's journey." At the end of my long association with this remarkable man whom I never met, I also make my own comparisons and ask what made his life a success. I have no doubt that Plaskett was very bright (Beals called him "shrewd and brainy as all get out"[36]), but I cannot say that his accomplishments show the spark of originality or creativity associated with real genius. He got his start on the road to fame by embarking on a field of research that was widely recognized as important but requiring more personnel and observing resources. Everything he did required persistence and effort. Partly, he inherited this determination from his mother, but as the oldest child of a large farm family, he had learned the need for hard work. From his father, he got his gregarious disposition, and with his siblings he experienced the pleasures of simple fun. In his service to the church as warden and on the building committee, he followed the example of his forebears. As part of the faith in which he was brought up, he believed that universal laws guiding the motion, structure, and evolution of the cosmos are part of a divine plan.

We can never know what different paths Plaskett would have followed had he been born a generation earlier or later, but we can be almost certain he would not have been Canada's first professional astrophysicist. Even if he had pursued his university studies right after high school, things would have been different. Without his experience in the machine shop, factory, and laboratory, or without his experiments in photography, he would not have been hired for his first observatory position. If his boss, W.F. King, had not had the foresight to see the value of astrophysics in a government institution, or if he had not recognized Plaskett as the right man to put that vision into practice, the history of astronomy in Canada would have been radically different and Canada's reputation in the world of science would have been much the poorer. The fact that Canada's first large, privately funded observatory did not open until 1935 proves the importance of government financing for such a large project. Officials did not ask for, or expect, immediate payback. But here we are a century later building on the legacy of pure astrophysical research, and Canadian high-tech industry is reaping benefits by designing and building sophisticated space-age equipment required by astronomers and others using satellite-borne technology.

Though Plaskett's ambition was driven by strong self-confidence, and his stellar success by sustained effort, he knew when he needed advice and help, and he was not afraid to ask those who were the best qualified to give it. Without that assistance, his research would not have attained respectability and his dream observatory would not have been built. Had he not cultivated the support and friendships of others, he would not have accumulated the high honours bestowed on him by the elite of his profession. Above all, he himself would have insisted that he could never have managed without his wife's affection and encouragement.[37] Amen.

Appendices

Appendix A lists all of JSP's published papers (not including newspaper articles). Close inspection shows that some appeared in several versions in different periodicals, revised to greater or lesser extents. In addition, JSP frequently submitted papers for scientific conferences that he did not attend. If these contributions or their abstracts appeared in print, they are listed in Appendix A but not in Appendix B. Given these considerations, arriving at a total number of publications is problematic, but we can nonetheless say with certainty that during Plaskett's active career, 1905–39, he wrote well over a hundred papers.

Some interesting trends are noticeable. The only years in this period when no papers appeared were 1916 and 1917, when JSP was busy supervising the construction of what became the world's largest operational telescope, and 1929, when he was preparing his massive synthesis papers on the rotation of the Galaxy. Phases of his scientific output waxed and waned over the years. The solar research on which he embarked soon after the Dominion Observatory opened in 1905 lasted until he left that inadequate and unproductive equipment behind when he moved from Ottawa in 1917. His interest in spectroscopic binaries, which began almost as early, survived his move to Victoria and flourished with the advantage of the much more powerful new telescope. A subset of this work was his study of eclipsing binaries during a period in 1919–23. All his binary research was over by 1925, three years after his discovery of the massive system that has come to be known as Plaskett's Star. The first program at the Dominion Astrophysical Observatory, cataloguing the radial velocities of hundreds of stars, many of which were beyond the reach of all but a couple of other large telescopes, signalled the beginning of a new phase in which he no longer concentrated on individual stars, but looked for the big picture. Interstellar matter, which first came to his attention in spectra of individual stars, became the focus of large-scale studies in 1929–32, though the results continued to be an important part of his work on the rotation of the Galaxy until 1939. JSP's constant interest in instrumentation is evident in a number of papers beginning with his first in 1899, reaching a climax with the opening of the DAO in 1918 and lasting up to 1939 with his consulting work on the telescope of the McDonald Observatory.

Several of his published papers (those marked with an asterisk in Appendix A) were first presented at meetings. The details of where and to whom Plaskett gave these talks can be found in Appendix B, along with all the known speeches, lectures, and presentations he made in person to scientific bodies and to a variety of clubs, associations, church groups, and public gatherings. His tally is impressive, averaging about four papers and six talks per year throughout a forty-one-year span.

Appendix A

PUBLISHED PAPERS BY J.S. PLASKETT

The following table provides a list of published papers (but not newspaper articles) written or co-written by JSP. The papers are arranged chronologically by the date of completion (shown in the first column), which is often different from the date of publication. (If the date of completion is not known, it is taken to be the date of publication.) Co-authors are included in the second column.

= Indicates that the article was reprinted without changes in another publication.

* Indicates that the article was also given as a talk. For the details about such talks, see Appendix B.

See pages 415–17 for a list of abbreviations used in this table.

Date completed	Title and co-author(s)	Publication information
1899, Dec.	Mathematics in the Mechanical Trades*	*Canadian Engineer* 7 (Jan. 1900), 241 = *Practical Engineer* (Feb.–March 1900)
1901, Feb.	Colour Values in Monochrome: A Lecture on Orthochromatic Photography*	44-page booklet, https://archive.org/stream/cihm_72137
1902, March	Photography in Natural Colours*	*Transactions of the Canadian Institute* 7 (1904), 371–90
1902	Landscape Photography	*Acta Victoriana* 26, no. 3 (1902), 162–68
1905	Plates and Filters for Monochromatic and Three Colour Photography of the Corona	*Transactions RASC* (1905), 89–107
1905	Observatory Building and Instrumental Equipment	*Report 1905* (1906), 198–211
1905, Oct.	Report on Solar Eclipse Expedition	*Report 1905* (1906), 213–35
1906, Oct.	Report … on Observatory Instruments and Astrophysical Work	*Report 1906* (1907), 43–93
1907, Jan.	The Character of the Star Image in Spectrographic Work (Part 1)	*ApJ* 25 (1907), 195–217
1907, Feb.	The Spectrum of Mira Ceti	*JRASC* 1 (1907), 45–59
1907, April	Adapting a Universal Spectroscope for Radial Velocity Determinations	*JRASC* 1 (1907), 104–21
1907, April	Meetings of the Society at Ottawa [written by JSP as secretary of RASC Ottawa Centre]	*JRASC* 1 (1907), 129–32
1907, June	Meetings of the Society at Ottawa [written by JSP as secretary of RASC Ottawa Centre]	*JRASC* 1 (1907), 199–203
1907, July	Astronomical and Astrophysical Work	*Report 1907* (1908), 41–216
1907, Oct.	The Star Image in Spectrographic Work*	*JRASC* 1 (1907), 297–322

1907, Dec.	The Character of the Star Image in Spectrographic Work (Part 2)	*ApJ* 27 (1908), 139–51
1908, Feb.	The Spectroscopic Binary ι Orionis (with W.E. Harper)	*ApJ* 27 (1908), 272–79
1908, March	Astrophysical Work	*Report 1908* (1909), 61–189
1908, May	The Design of Spectrographs for Radial Velocity Determinations*	*JRASC* 3 (1909), 190–209 = *Report 1909*, 152–63
1908, July	The Spectroscopic Binary ψ Orionis	*ApJ* 28 (1908), 266–73
1908, July	The Orbit of ι Orionis	*ApJ* 28 (1908), 274–77
1908, Aug.	Effect of Increasing the Slit-Width on the Accuracy of Radial Velocity Determination (Part 1)*	*ApJ* 28 (1908), 259–65
1908, Oct.	The Astronomical and Astrophysical Society of America	*JRASC* 2 (1908), 255–60
1908, Dec.	Notes from the Dominion Observatory: Astrophysics	*JRASC* 2 (1908), 324–25
1909, March	Astrophysical Work	*Report 1909* (1910), 143–224
1909, March	Camera Objectives for Spectrographs*	*ApJ* 29 (1909), 290–300
1909, May	The Spectroscopic Binary β Orionis	*ApJ* 30 (1909), 26–32
1909, May	The Spectroscopic Binary β Orionis	*Proc. Trans. RSC* 3 (1909), 215–26
1909, May	Slit Width and Errors of Measurement in Radial Velocity Determination	*Proc. Trans. RSC* 3 (1909), 209–14
1909, July	The Ottawa Spectrographs	*JRASC* 3 (1909), 287–305
1909, Aug.	Two Curiously Similar Spectroscopic Binaries (with Harper)	*ApJ* 30 (1909), 373–82
1910, March	Astrophysical Work	*Report 1910* (1911), 1: 81–130

(*Continued*)

(Continued)

Date completed	Title and co-author(s)	Publication information
1910, July	Slit Width and Errors of Measurement in Radial Velocity Determinations	*JRASC* 4 (1910), 333–44
1910, Aug.	Probable Errors of Radial Velocity Determinations [shorter account]	*ApJ* 32 (1910), 230–42
1910, Aug.	Probable Errors of Radial Velocity Determinations [full account]*	*Proc. Trans. RSC* 4 (1910), 77–88
1910, Aug.	The Collimation of the Correcting Lens	*ApJ* 32 (1910), 243–48
1910, Oct.	The Astronomical and Astrophysical Society of America	*JRASC* 4 (1910), 373–78
1910, Nov.	The Spectroscopic Binary ε Ursae Minoris	*JRASC* 4 (1910), 460–65
1911, Feb.	Some Interesting Developments in Astronomy*	*JRASC* 5 (1911), 245–65 = *Ann. Rep. Smithsonian Inst. for 1911* (1912), 255–70
1911, Feb.	Some Interesting Developments in Astronomy [abridged]	*Sci. Am. Supp.*, no. 1869 (28 Oct. 1911), 278–80
1911, April	Astrophysical Work	*Report 1911* (1912), 95–153
1911, June	Methods and Preliminary Results in Spectroscopic Determination of the Solar Rotation (with R. DeLury)*	*Proc. Trans. RSC* 5 (1911), 107–11
1911, Oct.	Review of W. Adams, *An Investigation of the Rotation Period of the Sun by Spectroscopic Methods*	*Science* 34 (1911), 606–8
1912, Feb.	The Evolution of the Worlds	*Ottawa Naturalist* 26 (1912), 29–34, 52–60
1912, April	The Spectrum of Nova Geminorum	*JRASC* 6 (1912), 27–36
1912, April	The Spectrum of Nova Geminorum	*PDO* 1 (1913), 159–69

1912, May	The Solar Rotation (with R.E. Delury)*	*Proc. Trans. RSC* 6 (1912), 23–70
1912, May	The Spectroscopic Binary θ2 Tauri*	*JRASC* 6 (1912), 231–39 = *Proc. Trans. RSC* 8 (1914), 3:157–66
1912, Sept.	The Solar Rotation in 1911 (with R.E. DeLury)*	*ApJ* 37 (1913), 73–104 = *Proc. Trans. RSC* 6 (1913), 3: 23–70
1912, Sept.	Astronomical and Astrophysical Society of America: Pittsburgh Meetings	*JRASC* 6 (1912), 272–80
1913, Jan.	The Plane Grating for Stellar Spectroscopy	*ApJ* 37 (1913), 373–79
1913, Sept.	The Solar Union	*JRASC* 7 (1913), 420–37
1913, Sept.	Spectroscopic Binaries under Investigation at Different Institutions [JSP's response to survey]	*ApJ* 38 (1914), 265–66
1913, Dec.	A Great Reflector for Canada	*JRASC* 7 (1913), 448–55
1913, Dec.	A New Prism for Spectrograph I [in Notes from the Dominion Observatory]	*JRASC* 7 (1913), 458–60
1914, Feb.	Experiments Regarding Efficiency of Spectrographs	*PDO* 1 (1914), 173–203
1914, Feb	Improvements in the Optical System of the Stellar Spectrograph	*ApJ* 40 (1914), 127–36
1914, June	The 72-inch Reflecting Telescope	*JRASC* 8 (1914), 180–87
1915, Jan.	The Sidereal Universe*	*JRASC* 9 (1915), 37–56 = *Sci. Am. Supp.*, nos. 2078, 274 and 2079, 299
1915, Feb.	The Spectroscopic Binary θ2 Tauri	*PDO 2* (1915), 61–85
1915	The Second Spectrum in ω Ursae Majoris and b Persei [in Notes from the Dominion Observatory]	*JRASC* 9 (1915), 80–81
1915, June	The Spectroscopic Determination of the Solar Rotation at Ottawa	*ApJ* 42 (1915), 373–93

(*Continued*)

(Continued)

Date completed	Title and co-author(s)	Publication information
1915, June	The Spectroscopic Determination of the Solar Rotation at Ottawa [abstract of *ApJ* paper]	*JRASC* 10 (1916), 170–74
1916, Jan.	Modern Views of the Sun*	*JRASC* 10 (1916), 101–20
1916, Aug.	W.F. King	*JRASC* 10 (1916), 267–74
1916, Aug.	William Frederick King [slightly abbreviated version of the preceding obituary]	*Obs.* 39 (1916), 338–44
1916, July	The 72-inch Reflecting Telescope	*JRASC* 10 (1916), 275–80
1916, July	Canada's 72-inch Reflecting Telescope Installation	*Canadian Machinery* 16 (1916), 55–58
1918, July	The 72-inch Reflecting Telescope	*JRASC* 12 (1918), 397–410
1918, July	Canada's Great Telescope	*Saturday Night* 31 (20 July 1918), 4
1918, Aug.	The 72-inch Reflecting Telescope of the Dominion Astrophysical Observatory, Victoria, B.C.	*PASP* 30 (1918), 267–82
1918, Sept.	Notes on the Spectrum of Nova Aquilae No. 3	*JRASC* 12 (1918), 350–56
1918, Aug.	Notes on the Spectrum of Nova Aquilae [abstract]	*PA* 27 (1919), 93 = *PAAS* 4 (1922), 41
1918, Aug.	The 72-inch Reflecting Telescope of the Dominion Astrophysical Observatory [abstract]	*PA* 27 (1919), 94 = *PAAS* 4 (1922), 42
1919, April	The 72-inch Reflecting Telescope of the Dominion Astrophysical Observatory, Victoria, B.C.	*PA* 27 (1919), 210–14
1918, Oct.	Notes from the Dominion Astrophysical Observatory, Victoria, B.C.: 12 Spectroscopic Binaries	*JRASC* 12 (1918), 460–63
1918. Nov.	Dominion Astrophysical Observatory at Victoria, B.C.	*Construction* 11 (1918), 364

1918, Dec.	Notes from the DAO, Second List of Spectroscopic Binaries (with R.K. Young)	*JRASC* 13 (1919), 59–64
1919, Jan.	An Interesting Double Star [Boss 6158 = Burnham 12675]	*PASP* 31 (1919), 39–40
1919, Jan.	Notes from the Dominion Astrophysical Observatory: An Interesting Double	*JRASC* 13 (1919), 146–47 = *PASP* 31 (1919), 39
1919, Jan.	The Stellar Spectrograph of the 72-inch Reflecting Telescope	*ApJ* 49 (1919), 209–23
1919, Feb.	The Spectroscopic Binary H.R. 8170	*JRASC* 13 (1919), 174–78
1919, Feb.	The Spectroscopic Binary H.R. 8170 [corrected version]	*PDAO* 1 (1922), 113–17
1919, Feb.	The Great Canadian Telescope	*Canadian Magazine* 52 (1919), 861–72
1919, March	Notes from the DAO: Third List of Spectroscopic Binaries (with R.K. Young)	*JRASC* 13 (1919), 191–98
1919, April	The Dominion of Canada's 72-in. Telescope	*Nature* 103 (1919), 105–8
1919, June	Description of Building and Equipment	*PDAO* 1 (1922), 1–103
1919, Aug.	The Spectroscopic Orbits of the Eclipsing Variables U Ophiuchi, RS Vulpeculae, and TW Draconis	*PDAO* 1 (1922), 137–51
1919, Aug.	The Spectroscopic Orbits of the Eclipsing Variables U Ophiuchi, RS Vulpeculae, and TW Draconis	*JRASC* 14 (1920), 1–15
1919, Sept.	The Spectroscopic Orbits and Dimensions of … U Oph, RS Vul, and TW Dra [abstract]	*PA* 27 (1919), 672–73 = *PAAS* 4 (1922), 111
1919, Sept.	Report of Progress of Work with the 72-inch Telescope [abstract]	*PA* 27 (1919), 674 = *PAAS* 4 (1922), 112

(*Continued*)

(Continued)

Date completed	Title and co-author(s)	Publication information
1919, Sept.	Notes from the DAO: Fourth List of Spectroscopic Binaries (with W.E. Harper and R.K. Young)	*JRASC* 13 (1919), 372–78
1920, Jan.	One Hundred Spectroscopic Binaries (with W.E. Harper, R.K. Young, and H.H. Plaskett)	*PDAO* 1 (1922), 163–85
1920, Feb.	The Dominion Astrophysical Observatory, Victoria, B.C. [annual report for 1919]	*MNRAS* 80 (1920), 392–93
1920, April	The Optical Parts of the 72-inch Telescope [extract from *PDAO* 1 (1922), 1]	*JRASC* 14 (1919), 177–99
1920, June	The Spectroscopic Orbit of U Coronae	*PDAO* 1 (1922), 187–91 = *Proc. Trans. RSC* 14 (1920), xxiv
1920, June	[letter]	*PASP* 32 (1920), 201
1920, Aug.	The Relative Surface Intensity of the Components of Eclipsing Variables [note]	*PASP* 32 (1920), 230–31
1920, Aug.	The Spectroscopic Orbit of TX Herculis	*PDAO* 1 (1922), 207–11
1920, Aug.	The Spectroscopic Orbit of Y Cygni	*PDAO* 1 (1922), 213–17
1920, Sept.	The Spectroscopic Orbits and Absolute Dimensions of the Eclipsing Variables TX Her and Y Cyg	*PA* 29 (1921), 26 = *PAAS* 4 (1922), 177
1920, Sept.	The Spectroscopic Orbits and Dimensions of the Eclipsing Variables U CrB, TX Her, and Y Cyg	*JRASC* 14 (1920), 409–24
1920, Nov.	The Spectroscopic Orbit and Dimensions of Z Vulpeculae	*PDAO* 1 (1922), 251–55

1920, Dec.	Notes from the Dominion Astrophysical Observatory, Victoria, B.C. Meetings of the Seminar	*JRASC* 14 (1920), 428–29
1920, Dec.	The Spectroscopic Orbit of Z Vulpeculae	*PA* 29 (1921), 227–28
1921, Feb.	The Dominion Astrophysical Observatory, Victoria, B.C. [annual report for 1920]	*MNRAS* 81 (1921), 290–91
1921, July	The Radial Velocities of 594 Stars (with Harper, Young, and H.H. Plaskett)	*PDAO* 2 (1924), 3–127
1921, Aug.	The Radial Velocities of 600 Stars [abstract] (with Harper, Young, and H.H. Plaskett)	*PASP* 33 (1921), 212
1921, Aug.	Temperature Regulation of Mirror and Spectrograph [abstract; see also *JRASC* 16 (1922), 91]	*PASP* 33 (1921), 212
1921, Aug.	The Spectroscopic Orbit and Dimensions of TV Cassiopeiae [see *PDAO* 2 (1924), 141–45]	*PA* 29 (1921), 639–40 = *PAAS* 4 (1922), 276
1921, Aug.	The Radial Velocities of 594 Stars (with Harper, Young, and H.H. Plaskett) (see *PDAO* 2 (1924), 3)	*PA* 30 (1922), 2 = *PAAS* 4 (1922), 277
1921, Nov.	The Dimensions of the Stars*	*PASP* 34 (1922), 79–93
1921, Nov.	Eighty-eight Spectroscopic Binaries (with Harper, Young, and H.H. Plaskett)	*PDAO* 1 (1922), 287–306
1922, Jan.	Temperature Regulation of Spectrograph and Mirror	*JRASC* 16 (1922), 91–100
1922, April	The Spectroscopic Orbit of TV Cassiopeiae	*PDAO* 2 (1924), 141–45
1922, May	The Spectroscopic Orbit of BD 6°1309	*PDAO* 2 (1924), 147–57
1922, May	A Very Massive Star [almost identical to *PDAO* 2 (1924), 147–57]	*MNRAS* 82 (1922), 447–57

(*Continued*)

(Continued)

Date completed	Title and co-author(s)	Publication information
1922, Aug.	The Star of Greatest Known Mass	*JRASC* 16 (1922), 284–93
1922, Sept.	The Ultra-violet Spectrograph of the 72-inch Telescope*	*PA* 31 (1923), 20–21 = *PAAS* 4 (1922), 381
1922, Dec.	The Spectroscopic Orbit of 44°3639	*PDAO* 2 (1924), 183–88
1923, Feb.	The Dominion Astrophysical Observatory, Victoria, B.C. [annual report for 1921–22]	*MNRAS* 83 (1923), 275–77
1923, May	A Remarkable Variable Spectrum [HD 45910, P Cygni type]	*PASP* 35 (1923), 145–49
1923, May	The Dominion Astrophysical Observatory [booklet]	Ottawa: F.A. Acland, 1923
1923, May	The Work of the Telescope [extract from booklet]	*JRASC* 18 (1924), 169–75
1923, July	The Optical Parts of the Victoria Spectrograph	*ApJ* 59 (1924), 65–75
1923, Aug.	The Spectroscopic Orbit of BD 56 ° 2617	*PDAO* 2 (1924), 269–74
1923, Nov.	The H and K Lines of Calcium in O-Type Stars	*MNRAS* 84 (1923), 80–93
1923, Dec.	A Great Astronomical Meeting	*JRASC* 17 (1923), 325–35
1924	American Astronomical Society Reports of Observatories, 1922–1923 [DAO]	*PA* 32 (1924), 96–97
1924	The Dominion Astrophysical Observatory, Victoria, B.C. [annual report for 1923]	*MNRAS* 84 (1923), 250–51
1924	Problems of the O-type Stars	*Probleme der Astronomie: Festschrift für H. v. Seeliger* (Berlin)
1924, Feb.	The O-type Stars	*PDAO* 2 (1924), 287–358
1924, April	James B. McDowell: An Appreciation	*JRASC* 18 (1924), 185–93
1924, May	The O-type Stars and Their Relation to the Stellar Evolutionary Sequence*	*JRASC* 18 (1924), 321–38

1924, Nov.	The Dominion Astrophysical Observatory, Victoria, B.C. [based on JSP's description of DAO]	*Obs.* 47 (1924), 329–35
1925	The Dominion Astrophysical Observatory, Victoria, B.C. [annual report for 1924]	*MNRAS* 85 (1924), 338–40
1925, April	The Orbits of Two Double-Lined B-Type Binaries	*PDAO* 3 (1927), 179–88
1925	Canada Prominent at Astronomical Union	*Natural Resources Canada* 4, no.10 (1925)
1926	The Dominion Astrophysical Observatory, Victoria B.C. [annual report for 1925]	*MNRAS* 86 (1925), 205–6
1926, Jan.	Three Spectroscopic Binary Orbits [21 Cas, Boss 3354, HD 191201]	*PDAO* 3 (1927), 247–64
1926, May	The Inauguration of the Buildings of UBC at Point Grey (with R.W. Brock)	*Proc. Trans. RSC* 14 (1926), xxiv
1926, Aug.	Three Peculiar Spectra [abstract under Notes from Pacific Coast Observatories]	*PASP* 38 (1926), 265
1926, Aug.	Three Peculiar Spectra [υ Sag, HD 50820, HD 45910]	*PDAO* 4 (1931), 1–26
1926	Interesting Data on a Remarkable Class of Stars	*Natural Resources Canada* 5, no. 7 (1926)
1926, Nov.	Note on Three Peculiar Spectra	*MNRAS* 87 (1926), 31–34
1927	Dominion Astrophysical Observatory, Victoria B.C. [annual report for 1926]	*MNRAS* 87 (1927), 279–81
1927, March	The Dominion Astrophysical Observatory	*PASP* 39 (1927), 88–96
1927, May	Sixty Years' Progress in Astronomy*	*JRASC* 21 (1927), 295–310
1927, June	The Orbit of the B-type Spectroscopic Binary, HD 193536 [abstract]*	*PASP* 39 (1927), 258

(*Continued*)

(Continued)

Date completed	Title and co-author(s)	Publication information
1927	Dominion Astrophysical Observatory, Victoria, B.C. [annual report for year ending 30 June 1927]	*PAAS* 6 (1931), 48–49
1927	Astronomers Seeking New Source of Energy	*Natural Resources Canada* 6, no. 4 (1927)
1928	Dominion Astrophysical Observatory, Victoria B.C. [annual report for 1927]	*MNRAS* 88 (1928), 279–81
1928, Feb.	Two Spectroscopic Orbits [HD 193536 and 56°2617] and Notes on υ Sagittarii	*PDAO* 4 (1931), 103–18
1928, Feb.	The Motion of the Stars*	*JRASC* 22 (1928), 111–34
1928, Feb.	Review of A. Beer, *The Characteristics of Spectroscopic Binary Stars*	*JRASC* 22 (1928), 296–98, 337–41
1928, March	The Rotation of the Galaxy	*MNRAS* 88 (1928), 395–403
1928, April	Lick Observatory Radial Velocities [review]	*PASP* 40 (1928), 118–25
1928	Dominion Astrophysical Observatory, Victoria B.C. [annual report for year ending 30 June 1928]	*PAAS* 6 (1931), 167–69
1928	Study of Atom Aids Modern Development	*Natural Resources Canada* 7, no. 2 (1928)
1929	Dominion Astrophysical Observatory, Victoria B.C. [annual report for 1928]	*MNRAS* 89 (1929), 347–49
1930	Dominion Astrophysical Observatory, Victoria B.C. [annual report for year ending 31 March 1929]	*PA* 38 (1930), 156–58 = *PAAS* 6 (1931), 296–98
1929, Aug.	The Dominion Astrophysical Observatory and its Work [abstract]	*PA* 38 (1930), 90 = *PAAS* 6 (1931), 276
1929, Aug.	The Galactic Rotation and the K-term for O- and B-type Stars (with J.A. Pearce) [abstract]	*PA* 38 (1930), 90 = *PAAS* 6 (1931), 277

1929, Aug.	The Motions of Interstellar Calcium (with J.A. Pearce) [abstract]	*PA* 38 (1930), 81 = *PAAS* 6 (1931), 277
1929	How the Universe Revolves in Space	*Natural Resources Canada* 8, no. 5 (1929)
1929, Dec.	The Motions of the B Stars*	*Science* 71 (1930), 225–30
1930	Dominion Astrophysical Observatory, Victoria B.C. [annual report for 1929]	*MNRAS* 90 (1930), 407–9
1930, Jan.	The Motions and Distribution of Interstellar Matter (with J.A. Pearce)	*MNRAS* 90 (1930), 243–68
1930, April	Diffuse Matter in Interstellar Space	*Proc. Am. Acad. Arts Sci.* 64 (1930), 335–46
1930, April	The Structure and Rotation of the Galaxy	*Proc. Am. Phil. Soc.* 69 (1930), 401–17
1930, May	The High-temperature Stars*	*MNRAS* 90 (1930), 616–35
1930, June	A Catalogue of the Radial Velocities of O- and B-Type Stars (with Pearce) [abstract]	*PASP* 42 (1930), 248
1930, Oct.	A Catalogue of the Radial Velocities of O and B Type Stars (with J.A. Pearce)	*PDAO* 5 (1935), 99–165
1930, Nov.	The Radial Velocities of 523 O and B Type Stars Obtained at Victoria, 1923–1929 (with J.A. Pearce)	*PDAO 5* (1935), 1–98
1931	Dominion Astrophysical Observatory, Victoria B.C. [annual report for year ending 31 March 1930]	*PA* 39 (1931), 87–89 = *PAAS* 6 (1931), 387–89
1931	Dominion Astrophysical Observatory, Victoria B.C. [annual report for 1930]	*MNRAS* 91 (1931), 358–60
1931, June	On the Motions and Distribution of Interstellar Calcium [abstract]*	*PASP* 43 (1931), 291–92
1931, Dec.	The Problems of the Diffuse Matter in the Galaxy (with J.A. Pearce)	*PDAO* 5 (1935), 167–237

(*Continued*)

(Continued)

Date completed	Title and co-author(s)	Publication information
1931	Dominion Astrophysical Observatory, Victoria B.C. [annual report for year ending 30 June 1931]	*PAAS* 7 (1933), 58–59
1932	Dominion Astrophysical Observatory, Victoria B.C. [annual report for 1931]	*MNRAS* 92 (1932), 288–90
1932	Fifty Years of Canadian Astronomy	*Fifty Years Retrospect* 117–22
1932, April	The Structure and Rotation of the Galaxy*	*PASP* 44 (1932), 141–66
1932	Report of Commission 30 (Radial Velocities)	*Trans. IAU* 4 (1932), 178–89
1932, June	A Review of the Progress of Astronomy*	*PASP* 44 (1932), 215–29
1932	Dominion Astrophysical Observatory, Victoria B.C. [annual report for year ending 31 March 1932]	*PAAS* 7 (1933), 134–36
1933	Dominion Astrophysical Observatory, Victoria B.C. [annual report for 1932]	*MNRAS* 93 (1933), 255–57
1933, March	The Expansion of the Universe*	*JRASC* 27 (1933), 235–52
1933, March	Observational Confirmation of the Rotation of the Galaxy	*Obs.* 56 (1933), 328–32
1933, June	Mean Parallaxes and Absolute Magnitudes of (with J.A. Pearce) [abstract]	*PASP* 45 (1933), 207
1933, June	Determination of the K-Term, the Solar Motion, and the Galactic Rotation (with J.A. Pearce) [abstract]	*PASP* 45 (1933), 207
1933, June	Determination of the K-term, Solar Motion, and Galactic Rotation (with J.A. Pearce)	*Proceedings of the Pacific Science Congress* 2 (1934), 1101–35

1933, June	The Work of the DAO	*Proceedings of the Pacific Science Congress* 2 (1934), 1135
1933, Nov.	The Distance and Direction to the Gravitational Centre of the Galaxy from the Motions of the O5 to B7 Stars (with J.A. Pearce)	*MNRAS* 94 (1934), 679–713
1934, March	Religion and Science	*Canadian Churchman* 61 (15 March 1934), 1–2
1934	Dominion Astrophysical Observatory, Victoria B.C. [annual report for 1933]	*MNRAS* 94 (1934), 308–10
1934	Dominion Astrophysical Observatory, Victoria, B.C. [annual report for year ending 31 March 1934]	*PAAS* 8 (1936), 67–68
1934, April	Notes added to "Structure and Rotation of the Galaxy" (see *PASP* 44 (1932), 141)	*Smithsonian Inst. Rep. and Proc. for 1933*, 189
1934, April	The Motions of the O and B Type Stars and the Scale of the Galaxy (with J.A. Pearce)	*PDAO* 5 (1935), 241–328
1935, June	The Dimensions and Structure of the Galaxy (Halley Lecture)*	*Five Halley Lectures* (Oxford 1936)
1935, June	The Dimensions and Structure of the Galactic System [summary of the Halley Lecture]	*JRASC* 30 (1936), 153–64
1935, June	The Galaxy	*Scientia* 60 (1936), 309–17
1936, Dec.	Edward P. Burrell	*PASP* 49 (1937), 141–43
1937, June	Edward P. Burrell	*Science* 85 (1937), 597
1937, June	Ambrose Swasey, Engineer, Scientist, Philanthropist	*JRASC* 31 (1937), 409–16

(*Continued*)

(Continued)

Date completed	Title and co-author(s)	Publication information
1937, Dec.	Obituary Notices – Swasey, Ambrose	*MNRAS* 98 (1938), 258–62
1937, Dec.	An Analysis of the K Term in the B-Type Stars	*MNRAS* 98 (1938), 518–27
1938, June	Modern Conceptions of the Stellar System*	*PA* 47 (1939), 239–55
1938, Nov.	The 82-inch Mirror of the McDonald Observatory	*ApJ* 89 (1939), 84-98
1939, May	Some Features of the New Mirror [booklet]	*Addresses Made at the Dedication … of the … McDonald Observatory* (1939)
1939, Oct.	The History of Astronomy in British Columbia*	*BCHQ* 4 (1939), 63–78
1983, June	The History of Astronomy in British Columbia [almost the same as the article above]	*JRASC* 77 (1983), 108–20

Appendix B

TALKS GIVEN BY J.S. PLASKETT*

See pages 415–17 for a list of abbreviations used in this table.

	Date	Title	Group	Location	Source (P: Published; S: Summarized; R: Referred to)
1899	Dec.	Mathematics in the Mechanical Trades	Mathematical and Physical Society	Toronto	R: *Varsity* 19 (6 Dec. 1899), 95; P: *Canadian Engineer* (Jan. 1900), 241
1901	16 Feb.	Colour Values in Monochrome	Canadian Institute	Toronto	P: https://archive.org/stream/cihm_72137
	5 Dec.	Colour Photography	Mathematical and Physical Society	Toronto	R: *UT Monthly* 2 (Dec. 1901), 84
1902	5 Feb.	Colour Photography and Colour Values in Monochrome	Conversazione	Toronto	R: *UT Monthly* 2 (Feb. 1902), 137
	March	Photography in Natural Colours	Camera Club; Canadian Institute	Toronto	P: *Transactions of the Canadian Institute* 7 (1904), 301
	2 April	Photography	ON Educational Association	Toronto	R: *Toronto Globe*, 3 April 1902, 9
1903	13 Jan.	Orthochromatic Photography	Toronto Astronomical Society	Toronto	S: *Toronto Globe*, 14 Jan. 1903, 9
	8 May	Colour Photography	St Margaret's College	Toronto	R: *Toronto Globe*, 9 May 1903, 5

(*Continued*)

(Continued)

	Date	Title	Group	Location	**Source** (P: Published; S: Summarized; R: Referred to)
1905	17 Feb.	[Progress of colour photography]	Normal School	Ottawa	R: LAC MG30, Series B13, vol. 3, file 21, Klotz diary, 17 Feb. 1905
	7 March	The Total Solar Eclipse of August 30 1905	RASC	Toronto	S: *Toronto Globe*, 8 March 1905, 14
1907	31 Jan.	The Star Image in Spectrographic Work	RASC	Ottawa	P: *JRASC* 1 (1907), 297; S: *JRASC* 1 (1907), 356
	5 March	Spectrographic Work at the Dominion Observatory	RASC	Toronto	R: *JRASC* 1 (1907), 66
	14 March	The Optics of the Telescope	RASC	Ottawa	S: *JRASC* 1 (1907), 131–32
1908	28 May	The Design of Spectrographs for Radial Velocity Determinations	RASC	Ottawa	P: *JRASC* 3 (1909), 190
	16 June	The Optics of the Telescope	RASC	Toronto	R: *JRASC* 2 (1908), 211
	25–27 Aug.	The Coelostat Telescope at the Dominion Observatory	AAS	Put-in-Bay, OH	R: *Science* 28 (1908), 854; S: *PAAS* 1 (1910), 303
	25–27 Aug.	Camera Objectives for Spectrographs	AAS	Put-in-Bay, OH	R: *Science* 28 (1908), 851; S: *PAAS* 1 (1910), 304
	25–27 Aug.	Effect on the Accuracy of Radial Velocity Determinations of Increasing Slit Width	AAS	Put-in-Bay, OH	R: *Science* 28 (1908), 851; S: *PAAS* 1 (1910), 311
1909	18–20 Aug.	The Width of Slit Giving Maximum Accuracy	AAS	Williams Bay, WI	S: *PA* 17 (1909), 464
	19–21 Aug.	Effect of Faulty Collimation of the Correcting Lens on the Star Image	AAS	Williams Bay, WI	S: *PA* 17 (1909), 526–27 = *Report 1910* (1911), 93–96

	Nov.	The Work of the Dominion Observatory	RASC	Toronto	R: *Toronto Globe*, 11 Nov. 1909, 14
1910	24 Feb.	The Optics of the Telescope	RASC	Ottawa	S: *JRASC* 4 (1910), 215
	24 March	The Constitution and Radial Motions of the Stars	RASC	Ottawa	S: *JRASC* 4 (1910), 135
	21 April	Stellar Evolution and Theories of World Building	RASC	Ottawa	R: *JRASC* 4 (1910), 388
	Aug.	Probable Errors of Radial Velocity Determinations	AAS	Cambridge, MA	R: *Report 1911* (1912), 152; S: *Science* 32 (1910), 876
	28 Sept.	Probable Errors of Radial Velocity Determinations	RSC	Ottawa	P: *Proc. Trans. RSC* 4 (1910), 77–88
1911	12 Jan.	Notes from Two Recent Astronomical Gatherings	RASC	Ottawa	R: *Report 1911* (1912), 152
	23 Feb.	Some Recent Interesting Developments in Astronomy	RASC	Ottawa	P: *JRASC* 5 (1911), 245–65
	17 May	Methods and Preliminary Results in the Spectroscopic Determination of the Solar Rotation	RSC	Ottawa	P: *Proc. Trans. RSC* 5 (1911), 107–121
	Aug.	Preliminary Measures of the Solar Rotation	AAS	Ottawa	S: *Science* 34 (1911), 530
	7 Dec.	Solar Rotation	RASC	Ottawa	S: *JRASC* 5 (1911), 429
	27–29 Dec.	Preliminary Measures of the Solar Rotation	AAS/AAAS	Washington, DC	S: *Science* 35 (1912), 710
1912	27 Feb.	The Evolution of the Worlds	Field-Naturalists	Ottawa	P: *Ottawa Naturalist* 26 (1912) 29, 52

(*Continued*)

(Continued)

	Date	Title	Group	Location	**Source** (P: Published; S: Summarized; R: Referred to)
	14–16 May	The Spectroscopic Binary θ2 Tauri	RSC	Ottawa	R: *Proc. Trans. RSC* 6 (1912), xxxi
	16 May	The Solar Rotation	RSC	Ottawa	P: *Proc. Trans. RSC* 6 (1912), 3: 23–70
	27–31 Aug.	The Solar Rotation	AAS	Pittsburgh	P: *ApJ* 37 (1913), 73; S: *Science* 37 (1913), 25
	27–31 Aug.	A Plane Grating Spectrograph for Stellar Work	AAS	Pittsburgh	P: *ApJ* 37 (1913), 373; S: *Science* 37 (1913), 25
	Dec.	Solar Rotation	AAS/AAAS	Cleveland	S: *Science* 37 (1913), 639
1913	11 April	[The new telescope]	Union Club	Cleveland	R: *Cleveland Plain Dealer*, 11 April 1913, 7
	27–29 May	Solar Rotation in 1912	RSC	Ottawa	R: *Proc. Trans. RSC* 7 (1913), xlix
	16 Oct.	Bonn Meeting of the Solar Union	RASC	Ottawa	R: *JRASC* 8 (1914), 54
1914	4 March	The Fixed Stars, Their Nature, and Constitution	Natural History Society	Victoria	S: *Victoria Daily Colonist*, 28 Feb. 1914, 2, and 5 March 1914, 4
	12 April	[Astronomy]	Alexandra Club	Victoria	R: *Dawson [YK] Daily News*, 13 April 1914
	26–28 May	On Prism Materials for Stellar Spectrographs	RSC	Montreal	S: *Proc. Trans. RSC* 8 (1914)
	26–28 May	The New 72-inch Reflecting Telescope for Canada	RSC	Montreal	S: *Proc. Trans. RSC* 8 (1914)
	25–28 Aug.	The Solar Rotation in 1912	AAS	Evanston, IL	S: *PA* 22 (1914), 637–38 = *PAAS* 3 (1918), 89–90
	Fall	Evanston Meeting of the AAS	RASC	Ottawa	R: *JRASC* 9 (1915), 66

	22 Oct.	The New Six-foot Reflecting Telescope	RASC	Ottawa	R: *JRASC* 9 (1915), 66; S: *Ottawa Citizen*, 23 Oct. 1918, 2
	1 Dec.	The New Reflecting Telescope of the Dominion Observatory	RASC	Toronto	R: *JRASC* 9 (1915), 18; *Toronto Globe*, 2 Dec. 1914, 6
	2 Dec.	Canada's New Telescope at Victoria and Its Uses	RASC	Guelph, ON	R: *JRASC* 8 (1914), 407
1915	12 Jan.	The Sidereal Universe [President's Address, annual meeting]	RASC	Toronto	P: *JRASC* 9 (1915), 37
	27 Jan.	Particulars of the Largest Telescope in the World, Now Being Made for Canada	RASC	Hamilton, ON	R: *JRASC* 9 (1915), 126
	25–27 May	Progress on the 72-inch Reflecting Telescope	RSC	Ottawa	R: *Proc. Trans. RSC* 9 (1915), xxxix
	25–27 May	The Solar Rotation	RSC	Ottawa	R: *Proc. Trans. RSC* 9 (1915), xxxix
	25 June	Jupiter and His Moons	RASC	Victoria	R: *JRASC* 10 (1916), 142; S: *Victoria Daily Colonist*, 26 June 1915, 5
	27 Oct.	Planets Jupiter and Saturn	RASC	Victoria	R: *JRASC* 10 (1916), 142; S: *Victoria Colonist*, 28 Oct. 1915, 9
1916	25-Jan.	Modern Views of the Sun [Retiring President's Address, annual meeting]	RASC	Toronto	P: *JRASC* 10 (1916), 101
	7 April	72-inch Telescope	RASC	Ottawa	S: *JRASC* 10 (1916), 323
	16–18 May	Progress on the 72-inch Reflecting Telescope	RSC	Ottawa	R: *Proc. Trans. RSC* 10 (1916), xxxiv

(*Continued*)

(Continued)

	Date	Title	Group	Location	**Source** (P: Published; S: Summarized; R: Referred to)
	10 Aug.	The Purpose of Modern Astronomy	Rotary Club	Victoria	S: *Victoria Daily Colonist*, 11 Aug. 1916), 7
	4 Oct.	The New Telescope	Canadian Society of Civil Engineers	Victoria	S: *Victoria Daily Colonist*, 5 Oct. 1916), 14
	21 Oct.	Observatory Visit	RASC	Victoria	R: *JRASC* 11 (1917), 105
	16 Dec.	A Journey through Space	Field-Naturalists	Ottawa	R: *Ottawa Naturalist* 31, no. 1 (1917), 10
	27–29 Dec.	The 72-inch Canadian Telescope and Its Dome	AAS/AAAS	New York	S: *PA* 25 (1917), 375 = *PAAS* 3 (1918), 275
1917	16 Feb.	Canada's Great Reflecting Telescope	RASC	Ottawa	R: *JRASC* 11 (1917), 242
	25 Feb.	A Journey through Space	People's Forum	Ottawa	R: *Ottawa Citizen* (26 Feb. 1917), 12
	21–24 May	The 72-inch Reflecting Telescope	RSC	Ottawa	R: *Proc. Trans. RSC* 11 (1917), I
	Aug.	A Trip through Space [tour of DAO]	Teachers' summer school	Victoria	R: *Victoria Daily Colonist* (26 Aug. 1917), 18
1918	25 Jan.	Polishing the Mirror of the New Telescope	RASC	Victoria	R: *JRASC* 13 (1919), 138; S: *Victoria Daily Colonist*, 26 Jan. 1918), 7
	29 Feb.	The Fixed Stars	Teachers' Assn	Victoria	S: *Victoria Daily Colonist*, 1 March 1918, 10

	13 May	Coming Eclipse of the Sun	RASC	Victoria	R: *JRASC* 13 (1919), 138; *Victoria Daily Colonist*, 13 May 1918, 7
	11 July	The Purpose of Astronomy and the DAO	Union of Canadian Municipalities	Victoria	S: *Victoria Daily Colonist*, 12 July 1918, 4
	18 Dec.	Performance and Work of the New Telescope	RASC	Victoria	R: *JRASC* 13 (1919), 244
1919	13 March	Modern Views of the Universe	Vancouver Institute	Vancouver	R: www.library.ubc.ca/archives/vaninsti.html
	6 May	Visit to the DAO	RASC	Victoria	R: *JRASC* 14 (1920), 72; S: *Victoria Daily Colonist*, 7 May 1919), 4
	20–22 May	The Performance and Work of the 72-inch Mirror	RSC	Ottawa	R: *Proc. Trans. RSC* 13 (1919), xxix
	20–22 May	Tests of the Figure of the 72-inch Mirror	RSC	Ottawa	R: *Proc. Trans. RSC* 13 (1919), xxix
	20–22 May	The Spectroscopic Binary and Eclipsing Variable U Ophiuchi	RSC	Ottawa	R: *Proc. Trans. RSC* 13 (1919), xxix
	19–22 June	The Spectroscopic Orbit and Absolute Dimensions of the Eclipsing Variable U Ophiuchi	ASP	Pasadena, CA	R: *PASP* 31 (1919), 187
	19–22 June	The Figure of the 72-inch Mirror under Constant and Changing Temperature Conditions	ASP	Pasadena, CA	R: *PASP* 31 (1919), 187
	1 Aug.	Our Solar System	Teachers' Summer School	Victoria	S: *Victoria Daily Colonist*, 2 Aug. 1919, 5
	9 Oct.	A Journey through Space	St Michael's, Royal Oak	Victoria	S: *Victoria Daily Colonist*, 14 Oct. 1919, 16

(*Continued*)

(Continued)

	Date	Title	Group	Location	**Source** (P: Published; S: Summarized; R: Referred to)
	16 Dec.	Development of Astronomical Research	RASC	Victoria	S: *JRASC* 14 (1920), 31; R: *JRASC* 15 (1921), 104
1920	25 March	The Purpose and Work of Present-Day Astronomy	Kiwanis Club	Vancouver	R: *Vancouver Daily Sun*, 26 March 1920, 3
	25 March	The Romance of Astronomy	Vancouver Institute	Vancouver	R: www.library.ubc.ca/archives/vaninsti.html
	12 May	The 72-inch Telescope	RASC	Winnipeg	R: *Winnipeg Evening Tribune*, 13 May 1920, 5; *JRASC* 15 (1921), 103
	19–21 May	The Orbit and Dimensions of U Coronae	RSC	Ottawa	R: *Proc. Trans. RSC* 14 (1920) xxiv
	17–19 June	The Orbit and Dimensions of U Coronae	AAAS/ASP	Seattle	S: *PASP* 32 (1920), 190
	23 Nov.	Wonders of the Starry Universe	YWCA	Victoria	S: *Victoria Daily Colonist*, 24 Nov. 1920, 4
1921	1 Feb.	The Romance of Astronomy	RASC	Victoria	R: *JRASC* 15 (1921), 132
	10 May	Achievements at the DAO	RASC	Toronto	R: *Toronto Mail*, 11 May 1921; *JRASC* 15 (1921) 253-4
	18–20 May	The Temperature Control of the Stellar Spectrograph	RSC	Ottawa	R: *Proc. Trans. RSC* 15 (1921), xxvii
	18–20 May	The Orbit and Dimensions of TV Cassiopeiae	RSC	Ottawa	R: *Proc. Trans. RSC* 15 (1921), xxvii

	18–20 May	The Radial Velocities of 594 Stars	RSC	Ottawa	S: *Proc. Trans. RSC* 15 (1921), 141-142
	26 May	Modern Ideas of the Universe	RASC	Montreal	S: *JRASC* 15 (1921), 258
	3 Nov.	The Dimensions of Stars	RASC	Victoria	R: *JRASC* 16 (1922), 27
	11 Nov.	The Dimensions of the Stars	ASP	San Francisco	P: *PASP* 34 (1922), 79
1922	7 Feb.	The Observer's Handbook	RASC	Victoria	R: *JRASC* 16 (1922), 155
	27 Feb.		BC Academy of Science	Vancouver	R: UBC Library Special Collections
	17–19 May	The Most Massive Star Known	RSC	Ottawa	R: *Proc. Trans. RSC* 16 (1922), xxxvi
	5–8 Sept.	The Ultra-violet Spectrograph of the 72-inch Telescope	AAS	Williams Bay, WI	S: *PA* 31 (1923), 20–21
	26 Oct.	The Personality and Work of Astronomers Today	RASC	Victoria	R: *JRASC* 17 (1923), 45
1923	1 May	The O-type Stars	RASC	Victoria	R: *JRASC* 17 (1923), 414
	16 May	Convocation address	University of Alberta	Edmonton	S: *Edmonton Journal*, 17 May 1923, 3
	22–24 May	A Study of the O-type Stars	RSC	Ottawa	R: *Proc. Trans. RSC* 17 (1923), xxxiv
	22–24 May	A Remarkable Stellar Spectrum	RSC	Ottawa	R: *Proc. Trans. RSC* 17 (1923), xxxiv
	30 Aug.	Some Popular Misconceptions about Astronomy	Rotary Club	Victoria	S: newspaper clippings in Rotary Club file, Victoria City Archives
	17–20 Sept.	Two New Camera Lenses for Spectrographs	AAS	Pasadena, CA	S: *PA* 31 (1923), 659–60 = *PAAS* 5 (1927), 78

(*Continued*)

(Continued)

	Date	Title	Group	Location	**Source** (P: Published; S: Summarized; R: Referred to)
	17–20 Sept.	The Absorption Line O-type Stars	AAS	Pasadena, CA	S: *PA* 31 (1923), 660–61 = *PASS* 5 (1927), 79
	20 Nov.	Recent Astronomical Events in California	RASC	Victoria	R: *JRASC* 18 (1924), 69; S: *Victoria Daily Times*, 21 Nov. 1923
1924	7 Feb.	The Evolution of Stars	Vancouver Institute	Vancouver	www.library.ubc.ca/archives/vaninsti.html
	21 May	O-type Stars and Their Relation to the Stellar Evolutionary Sequence	RSC	Quebec	P: *JRASC* 18 (1924), 321–38; R: *Proc. Trans. RSC* 18 (1924), xxxv
	25 Nov.	The Evolution of Stars	RASC	Victoria	S: *JRASC* 19 (1925), 27
1925	19–21 May	The Orbits of Three Spectroscopic Binaries	RSC	Ottawa	R: *Proc. Trans. RSC* 19 (1925), xli
	24 July	[Report on latest work at DAO]	RAS	London, UK	S: *Obs.* 48 (1925), 280
	July		BAA	London, UK	R: *JRASC* 19 (1925), 253
	25 Sept.	Account of JSP's trip to England	Canadian Club	Victoria	P: *JRASC* 19 (1925), 247–54
	2 Oct.	The Cambridge Meeting of the International Astronomical Union	RASC	Victoria	S: *JRASC* 19 (1925), 303–4
	3 Nov.	Impressions of the Old Country	Royal Society of St George	Victoria	R: *Victoria Daily Times*, 2 Nov. 1925, 6
1926	19 Feb.	Astronomy as a National Asset	Canadian Club	Vancouver	R: JSP to W.W. Cory, LAC, RG 48, vol. 47, 11 Feb. 1926
	19 Feb.	[Title unknown]	Canadian Mining Institute	Vancouver	R: *Vancouver Evening Sun*, 20 Feb. 1926, "Society and Clubs"

	21 May	The Observatory: Its Origin, Installation, and Work	RASC	Victoria	R: *JRASC* 20 (1926), 307
1927	14 Jan.	Three Peculiar Spectra	DAO seminar	Victoria (DAO)	R: *JRASC* 22 (1928), 49
	11 Feb.	On Corrections to Proper Motion	DAO seminar	Victoria (DAO)	R: *JRASC* 22 (1928), 49
	11 March	Extra-Galactic Nebulae [review of Hubble paper]	DAO seminar	Victoria (DAO)	R: *JRASC* 22 (1928), 49
	5 April	The Galaxy and Beyond	RASC	Victoria	R: *JRASC* 22 (1928), 70
	15 April	On the Masses of Stars [review of Gerasimovic paper]	DAO seminar	Victoria (DAO)	R: *JRASC* 22 (1928), 49
	12 May	Parallaxes of the Helium Stars [review of Kapteyn's work]	DAO seminar	Victoria (DAO)	R: *JRASC* 22 (1928), 49
	12 May	Remarkable Range of Astronomical Research	RASC	Victoria	R: *JRASC* 21 (1927), 254
	17 May	The Galaxy and Beyond	RASC	Winnipeg	R: *JRASC* 22 (1928), 69; *Winnipeg Free Press*, 18 May 1927, 12
	25 May	Sixty Years' Progress in Astronomy	RSC	Ottawa	P: *JRASC* 21 (1927), 295–310
	24–26 May	The B-type Stars	RSC	Ottawa	R: *Proc. Trans. RSC* 21 (1927), lxxxvi
	17 June	The Period-Luminosity Law of Cepheid Variables	DAO seminar	Victoria (DAO)	R: *JRASC* 22 (1928), 49
	22–24 June	The Orbit of B-type Spectroscopic Binary HD 193536	ASP/AAAS Pac Div	Reno, NV	P: *PASP* 39 (1927), 254
	5 Aug	The Rotation of the Galaxy (Lindblad and Oort)	DAO seminar	Victoria (DAO)	R: *JRASC* 22 (1928), 49

(*Continued*)

(Continued)

	Date	Title	Group	Location	**Source** (P: Published; S: Summarized; R: Referred to)
	16 Sept.	The Stationary H and K lines of Calcium and Sodium	DAO seminar	Victoria (DAO)	R: *JRASC* 22 (1928), 49
	28 Oct.	The Rotation of the Galaxy from Class B Stars at Victoria	DAO seminar	Victoria (DAO)	R: *JRASC* 22 (1928), 49
	25 Nov.	Distribution of the Absolute Magnitudes of the Stars (Malmquist)	DAO seminar	Victoria (DAO)	R: *JRASC* 22 (1928), 49
	6 Dec.	Journey through Space	Strathcona Lodge	Shawnigan Lake, BC	R: *JRASC* 22 (1928), 202
	30 Dec.	The Masses and Luminosities of the Eclipsing Variables (McLaughlin)	DAO seminar	Victoria (DAO)	R: *JRASC* 22 (1928), 50
1928	27 June	[Rotation of the Galaxy]	British Astronomical Association	London, UK	S: *Obs.* 51 (1928), 263
	3 Feb.	Statisitical Discussion of Spectroscopic Binaries Stars (Beer)	DAO seminar	Victoria	P: *JRASC* 22 (1928), 296, 337
	21 Feb.	The Motions of the Stars	RASC	Victoria	P: *JRASC* 22 (1928), 111; S: *JRASC* 22 (1928), 201
	13 Nov.	The Leiden Meeting of the IAU	RASC	Victoria	S: *JRASC* 23 (1929), 115–16
1929	30 April	The History of the DAO Telescope	RASC	Victoria	R: *JRASC* 23 (1929), 341–42; *Victoria Daily Colonist*, 1 May 1929, 2

	20–21 June	The Rotation of the Galaxy	AAAS/ASP	Berkeley, CA	S: *PASP* 41 (1929), 251
	4 Nov.	The Rotation of the Galaxy	Canadian Club	Toronto	P: https://www.canadianclub.org/Events/EventDetails.aspx?id=786
	Nov.	[Title unknown]		Various cities in NY, CT, OH, AZ, and CA	R: LAC, RG48, vol. 47, file 4-11, JSP to R.M. Stewart, 9 Dec. 1929
	2 Dec.	The Rotation of the Galaxy	Vancouver Institute	Vancouver	R: www.library.ubc.ca/archives/vaninsti.html
	10 Dec.	The Progress of the 200-inch Telescope	RASC	Victoria	R: *JRASC* 24 (1930), 46; *Victoria Daily Colonist*, 11 Dec. 1929, 5
	31 Dec.	The Motions of the B Stars	AAAS	Des Moines, IA	P: *Science* 71 (1930), 225–30; R: *Nature* 125 (18 Jan. 1930), 103
1930	28 Jan.	The Rotation of the Galaxy	RASC	Victoria	R: *JRASC* 25 (1931), 105; *Victoria Daily Colonist*, 2 Feb. 1930, 34
	9 April	Diffuse Matter in Interstellar Space	American Academy of Arts & Sciences	Boston	P: *Proc. Am. Acad. Arts Sci.* 64 (1930), 335– 46
	26 April	The Motions of the B Stars	American Philosophical Society	Philadelphia	P: *Proc. Am. Phil. Soc.* 69 (1930), 401–17
	9 May	The High-temperature Stars [George Darwin Lecture]	RAS	London, UK	P: *MNRAS* 90 (1930), 616; S: *Obs.* 53 (1930), 223–29
	28 May	The Structure and Rotation of the Galaxy	British Astronomical Association	London, UK	R: *Obs.* 53 (1930), 201

(*Continued*)

(Continued)

	Date	Title	Group	Location	**Source** (P: Published; S: Summarized; R: Referred to)
	17 July	History of Astronomy in Canada	RASC; Scientific Club	Winnipeg	R: *JRASC* 25 (1931), 106; *Winnipeg Tribune*, 18 July 1930, 3
1931	24 Feb.	Reminiscences of Some Famous Astronomers	RASC	Victoria	R: *JRASC* 25 (1931), 182; *Victoria Daily Colonist*, 25 Feb 1931, 2
	14 May	The Structure and Rotation of the Galaxy	RASC	Winnipeg	R: *JRASC* 26 (1932), 118
	21 May	The Motions of the Diffuse Gaseous Matter	RSC/RASC	Toronto	R: *Toronto Globe*, 22 May 1931; *Proc. Trans. RSC* 25 (1931), c
	21 May	Investigations of the Motions of the O- and B-type Stars at Victoria	RSC	Toronto	R: *JRASC* 25 (1931), 269; *Proc. Trans. RSC* 25 (1931), ci
	21 May	The Structure and Rotation of the Galaxy [popular lecture]	RSC	Toronto	S: *Toronto Globe*, 22 May 1931, 13
	June	On the Motions and Distribution of Interstellar Calcium	AAAS/ASP	Pasadena, CA	S: *PASP* 43 (1931), 291
1932	29 Jan.	Diffuse Gaseous Matter in Space	RASC	Victoria	S: *JRASC* 26 (1932), 239; *Victoria Daily Colonist*, 30 Jan. 1932, 2
	15 April	[Astronomy of the Early Explorers of British Columbia]	BCHA	Victoria	S: *Victoria Daily Colonist*, 16 April 1932, 2
	20 April	Rotation and Structure of the Galaxy	ASP	San Francisco	P: *PASP* 44 (1932), 141–66
1933	17 Feb.	Diffuse Matter in Space	RASC	Vancouver	S: *JRASC* 27 (1933), 217

	17 Feb.	The Absorption of Light in Interstellar Space	UBC Physics Club	Vancouver	R: *Ubyssey* 15 (17 Feb. 1933), 1
	18 Feb.	The Expanding Universe	Vancouver Institute	Vancouver	R: www.library.ubc.ca/archives/vaninsti.html
	10 March	Expansion of the Universe	RASC	Victoria	P: *JRASC* 27 (1933), 235–52
	June	The Work of the DAO	Pacific Science Congress	Vancouver	P: *Proc. Pac. Sci. Cong.* 2 (1934), 1135
	June	Determination of the K-Term, Solar Motion, and Galactic Rotation	AAAS/ASP	Salt Lake City	S: *PASP* 45 (1933), 207
	29 Dec.	The Structure and Rotation of the Galaxy	AAAS/AAS	Boston	S: *PAAS* 8 (1936), 19
	29 Dec.	The Expansion of the Universe	AAS	Boston	S: *PAAS* 8 (1936), 19
1934	March	Religion and Science	Christ Church Cathedral	Victoria	P: *Canadian Churchman* 61 (15 March 1934), 1–2
	21–24 May	The Dimensions of the Galaxy ... from the Motions of Class B Stars	RSC	Quebec	S: *Proc. Trans. RSC* 28 (1934), xcix
	29 May	The Nature of the Galactic System	RASC	Winnipeg	R: *Winnipeg Free Press*, 30 May 1934, 6; *JRASC* 28 (1934), 463–64
1935	24 April	[Summary of work at the DAO]	British Astronomical Association	London, UK	R: *Obs.* 58 (1935), 188
	5 June	The Distance and Direction to the Galactic Centre from the Rotational Constants …	Oxford University	Oxford	S: *JRASC* 30 (1936), 153–64
1936	6 Feb.	Some Sidelights of Astronomy in Europe	RASC	Victoria	S: *JRASC* 30 (1936), 261 7; *Victoria Daily Colonist*, 7 Feb. 1936, 3

(*Continued*)

(Continued)

	Date	Title	Group	Location	**Source** (P: Published; S: Summarized; R: Referred to)
	9 Oct.	Some Phases of Road Transportation in the Early Days	BCHA	Victoria	R: *Victoria Daily Colonist*, 10 Oct. 1936
1937	24 June	Progress on the 82-inch McDonald Telescope (with W. Seely)	AAAS/ASP	Denver, CO	R: *PA* 45 (1937), 363
	6 Oct.	Special Features of the New 200-inch and 82-inch Telescopes	RASC	Victoria	S: *JRASC* 31 (1937), 443
1938	June	Modern Conceptions of the Stellar System	AAAS	San Diego	P: *PA* 48 (1940), 239
1939	5 May	The 82-Inch Telescope	Dedication event	Alpine, TX	S: *Addresses Made at the Dedication of McDonald Observatory* (1939)
	5 May	Some Features of the New Mirror	McDonald Observatory	Alpine, TX	S: *Addresses Made at the Dedication of McDonald Observatory* (1939)
	13 Oct.	History of Astronomy in Canada	BCHA	Victoria	P: *BCHQ* 4 (1939), 63; *Victoria Daily Colonist*, 15 Oct. 1939, 5
	13 Dec.	History of the Founding of the DAO at Victoria	RASC	Victoria	S: *JRASC* 34 (1940), 164; *Victoria Daily Colonist*, 14 Dec. 1939, 2

* This list does not include papers written by JSP but presented by others.

Notes

Abbreviations

Archival Sources

ADBCA Anglican Diocese of British Columbia Archives, Victoria, BC
AIP American Institute of Physics, Niels Bohr Library, College Park, MD, USA
AO Archives of Ontario, Toronto Arnprior Arnprior McNab/Braeside Archives, Arnprior, ON
BCA British Columbia Archives, Victoria
BHL Bentley Historical Library, University of Michigan, Ann Arbor
CCL Citrus College Library, Glendora, CA
CITA California Institute of Technology Archives, Pasadena, CA
CWRU Case Western Reserve University, Kelvin Smith Library, Cleveland, OH
GAU Georg-August–Univesität, Göttingen, Germany
HL Huntington Library, San Marino, CA
HUA Harvard University Archives, Cambridge, MA
LAC Library and Archives Canada, Ottawa
LOA Lowell Observatory Archives, Flagstaff, AZ
MLSA Mary Lea Shane Archives of the Lick Observatory, University of California at Santa Cruz
NBL Niels Bohr Library, American Institute of Physics, College Park, MD
PUL Princeton University Library, Princeton NJ
RASCA Royal Astronomical Society of Canada Archives, Toronto

TFL RCI	Royal Canadian Institute papers, Thomas Fisher Library, University of Toronto
UCL	Special Collections, University of Chicago Library, Chicago
UCSC	University of California Santa Cruz, Special Collections
UPA	University of Pittsburgh Archives, Pittsburgh, PA
UTA	University of Toronto Archives, Toronto

Periodicals

The following list does not include newspapers or those periodicals that are cited only once.

Ann. Rep. Smithsonian Inst.	*Annual Reports of the Smithsonian Institution*
AN	*Astronomische Nachrichten*
ApJ	*Astrophysical Journal*
BAN	*Bulletin of the Astronomical Institutes of the Netherlands*
Biogr. Mems. Fell. R. Soc.	*Biographical Memoirs of Fellows of the Royal Society*
BCHQ	*British Columbia Historical Quarterly*
JAHH	*Journal of Astronomical History and Heritage*
JHA	*Journal for the History of Astronomy*
JRASC	*Journal of the Royal Astronomical Society of Canada*
LicOB	*Lick Observatory Bulletin*
MNRAS	*Monthly Notices of the Royal Astronomical Society*
Obit. Not. Fell. R. Soc.	*Obituary Notices of Fellows of the Royal Society*
Obs.	*The Observatory*
PA	*Popular Astronomy*
PAAS	*Publications of the American Astronomical Society*
PASP	*Publications of the Astronomical Society of the Pacific*
PDAO	*Publications of the Dominion Astrophysical Observatory*
PDO	*Publications of the Dominion Observatory*
Phil. Trans. Roy. Soc.	*Philosophical Transactions of the Royal Society of London*
PLO	*Publications of the Lick Observatory*
Proc. Am. Acad. Arts Sci.	*Proceedings of the American Academy of Arts & Sciences*

Proc. Am. Phil. Soc.	*Proceedings of the American Philosophical Society*
Proc. NAS	*Proceedings of the National Academy of Sciences*
Proc. Trans. RSC	*Proceedings and Transactions of the Royal Society of Canada*, Series 3
QJRAS	*Quarterly Journal of the Royal Astronomical Society*
Report	*Report of the Chief Astronomer for the Year* [Dominion Observatory, Ottawa; these reports, though circulated separately, are part of Canada, *Sessional Papers*, Department of the Interior]
Sci. Am. and *Sci. Am. Supp.*	*Scientific American* and its *Supplement*
S&T	*Sky & Telescope*
Trans. IAU	*Transactions of the International Astronomical Union*
Trans. IUCSR	*Transactions of the International Union for Cooperation in Solar Research*
UT Monthly	*University of Toronto Monthly*
VDT	*Victoria Daily Colonist*

Preface

1 Margaret E. Firth, *Handbook of Scientific and Technical Awards – United States and Canada, 1900–1952* (New York: Special Library Association, 1956) has Plaskett receiving four awards. It was not difficult to check her index for the names of those who got five or more. There were no Canadians among them. George Ellery Hale, the founder of three major U.S. observatories, received six, as did Ambrose Swasey, an American engineer, entrepreneur, and builder of telescopes.

2 See http://starchild.gsfc.nasa.gov/docs/StarChild/whos_who_level2/sagan.html.

3 *A Brief History of Time*, published in 1988, has more than ten million copies in print globally, and Dickinson's first three editions of *Night Watch* sold more than 800,000 copies, making it the top-selling stargazing guide in the world for the past twenty years.

4 Quoted in Christine Clement and Peter Broughton, "Helen Sawyer Hogg, 1905–1993," *JRASC* 87 (1993): 355.

5 Richard A. Jarrell, "The Value of Astronomy for a Civilized Society," *JRASC* 84 (1990): 314.

6 James B. Kaler, *Stars and Their Spectra: An Introduction to the Spectral Sequence*, 2nd ed. (Cambridge: Cambridge University Press, 2011).

7 J. Murray Clark, "Canada's Leaders in Science and Research," *Current History* 20 (1924): 101–8 named twelve scientists whose work had accorded Canada a high place in scientific activity at the time. Ten were at McGill or the University of Toronto. Other than JSP, only J.C. McLennan, F. Banting, and J.C. Fields have book-length biographies.

8 An exception, perhaps, is an article about JSP under the heading "Died This Day," *Toronto Globe and Mail*, 16 October 2004.

9 J.H. Hodgson, *The Heavens Above.*

10 R. Jarrell, *The Cold Light of Dawn*; Jarrell's papers that deal with aspects of JSP's career will be cited throughout the book.

11 T.S. Plaskett, *Plaskett–Stanley Family History.*

12 John Lankford, *American Astronomy*. His investigation of the community's demographics, institutional development, forms and uses of power, and reward systems, as well as the changing nature of scientific careers and the status of women within the community, is helpful in considering how Plaskett fitted into the American pattern.

13 This is a bit of an inside joke for Canadians. Starting in the 1980s, efforts were made to recognize Quebec as a distinct society within Canada, leaving many to wonder if the rest of Canada was also distinct.

14 Lankford, *American Astronomy*, 370

15 Chant's manuscript autobiography is in UTA, A1974-0027, box 10. Portions of it were published as Clarence A. Chant, *Astronomy in the University of Toronto.*

1 Rural Roots, to 1890

1 Records at the Woodstock Registry Office for East Zorra, Lot 20, Concession 12, show Timothy Plaskett purchased 100 acres in October 1859.

2 Doug Symons, *Giants of Oxford: Women and Men Who Changed Our World* (Woodstock, ON: Oxford Historical Society, 2001).

3 Many of the family-related details in this chapter come from Thomas S. Plaskett, *Plaskett-Stanley Family History*. A portion of this history was published in *JRASC* 82 (1988): 317–27. For the marriage of JSP's maternal grandparents, see the International Genealogical Index (IGI) database, England Marriages, 1538–1973, index, John Stanley and Mary Thwaite, 22 August 1822, https://familysearch.org/pal:/MM9.1.1/N2LQ-Z43. The name is spelled Thwaite, Thwaits, and Thwaites in various documents. For the birth of JSP's mother, see the IGI database, https://familysearch.org/search/collection/igi; also birthdate from the Province of British Columbia death certificate 1922-09-306642.

4 According to the 1901 Canadian census for Toronto, Annie Plaskett immigrated to Canada in 1848. The 1861 census for Canada West, Oxford County, gives the names and ages of Joseph Thwaites, Mary Stanley, and three of her children. The location and size of the farm are shown in the *Illustrated Historical Atlas of Oxford County, Ontario.*

5 According to Plaskett, *Plaskett-Stanley Family History*, 6, the Thwaites' farm was known as Maple Grove. The Stanley tombstone (see http://freepages.genealogy.rootsweb.ancestry.com/~dcoop/huntingford/hfc54.htm) states that Timothy Stanley (1833–95) died at Maple Grove and that his wife, Betsy Millman, died in 1927 at age 83, which fits with the account in *Plaskett-Stanley Family History*, 7, that Timothy, the eldest of Mary Stanley's boys, inherited the farm from his bachelor uncle, Joseph Thwaites.

6 The 1851 Canadian census for Zorra Township includes a household on Concession 12, Lot 11, comprising Joseph Thwaites, farmer, age 55, Mary Stanley, farmer, age 48, and her three children Timothy (age 18), Joseph (14), and Anne (12), "labourers." For Annie's marriage, see IGI database, entry for Annie Stanley, at https://familysearch.org/search/collection/igi.

7 Though I have not found the registration of JSP's birth, the date 17 November 1865, which must have been supplied by JSP himself, is in every biographical source published during his lifetime.

8 Plaskett, *Plaskett-Stanley Family History*, 1; England Marriages, 1538–1973 index, Family Search, IGI database, at https://familysearch.org/search/collection/igi, Wilford Plaskett [*sic*] and Anne Mounsey, 26 March 1797.

9 Plaskett, *Plaskett-Stanley Family History*, 4; the 1861 census for Canada West, East Zorra, lists Timothy and Sarah Plasket [*sic*] and their five children along with their ages.

10 Plaskett, *Plaskett-Stanley Family History*, 3. The 1851 census for Zorra Township includes a John Dawson, labourer, born in Ireland; Sarah Dawson was listed in the 1871 census as a widow, and in the 1901 census as a widow living in Woodstock who was born in England and immigrated to Canada in 1847. See also "Obituary of Mrs. Robert H. Little," *Woodstock Daily Sentinel-Review*, 22 September 1924, 1. This and other early newspapers were searched at the website https://paperofrecord.hypernet.ca/default.asp at the Woodstock Public Library.

11 Plaskett, *Plaskett-Stanley Family History*, 3–4. The details are confirmed in *Illustrated Historical Atlas of Oxford County* and records for East Zorra, Lot 20, Concession 12, in the land registry office in Woodstock.

12 The date of the fire is noted in the *Hickson Women's Institute, Tweedsmuir*, Book 1, at the Oxford Historical Society, Woodstock.

13 Plaskett, *Plaskett-Stanley Family History*, 4.

14 AO, Wills, Oxford County, G.S. Ont. 1-123, file 513. I have not found any standard way to convert the value of Canadian dollars in the 1870s to the present day but a cursory look at prices in newspapers for 1872 suggests to me that a factor of 50 is not unreasonable.
15 Plaskett, *Plaskett-Stanley Family History*, 4, 5–6. Tom said that his father was reeve, trustee on the school board, and commander of the village militia, though none of those offices has been independently confirmed. *Woodstock Review*, 30 January 1874, 1 and 18 September 1874, 1. *Volunteer Review and Military and Naval Gazette* 3 (3 May 1869), 297; *Woodstock Review*, 22 January 1875, 1; 12 February 1875, 1; 5 March 1875, 1; 17 September 1875, 4; *Weekly Sentinel Review*, 23 January 1880, 5; 2 April 1880, 1.
16 I did not find instructions in the *Boy's Own Annual* for building the type of electrical machine with a circular glass plate, but there were instructions for making and working an electric telegraph (2: 174) and instructions for making a Leyden jar (7: 318), as well as information on electric lights (3: 664), the curiosities of electricity (5: 650 and 763), and photography (8: 503).
17 I am grateful to Randall Rosenfeld, who pointed out the similarity to Gilbert and Sullivan's patter song, "I am the very model of a modern major general," also dating from 1879.
18 Woodstock Registry Office, East Zorra Lots 19 and 20, Concession 12.
19 AO, MS 937, reel 6, death record. As Richard Gwyn points out in *Nation Maker: Sir John A. Macdonald – His Life, Our Times* (Toronto: Random House, 2011), 29, heavy drinking "wasn't in the least unusual" at the time. Tom acknowledged that his father was fond of his liquor and thought that liquor may have been partly responsible for his early death, *Plaskett-Stanley Family History*, 5). AO, Wills, Oxford County, G.S. Ont. 1-123, file 1175.
20 "The Late Joseph Plaskett," *Woodstock Sentinel Review*, 25 November 1881, 1.
21 In adult life, JSP identified himself as a Conservative, *Canadian Who's Who* 2 (1936–7), 879.
22 Ralph Connor, *Glengarry School Days*, 207.
23 Plaskett, *Plaskett-Stanley Family History*, 8. According to the Census of Canada, 1881, for Zorra East, the ten children (and their ages at the time) were John Stanley (15), Robert Henry (14), Wilfred Stanley (12), Joseph (11), Tom Stanley (10), Frank (8), Charles Wm (7), Harry (5), Frederick (4), Annie J[osephine] (2).
24 J. George Hodgins, *Documentary History of Education in Upper Canada, from the Passing of the Constitutional Act of 1791, to the Close of Rev. Dr. Ryerson's Administration of the Education Department in 1876*, vol. 22 (1869–1871) (Toronto 1908), 213.
25 The Plaskett children's progress through elementary school can be traced through several reports sprinkled through the *Weekly Sentinel Review*

(Woodstock) from 1883 to 1890. JSP's name was listed under S.S. No. 7, East Zorra; the others attended S.S. No. 6, Strathallan and later the Hickson School.

26 W.A. MacKay, *Zorra Boys at Home and Abroad; or How to Succeed* (Toronto: Briggs, 1900), 42.

27 *Woodstock Review*, 11 January 1878, 6. A wonderful evocative telling, though fictional, of a contemporaneous examination day in another part of Ontario is in Connor, *Glengarry School Days*, chap. 3.

28 *Woodstock Review*, 1 February 1878, 8.

29 Although the maximum score was not stated in the report just cited, it was given in a similar report in the *Woodstock Review*, 7 August 1874.

30 As recalled by JSP's brother Frank in a CBC recording, BCA, T4022:56/0004, Voices of B.C., part of Series GR-3378 – Provincial Educational Media Centre, school broadcasts.

31 Art Williams and Edward Baker, *Woodstock: Bits & Pieces* (Erin ON: Boston Mills Press, 1990), 17.

32 1881 census for Zorra East, Oxford North.

33 Plaskett, *Plaskett-Stanley Family History*, 8.

34 Ibid., 9.

35 This first public incandescent station was in Victoria, BC, according to W.G. Richardson "Electric-Power Development," *The Canadian Encyclopedia*, http://www.thecanadianencyclopedia.com/articles/electricpower-development

36 *Woodstock Weekly Sentinel Review*, 1 April 1881, 3.

37 Philip Mozel, "The Woodstock College Observatory," *JRASC* 76 (1982): 168-80. The day of the transit, 6 December 1882, was not clear at Woodstock, though sporadic sightings of Venus on the Sun were made through breaks in the clouds.

38 Plaskett, *Plaskett-Stanley Family History*, 11.

39 Ibid., 11, 4.

40 Ibid., 12; Warren Elofson, *Somebody Else's Money: The Walrond Ranch Story* (Calgary: University of Calgary Press, 2009); P. David Green, *Visionary Veterinarian: The Remarkable Exploits of Dr. Duncan McNab McEachran* (Victoria: Anconalces, 2012).

41 Plaskett, *Plaskett-Stanley Family History*, 6; Philip M. Teigen, "The Establishment of the Montreal Veterinary College," *Canadian Veterinary Journal* 29 (February 1988): 185; Green, *Visionary Veterinarian.*

42 C.O. Pherrill, *150th Anniversary, Christ Church (Anglican) Huntingford, Ontario, 1839–1989* (n.p. 1954), 13–14.

43 Ruth Ellis, *Guide to Cemeteries and Early Churches in Oxford County* (Woodstock, ON: Ontario Genealogical Society, Oxford Branch, 1991).

44 Marriage Registers for Oxford County, AO, RG-8, Series 1-6-B, vol. 26.
45 Pherrill, *150th Anniversary*, 9, 12.
46 J.C. Farthing, *Recollections of the Right Rev. John Cragg Farthing, Bishop of Montreal, 1909–1939* (Toronto: Church House, 1945), 68.
47 JSP, "Small Dynamo," *English Mechanic and World of Science*, no. 1076 (6 November 1885): 210; no. 1081 (11 December 1885): 304.
48 UTA, 75-0026/003(05), Robert Alexander Ross file. During his subsequent career Ross acted, at one time or another, as engineering consultant to most of the larger cities in Canada and designed many important hydroelectric installations in Asia, Europe, and the United States. A founding member of National Research Council, he was its third chair, serving from August 1921 until May 1922 (Thistle, *The Inner Ring: The Early History of the National Research Council of Canada* (Toronto 1966), 111) and was awarded the Engineering Institute of Canada's highest honour, the Sir John Kennedy Medal in 1935. Ross died 23 September 1936 in Montreal.
49 C.A. Chant, "John S. Plaskett at the University of Toronto," *JRASC* 35 (1941): 412–14, and 36 (1942): 121.

2 Toronto, 1890–1903

1 Martin L. Friedland, *The University of Toronto*; Robert Craig Brown, *Arts and Science at Toronto: A History 1827-1990* (Toronto: University of Toronto Press, 2013).
2 Richard White, *The Skule Story: The University of Toronto Faculty of Applied Science and Engineering* (Toronto: University of Toronto Press, 2000), 58, shows that scientists at the university were engaged in research and publishing but the faculty members at the School of Practical Science did very little.
3 Yves Gingras, *Physics and the Rise of Scientific Research in Canada*, (Montreal and Kingston: McGill-Queen's University Press, 1991), 26.
4 Elizabeth J. Allin, *Physics at the University of Toronto, 1843–1980* (Toronto: University of Toronto Physics Department, 1981). For comparison with chemistry, see Adrian G. Brook and W.E.A. McBryde, *Historical Distillates: Chemistry at the University of Toronto since 1843* (Toronto: Dundurn, 2007), especially 40–1.
5 C.A. Chant, "Recollections of Astronomy in Canada," *JRASC* 35 (1941): 100.
6 A detailed list of equipment and the duties of the "skilled mechanician" are found in calendars of the period, including *University of Toronto Calendar for 1896–97*, 31.
7 Ari Gross, "In Praise of Small Instruments: J.S. Plaskett, the Physical Laboratory Workshop, and the Humble Resistance Box," University of

Toronto Scientific Instruments Collection, blog post (2010), available at https://utsic.escalator.utoronto.ca/home/blog/category/blogs/page/2/.

8 *Varsity* 22 (11 February 1903): 245.

9 UTA, A1974-0027/010, Chant's autobiography; see p.166 for information about projection lanterns.

10 *Toronto Globe*, 15 February 1890, 13.

11 W.J. Loudon, *Studies of Student Life*, 5: 169–70, 159–60.

12 C.A. Chant, "J.S. Plaskett at the University of Toronto," *JRASC* 35 (1941): 412–14.

13 JSP is described as "Mechanical Assistant and Superintendent of the Electrical Plant" with an annual salary of $800 in Ontario Legislative Assembly, *Sessional Papers* 25 (6) (1893), section 21, 12. The report of the building restoration fund is in 25 (9) (1893), 57. The cost of living in Canada was first calculated in 1913, so there are no exact ways to compare prices and wages prior to that date. However, to adjust for inflation, and bearing in mind that income tax was not introduced until 1917, a factor of about 30 is probably appropriate to convert to present-day dollars.

14 AO, MS935, reel 62.

15 The 1891 census for East Zorra, though supposedly conducted on 20 April 1891, lists the members of the Timothy Stanley household, including his mother, Mary, age eighty-eight. The same census for Woodstock lists Annie Plaskett and all ten children, from which I infer that JSP was there, temporarily.

16 T.S. Plaskett, *Plaskett-Stanley Family History*, 23–4.

17 BCA, T4022:56/0004, Voices of B.C., part of Series GR-3378 – Provincial Educational Media Centre, school broadcasts. JSP's brother Frank recalls Reba's "low, sweet voice."

18 The dates of birth and immigration are found in the Canadian census for 1901: http://data2.collectionscanada.ca/1901/z/z003/pdf/z000118035.pdf

19 Judy Tuck, *History of Harriston*, 25. The 1881 Canadian census gives names and ages, religious affiliation, and occupation.

20 Both families appear in the 1881 and 1891 Canadian census for Harriston. JSP's brother Tom also referred to the fact that his uncle lived in Harriston (Plaskett, *Plaskett-Stanley Family History*, 25).

21 In the 1891 census for Harriston, Rebecca H. Hemley, age twenty-three, gave her occupation as nurse. She and her three younger sisters and one younger brother were living with their parents, Alexander and Elizabeth.

The only Hemley in the city of Toronto directories of the early 1890s was "Wm. Hemley, mldr 389 King w," who made a one-year appearance in 1892.

22 AO, MS932, marriage records.

23 George Cheyne's successful entrance to Woodstock High School is noted (along with that of R.A. Ross) in the *Woodstock Review*, 2 August 1878, 1, half a year after JSP's. The Canadian census for Woodstock lists Cheyne's occupation as iron finisher in 1891 and machinist in 1901. He died of influenza and meningitis at age forty-two, on 6 July 1907.

24 Tuck, *History of Harriston*, 26.

25 U.S. Census for 1900, accessed through www.ancestry.com

26 Ontario Land Registry Office, no. 66, A73 – Lot 5, Plan 349.

27 The Plaskett's oldest son, Harry, recalled his happy childhood in the small house with the huge hickory tree, according to his biographer, W.H. McCrea, *Biogr. Mems. Fell. R. Soc.* 27 (1981): 446.

28 City of Toronto Archives, Assessment Roll for Ward 4, Division 3, 1893. JSP's salary for the first six months of 1890 was $350. Ontario Legislative Assembly, *Sessional Papers* 23 (7) (1891), 5.

29 The register, Easter reports, and yearbooks of the Church of the Redeemer are at the Anglican Archives of the Diocese of Toronto.

30 Edith's son, Peter Millman, told me of this family connection many years ago.

31 Plaskett, *Plaskett-Stanley Family History*, 23; Gayle M. Comeau-Vasilopoulos, "Oronhyatekha," *Dictionary of Canadian Biography* 13, available at http://www.biographi.ca/en/bio/oronhyatekha_13E.html.

32 *Toronto Globe*, 17 March 1898, 2.

33 Josephine A. Plaskett [first cousin of JSP], "Muskoka History: A Personal Account," *Early Canadian Life* 1, no. 5 (September 1977): 6, and 1, no. 6 (October 1977): 27.

34 Plaskett, *Plaskett-Stanley Family History*, 35.

35 According to records in the Ontario Land Registry Office, Bracebridge, JSP purchased Rock Island (West Twin Sister) and Thomas S. Plaskett purchased Fairy Island (also known as Snake Island or East Twin Sister). Each island comprised 0.8 hectares, the price of each was $100, and the deeds were registered in November 1898. Fairy Island was purchased a century later for $200 000.

36 William J. Loudon and John C. McLennan, *A Laboratory Course in Experimental Physics* (New York: Macmillan, 1895).

37 This lecture was presented to the Astronomical and Physical Society of Toronto and was written up in the *Toronto Daily Mail and Empire*,

3 December 1895. The demonstration was described, with an illustration of the apparatus, in *Canadian Engineer* 3, no. 9 (January 1896): 244. Again with Plaskett's help, Chant gave a similar lecture a couple of months later. A young reporter, William Lyon Mackenzie King, wrote it up in the illustrated supplement of the *Toronto Globe*, 15 February 1896 (Chant's autobiography, UTA, 1974-0027/010, 177–8).

38 Chant, "J.S. Plaskett at the University of Toronto," 414.

39 C.S. Beals, "John Stanley Plaskett," *JRASC* 35 (1941): 402, says JSP spoke "many times with deepest appreciation of the encouragement and inspiration given by his wife during his early efforts to complete his education as well as later in his astronomical work." Similar thoughts were echoed by R.M. Petrie, "John Stanley Plaskett, 1865–1941," *PA* 50 (1942): 3.

40 JSP "graduated from University without attending lectures," according to his brother, Plaskett, *Plaskett-Stanley Family History*, 25.

41 Details of the courses and instructors are found in university calendars. JSP's transcript is in UTA, A89-0011/74.

42 Prescribed textbooks for each course, as well as recommended reference works, are listed in the calendars; see, for example, *University of Toronto Calendar* for 1899–1900, 138. Aside from U of T's Loudon and McLennan, thirteen of the seventeen authors were British, the exceptions being Rudolf Clausius (Germany), Adolphe Ganot and Jules Jamin (France), and Mansfield Merriman (United States), although the Clausius and Ganot texts were in English editions.

43 Richard T. Glazebrook, *Light: An Elementary Text-Book, Theoretical and Practical* (Cambridge: Cambridge University Press, 1895), preface.

44 The text was Adolphe Ganot, *Elementary Treatise on Physics, Experimental and Applied, for the Use of Schools and Colleges,* edited by Arnold W. Reinold and translated by Edmund Atkinson (London: Longmans, 1893)., The uses of the spectroscope and the properties of the spectrum, including the Fraunhofer lines, are described on pp. 556–67. Evidence for the wave nature of light and diffraction gratings are found on pp. 632–5. For more information on this influential book, see Josep Simon, *Communicating Physics: The Production, Circulation and Appropriation of Ganot's Textbooks in France and England, 1851–1887* (London: Pickering and Chatto 2011.

45 William H. Wollaston, "A Method of Examining Refractive and Dispersive Powers, by Prismatic Reflection," *Phil. Trans. Roy. Soc.* 92 (1802): 365–80; Joseph Fraunhofer, "Bestimmung des Brechungs- und des Farben-Zerstreuungs - Vermögens verschiedener Glasarten, in Bezug auf die Vervollkommnung achromatischer Fernröhre," *Denkschriften der Königlichen Akademie der Wissenschaften zu München* (Memoirs of the

Royal Academy of Sciences in Munich) 5 (1814–15): 193–226. For an English translation, see Joseph Frauenhofer [*sic*], "On the Refractive and Dispersive Power of Different Species of Glass, in Reference to the Improvement of Achromatic Telescopes, with an Account of the Lines or Streaks which Cross the Spectrum," *Edinburgh Philosophical Journal* 9 (1823): 288–99 and 10 (1824): 26–39, especially the last three pages.

46 A.E. Becquerel, "Mémoire sur la constitution du spectre solaire," *Bibliothèque Universelle de Genève* 40 (1842): 341-67 cited in Malcolm S. Longair, *The Cosmic Century*, 9-10; H. Draper, 9 August 1872, Research Notebook XI, Museum of American History, Smithsonian Institution, cited by Barbara Becker, *Unravelling Starlight*, 190.

47 See, for example, another textbook in the 1890 calendars: Thomas Preston, *The Theory of Light*, 2nd ed. (London: Macmillan, 1895), 233–42.

48 See, for example, Becker *Unravelling Starlight*, 4

49 See, for example, Reinold, *Elementary Treatise*, 566.

50 See www-history.mcs.st-and.ac.uk/history/Biographies/Doppler.html.

51 Becker, *Unravelling Starlight*, 109.

52 The Doppler formula is $(\lambda-\lambda_0)/\lambda_0 = v/c$ where λ and λ_0 are the shifted and unshifted wavelengths, and v and c are the velocity of the source and of light, respectively. For example, if a star is approaching us with a typical speed of 30 km/s, $v/c = 0.0001$. Therefore a spectral line at 500 nm will be observed at 499.95 nm.

53 Becker, *Unravelling Starlight*, 115–20.

54 Simon Newcomb and Edward S. Holden, *Astronomy for High Schools and Colleges*, 6th ed. (New York: Holt, 1889).

55 For information about the Toronto observatory, see Morley Thomas, *The Beginnings of Canadian Meteorology*, Malcolm M. Thomson, *The Beginning of the Long Dash*, 6–7, and Brian Beattie, "The 6-inch Cooke Refractor in Toronto," *JRASC* 76 (1982): 109–28.

56 The arrangement was not very satisfactory. See the *Toronto Globe*, 9 June 1853; 3 January 1854, 12; 8 June 1855, 2; 23 April 1860, 4; 28 January 1863, 2.

57 "The correct time for Ontario is given from this observatory, and for seven years the time has been given to Toronto by striking all the fire alarm bells at a fixed instant, 11:55 a.m." "Notes from the Capital: Marine and Fisheries Report," *Toronto Globe*, 20 February 1879, 1.

58 J.C. McLennan, "Electrical Conductivity in Gases Traversed by Cathode Rays," *Phil. Trans. Roy. Soc.*, series A, 195 (1900): 49–77.

59 McLennan acknowledged Plaskett's assistance in his paper "Induced Radioactivity Excited in Air at the Foot of Waterfalls," *Philosophical Magazine* 5, Series 6, (1903): 419–28.

60 LAC, RG48, vol. 34, file 676.13, JSP to W.F. King, 19 February 1916.
61 Plaskett, *Plaskett-Stanley Family History*; Hugh H. Langton, *Sir John Cunningham McLennan*.
62 Chant, "J.S. Plaskett at the University of Toronto," 414; UTA, A89-0011/74, JSP's transcript confirms his excellent marks but gives very little information on the courses he studied, listing only maths and physics in third year and physics in fourth year.
63 All JSP papers and talks are listed in Appendix A and B.
64 H. Spencer Jones, *Obit. Not. Fell. R. Soc.* 4 (1942): 68 alludes to McLennan's attitude; LAC, RG48, vol. 34, file 676.13, JSP to W.F. King, 19 February 1916, said that there was no chance of advancement.
65 UTA, B72-0031/014(03), draft letter from J. Loudon to O.M. Stewart, University of Missouri, 11 August 1903.
66 LAC, MG 28 I 181, Toronto Camera Club fonds, vols. 1 and 2, give information about membership, general activities of the club, exhibitions, and so on. I have not found the actual date that JSP joined, but the earliest occasion on which he seemed to attend was the club's eleventh annual meeting, 6 November 1899.
67 Andrew Oliver, *The First Hundred Years: an Historical Portrait of the Toronto Camera Club* (Gormley, ON: Aurora Nature Photography and Pub., 1988).
68 Lilly Koltun, ed., *Private Realms of Light*. In the chapter "Art Ascendant, 1900–1914," Koltun credits JSP with the first-known attempts at colour photography in Canada (p. 59) and documents some of his specific activities with the Toronto Camera Club (p. 116).
69 JSP's award was noted in *American Amateur Photographer* 13 (1901): 40.
70 In the preface to *The University of Toronto and Its Colleges*, the author (W.J.A.) specifically mentioned "Dr. A.H. Abbott and Mr. J.S. Plaskett, for services in connection with the views of the buildings." See also JSP's photo of cannons on the campus in *UT Monthly* 1 (December 1900): 120, available online at https://archive.org/details/universityoftoro01univuoft. *The Photographic Art Club of Ottawa: Sixth Exhibition* (1917) lists fourteen prints by JSP, two of them being the main door and the rotunda of University College (Arnprior Archives).
71 JSP, "Landscape Photography," *Acta Victoriana* 26 (1902): 162–8.
72 See Koltun, *Private Realms of Light*, 59.
73 J.C. Maxwell, "Experiments on Colour, as Perceived by the Eye, with Remarks on Colour-Blindness," *Transactions of the Royal Society of Edinburgh* 21 (1855): 275–98.
74 TFL, RCI Papers, MS 193, box 39, Minutes 1894–1910. JSP lectured to sixty-five members of the Canadian Institute on 16 February 1901 on

the topic "Colour Values in Monochrome." Perhaps he also spoke to the Toronto Camera Club, for at their annual meeting in November 1901, the president commended Plaskett and Carl Lehman for their lectures on colour work (LAC, MG 28 I 181, Toronto Camera Club fonds). The club included three of Plaskett's three-colour slides in its submission to the American Lantern Slide Interchange in 1902–3.

75 I assume, given the identical title "Photography in Natural Colours," that the lectures to the Toronto Camera Club on 24 March 1902 and the Canadian Institute on 15 March 1902 were much the same. The institute published the lecture in *Transactions of the Canadian Institute* 7 (1902): 371–90.

76 *Varsity* 22 (11 February 1903): 245; Fraser's lecture notes, including indications of the slides that were shown and the dates of presentation, are in UTA, B1995-0044/004. The slides themselves have not been found.

77 The eight occasions with high attendance were George M. Dawson speaking on Canadian mines (200 in attendance), Chant on wireless telegraphy (250), the fiftieth-anniversary conversazione (1000), all in 1899, then a very large joint meeting with the Canadian Forestry Association in 1901, Joseph-Elzéar Bernier on his plans for a Canadian polar expedition (200), also in 1901, and three illustrated lectures in 1902: one on the relation of geology to scenery (150), another on exploration in the Rocky Mountains (160), and Plaskett on colour photography (120). TFL, RCI Papers, MS 193, box 39, Minutes 1894–1910.

78 Ibid., minutes for 1 February 1902. Abbott's 1898 paper "Recent Views on Colour" was published in *Proceedings of the Canadian Institute*, n.s. 1 (1895–8): 107.

79 *Transactions of the Canadian Institute* 7 (1904): 371–90. In this paper, JSP acknowledged Abbott's assistance.

80 TFL, RCI Papers, MS 193, box 20, Council minutes, 1887–1908, for 16 December 1899, 6 March 1901, and 25 April 1902.

81 UTA, P78-0737 through P78-0740, *Papers Read Before the Mathematical and Physical Society of the Toronto University* (Toronto n.d.).

82 *UT Monthly* 2 (December 1901): 84.

83 *Toronto Globe*, 14 January 1903, 9.

84 Ernest J. Chambers, *The Canadian Marine; a History of the Department of Marine and Fisheries* (Toronto, [1905]); see also Thomas, *The Beginnings of Canadian Meteorology* and Thomson, *The Beginning of the Long Dash*.

85 For the history of the Toronto observatory and its telescope, see Beattie, "The 6-inch Cooke Refractor in Toronto"; for the Quebec observatory, see Paulette Smith-Roy, *L'Observatoire astronomique de Québec: Histoire*

et reminiscences, 1850–1936 (Quebec: Commission des Champs de Bataille Nationaux, 1983); R. Jarrell, "Origins of Canadian Government Astronomy*JRASC* 69 (1975): 77–85.

86 R.M. Stewart, "The Early History of Astronomical Activity in the Canadian Public Service," *JRASC* 65 (1971): 206–16, adapted from a report he wrote in 1930: "Activities of the Astronomical Branch, Department of the Interior" and used by John H. Hodgson, *The Heavens Above*, 4–8.

87 Order-in-council, dated 19 June 1889, LAC, available at http://www.collectionscanada.gc.ca/databases/orders/

88 J.H. Taylor, *Ottawa*, 148.

89 Hodgson, *Heavens Above*, 2–23, describes the background to, and founding of the Dominion Observatory, as well as the rivalry between King and Klotz, and on pages 10–12 quotes the memorandum in full.

90 "The new astronomy" was used by the pioneering American solar astrophysicist, Samuel P. Langley, as the title of his popular book first published in 1884.

91 Hodgson, *Heavens Above*, 10.

92 LAC, MG30, series B-13, Klotz diaries, 23 May 1901, quoted by Hodgson, *Heavens Above*, 14.

93 Hodgson, *Heavens Above*, 12–15; also LAC, MG30, series B-13, Klotz diaries.

94 RASCA, W.F. King to his relation, Robert King, a member of the Toronto Astronomical Society, 24 June 1901. The letter, read to the society the next day (*Transactions of the TAS* for 1901, 69), referred in detail to the equipment that W.F. King had ordered from Warner & Swasey and was inserted in a copy of its publication entitled *A Few Astronomical Instruments*.

95 See John Lankford, *American Astronomy*, especially chap. 3, "The New Astronomy: Identity and Conflict."

96 The source of the two quotes attributed to Fisher and Mulock are in a report of a visit by newspapermen to the observatory, *Saint John Daily Sun*, 4 May 1905.

97 Canada, *House of Commons Debates* 1901 (2), 5783, 21 May 1901.

98 Warner & Swasey's invoice totaling $14 637 for the telescope and auxiliary equipment refers to a letter of 14 June 1901: CWRU, Warner & Swasey Collection, box 26, folder 7. The original estimated cost of the building, $16 075, appears in the 1898 memorandum prepared by King for Sifton. For the approval of $75 000 for the building, see LAC Orders-in-Council, dated 12 June 1902, at http://www.collectionscanada.gc.ca/databases/orders/. However, during a visit to the completed

observatory in 1905, members the press were told the cost of the building was $125 000 and of the telescope was $14 625 (*Ottawa Evening Journal*, 1 May 1905, 6).

99 Annual Reports of the Department of the Interior, Canada, *Sessional Papers of Parliament*, 36, (no. 10) (1902), pp. xxxvi, 1, for the year ended 30 June 1901,; 37 (no. 10) (1903), p. i for the twelve months ended 30 June 1902; vol. 38 (no. 10) (1904), p. liii for the twelve months ended 30 June 1903.

100 Reference to the "Royal Observatory" are found in Annual Reports of the Department of Public Works, Report of the Chief Architect, Canada, *Sessional Papers of Parliament*, vol. 37 (no. 19), each year from 1902 to 1906.

101 W.J. Loudon, *Studies of Student Life*, 4: 37–8, 171 ff., 227–8, wrote at length about Billy King, in particular his work on the mathematical theory of solitaire.

102 LAC, RG48, vol. 34, file 676.13, JSP to W.F. King, 31 October 1902.

103 "One of Plaskett's teachers, Alfred Baker, wrote to Sifton, who in turn told King that he 'would favour this man's appointment'"; Richard A. Jarrell, *The Cold Light of Dawn*, 55, citing LAC, MG27 II, D15, vol. 250, letter from Sifton to King, 19 February 1903. The *Manitoba Free Press* (Winnipeg), 17 March 1903, 7, reported from Ottawa that the Commons sat for one hour and a half on the previous day, when "John S. Plaskett, the electrician and mechanical expert of Toronto university [was] appointed by Hon. Mr. Sifton mechanical superintendent of the new astronomical observatory at Ottawa." A similar announcement in the *Toronto Globe*, 16 March 1903, 4, added that JSP would "undertake work in spectrum analysis and stellar spectroscopy" and went on to describe his qualifications.

104 LAC, RG48, vol. 34, file 676.13, JSP to W.F. King, 24 April 1903.

105 Ibid., W.F. King to Acting Deputy Minister T.G. Rothwell, 5 March 1903; JSP first appears on the payroll as a mechanician earning $1100 per annum, see Canada, *Sessional Papers of Parliament*, vol. 39, Auditor General's Report for 1903–4, vol. 1, part 1, L6.

106 JSP's expenses for inspecting boundary monuments were paid in 1904 for May–June, August, and November–December and for unspecified dates in 1905–6; see Canada, *Sessional Papers of Parliament*, vol. 39, Auditor General's Reports for 1903–4, L58; for 1904–5, L56; and for 1905–6, L51.

107 Hodgson, *Heavens Above*, 23n.66.

108 Much earlier, Allan F. Miller, an important figure in the astronomical society in Toronto, reported using a spectroscope to examine the bright line spectra of the Ring Nebula and the Orion Nebula: *The English Mechanic and World of Science*, 13 February 1885): 518. As I described elsewhere, he continued to do pioneering work with a spectroscope

attached to his 10 cm refractor over the next decade, but his efforts cannot be described as systematic: Broughton, *Looking Up*, 137. Another pioneer was Edward D. Ashe, director of the Quebec observatory, who did some solar research in the 1860s. In the *Dictionary of Canadian Biography*, vol. 12, the article on Ashe, Richard Jarrell calls Ashe "Canada's first astrophysicist."

109 *Toronto Star*, 20 October 1897, 2; 21 July 1903, 6; 21 June 1906; *UT Monthly* 4, no. 9 (June–July 1904): 252; for petitions from Toronto to the federal government, see http://www.collectionscanada.gc.ca/databases/orders/, numbers 1906-0480, 1.

110 *UT Monthly* 4, no. 9 (June–July 1904): 276.

111 UTA, Senate minutes 7 (1900), 276.

112 LAC, MG30, series B-13, Klotz diaries, 10 June 1904.

113 *UT Monthly* 4, no. 9 (June–July 1904): 278–9.

114 C.A. Chant, *Varsity* 24 (16 February 1905): 278.

115 Friedland, *University of Toronto*, chap. 18; Charles W. Humphries, *"Honest Enough to Be Bold": The Life and Times of Sir James Pliny Whitney* (Toronto: University of Toronto Press, 1985), 126–8.

116 C.A. Chant, *Astronomy in the University of Toronto*, 5.

3 Ottawa Advancement, 1903–1907

1 Hodgson, *Heavens Above*, 8.

2 W.H. McCrea, "Harry Hemley Plaskett," *Biogr. Mems. Fell. R. Soc.* 27 (1981): 446; AO, RG2–256, vols. 256 and 458, Ottawa Model School, shows that Harry was an above-average, but not outstanding, student. He passed his entrance exams in June 1906 but did not go to university until June 1912, suggesting that he spent six years in high school instead of the normal five. His high school records are not available.

3 OA, Birth records, Registration No. 011941, MS 929, reel 167; Anglican Diocese of Ottawa Archives, Register 377, p. 7.

4 C.H. Little, *All Saints' Church: Sandy Hill, Ottawa – The First Ninety Years* (Ottawa: Anglican Book Society, 1990).

5 Anglican Diocese of Ottawa Archives, Christ Church, Sandy Hill, minute book, 1900–20.

6 JSP was rector's warden at St Matthias, Hintonburg, according to the *Ottawa Journal*, 5 April 1910, 10. The church, at 17 Fairmont Avenue, was about a kilometer north of the Plasketts' home. The church hall later served Crawley Films as a studio and is now used by the Orpheus Society for musical theatre productions.

7 Ottawa City Directories, 1904–16 and 1911 census.
8 W.F. King, *Report … Ending June 30, 1905*, 3.
9 Ibid., Appendix 1, 13.
10 Hodgson, *Heavens Above*, 21, citing Klotz's diary.
11 Canada, *Sessional Papers*, vol. 41, Auditor General's Report for 1905–6, vol. 1, part 1, L3.
12 Otto J. Klotz, "Transpacific Longitudes between Canada and Australia and New Zealand, Executed during the Years 1903 and 1904," *Report … Ending June 30, 1905*, 35–197.
13 JSP, "Observatory Building and Instrumental Equipment," *Report … Ending June 30, 1905*, 198–212.
14 John Macara's wife was Emma King, the sister of W.F. King. I believe he was the John Macara (1855–1926) who identified himself as a civil servant in Ottawa in the 1901 census.
15 *Report … Ending June 30, 1906*, 44.
16 Hodgson, *Heavens Above*, 27.
17 JSP, "Observatory Building and Instrumental Equipment," *Report … Ending June 30, 1905*, 198–212; Some of the original equipment, including the governor used in the drive clock, the filar micrometer, photometer, and the objective lens are part of the collection of the Canadian Science and Technology Museum. The museum is also home to the Helen Sawyer Hogg Observatory, which houses the 38 cm telescope with a replacement objective installed in 1958.
18 RASCA, W.F. King to Robert King, 24 June 1901, inserted in Warner & Swasey's publication entitled *A Few Astronomical Instruments* (1900). W.F. King says, "Plate IV shews the general pattern, except that mine will be 15 inch, instead of 12."
19 J.B. Hearnshaw, *The Analysis of Starlight* provides a valuable historical interpretation of the connection between astronomy, spectroscopy, and the technical developments in both areas.
20 Matthew Stanley, "Spectroscopy – So What?" *JAHH* 13 (2010): 105-11.
21 A starting point in understanding how a binary star's orbital velocity is related to its mass is Kepler's third law: $P^2 (m_1 + m_2) = a^3$ where P is the period or the time required for one revolution of the binary system in years, $m_1 + m_2$ is the sum of the stars' masses (in solar masses) and a is the separation of the components in astronomical units (i.e., the mean distance from Earth to the Sun). Try substituting 1 for a and for each m, to get $P = 0.7$ years and then find the circular orbital velocity to be 21 km/s.
22 Victor Gaizauskas, "The Grand Schism in Canadian Astronomy, (Part 2)," *JRASC* 106 (2012): 195.

23 JSP, "Total Solar Eclipse," *Report … Ending June 30, 1905*, 212.
24 Edward D. Ashe, "The Proceedings of the Canadian Eclipse Party," *Transactions, Literary and Historical Society of Quebec* 4 (1870): 1–23.
25 Lindsay joined the Yerkes expedition and Lumsden joined the team from Lick Observatory, *Toronto Daily Star*, 28 May 1900, 1.
26 W.F. King, *Report … Ending June 30, 1905*, 6–7 quotes the letter from the secretary of the RASC to the prime minister.
27 The grant, dated 19 December 1904, is in www.collectionscanada.gc.ca/databases/orders/, 1904–2281. For Plaskett's report (dated 31 October 1905) on the total solar eclipse, see *Report … Ending June 30, 1905*, 213–35 (including 10 figures). Whether Labrador was part of Quebec (and therefore part of Canada) or Newfoundland was not resolved until 1917.
28 Jeffrey Crelinsten, *Einstein's Jury* 51.
29 W.F. King, *Report … Ending June 30, 1905*, 7.
30 Canada, *Sessional Papers of Parliament*, vol. 41, Auditor General's Report for 1905–6, vol. 1, part 1, L46–L47 lists nineteen people on the eclipse expedition along with their expenses that were reimbursed by the government. The total cost to the government was $9356 including $5490 for the return trip from Quebec to Labrador, but not including the coelostat.
31 A.T. DeLury, "The Eclipse Expedition to Labrador, August, 1905," *RASC Transactions for 1905*, 57, lists twenty-three people, but he names a total of thirty-one elsewhere in the article and in the accompanying group photo. One of the party, J.A. Russell, was Annie Maunder's brother, according to an account in F.W. Levander, ed., *The Total Solar Eclipse of 1905* (London: Eyre and Spottiswoode for the British Astronomical Association, 1906).
32 JSP described the instruments and proposed observations in *Report … Ending June 30, 1905*, 213 ff.; King described the equipment and plans of the others in the same report, 8–10.
33 "The Eclipse Expedition," *Toronto Globe*, 8 March 1905, 11 and *RASC Transactions for 1905*, 26.
34 For Plaskett's work and salary, see *Report … Ending June 30, 1906*, 44 and Canada, *Sessional Papers of Parliament*, vol. 41, Auditor General's Report for 1905–6, vol. 1, part 1, L3, L7–L8 and L51. JSP seems to have shed the more general responsibilities even before a series of steps, beginning in 1909, separated the surveying from the astronomical work.
35 W.F. King, *Report … Ending June 30, 1905*, 7.
36 Hodgson, *Heavens Above*, 21, quoting Klotz's diary. This seems to have been a very good salary. According to Bartusiak, *The Day We Found the Universe*, 117 and 181, Harlow Shapley, just completing his PhD at Princeton, was offered $1080 plus free board at Mount Wilson in 1912;

Edwin Hubble, having earned his PhD and returned from war service, got $1500 in 1919. An inflation factor of about 25 is appropriate to convert to today's dollars (see www.measuringworth.com/ppowerus for U.S. figures and http://data.bls.gov/cgi-binnl/cpicalc.pp for Canada) but remember that Canada did not introduce income tax until 1917.

37 John Lankford, *American Astronomy*, 76–9.

38 LAC, RG48, vol. 34, file 676.13, JSP openly expressed his admiration for King in a letter to him, 19 February 1916, and in his obituary of King, which appeared in both *JRASC* 10 (1916): 267 and *Obs* 39 (1916): 339.

39 JSP, "Plates and Filters for Monochromatic and Three-Color Photography of the Corona," *RASC Transactions for 1905*, 89.

40 Ibid.

41 The green coronal line had been noticed as early as 1869, but its origin in highly ionized iron was not established until about 1940.

42 JSP, *Report ... Ending June 30, 1905*, 213–35.

43 Plaskett called it a coelostat, which strictly speaking tracks the stars, rather than a heliostat, which is designed specifically to follow the Sun. However, for the short duration of a total eclipse, the difference in tracking rates is unimportant.

44 The details of all the eclipse expenses, including the coelostat, are found in Canada, *Sessional Papers of Parliament*, vol. 41, Auditor General's Report for 1905–6, vol. 1, part 1, L46–L49.

45 UTA, A1974-0027/010, C.A. Chant's autobiography, 300.

46 The *King Edward* was about 46 m long by 6.7 m wide, according to C.A. Chant, "The Labrador Eclipse Expedition," *Acta Victoriana* 29 (1905): 137. The nature of the accommodation was described in Anthony Kinder, "E.W. Maunder and the Labrador 1905 Solar Eclipse," *Journal of the British Astronomical Association* 111 (2001): 272.

47 At least five versions of the album exist – at the LAC, at the Canada Science and Technology Museum, at the Hamilton Public Library (digitized at http://preview.hpl.ca:8080/Sites/), at the Musées de la Civilisation Québec, and King's album now at the RASC. The last, available online at http://www.rasc.ca/labrador-expedition-1905, is the source for the four photos reproduced here.

48 Recalled by Frank Plaskett on a CBC recording, BCA, T4022:56/0004, Voices of B.C., part of Series GR-3378 – Provincial Educational Media Centre, school broadcasts.

49 Letters written from the eclipse camp by E.W. Maunder to his children were edited and published by Kinder, "E.W. Maunder and the Labrador 1905 Solar Eclipse," 272.

50 D.B. Marsh and G. Parry Jenkins, "With the Canadian Government Eclipse Expedition to Labrador, 1905," *Journal and Proceedings of the Hamilton Association* 22 (1905–6): 50–1.
51 C.A. Chant, "W.F. King, Late Chief Astronomer for Canada," *PA* 24 (1916): 337.
52 JSP, "Report on Solar Eclipse Expedition," *Report ... Ending June 30, 1905*, 213–35. This quote and those in the next two paragraphs are found on 234–5. It is an anomaly that the eclipse was covered in a report that supposedly ended two months before the eclipse occurred.
53 Ibid.
54 Ibid.
55 John C. Pearson and Wayne Orchiston, "The 40-Foot Solar Eclipse Camera of the Lick Observatory," *JAHH* 11, no. 1 (2008): 25–37.
56 UTA, A1974-0027/010, C.A. Chant's autobiography, 317.
57 The title of C.A. Chant's talk, "The Expedition to Labrador to Observe the Total Solar Eclipse of August 30, 1905," is noted in *RASC Transactions for 1905*, 26.
58 *Register of the University of Toronto for the Year 1920*, lists all graduates up to that time.
59 Because his account, dated 31 October 1905, reports on the eclipse of August 1905, it seems out of place in the *Report ... Ending June 30, 1905*, but that is where it was published. It immediately followed his report, also dated 31 October 1905, on the observatory building and instrumental equipment.
60 *Report ... Ending June 30, 1906*, 43.
61 This description is based on Plaskett's detailed description in *Report ... Ending June 30, 1905*, 204–5.
62 Plaskett provides details for reducing the measurements to wavelength and removing the effects of the Earth's motion in *Report ... Ending June 30, 1906*, 64.
63 H.F.G., "The Press Men as Star Gazers," *Toronto Star*, 2 May 1905, 2. LAC, MG30, vol. 3, Klotz diaries, 29 April 1905 is the source of the information that King provided mineral water and scotch. A somewhat more sober account of the press gallery's visit is in *Toronto Globe*, 1 May 1905, 1.
64 LAC, MG30, vol. 3, Klotz diaries, 16 March 1906; for a photo of Lady Laurier in a car at the observatory, see Hodgson, *The Heavens Above*, 63.
65 *Report ... Ending June 30, 1906*, 43.
66 *Ottawa Citizen*, 10 February 1908, 13.
67 H.D. Curtis, "Astronomical Discovery," in *Adolfo Stahl Lectures*, edited by R.G. Aitken, 111.
68 The details in the ensuing paragraphs are all drawn from the *Reports* ... for 1906 and subsequent years.

69 LAC, RG48, vol. 48, file 5–12, correspondence in 1906 between JSP and Hale, and vol. 49, file 6–3, between JSP and C.G. Abbott regarding coelostat design.

70 Government approval in order-in-council, 25 March 1907, www.collectionscanada.gc.ca/databases/.

71 JSP, "The Coelostat Telescope," *Report ... Ending March 31, 1909*, 207–9.

72 E-mail from V. Gaizauskas to Paul Herzberg, 22 August 2006, quoted in Paul A. Herzberg, *Luise Herzberg*, 98.

73 JSP, *Report ... Ending June 30, 1906*, 50.

74 JSP, *Report... Ending June 30, 1906*, 51–60.

75 LAC, RG48, vol. 49, file 6-13, correspondence between JSP and E.B. Frost; UCSC, Directors' files, UA 36 Ser.03, correspondence between JSP and W.W. Campbell. See also Crelinsten, *Einstein's Jury*, 14, who points out that "Campbell's accurate and systematic program on radial velocities inspired many other observatory directors to put similar programs on their research agendas. Campbell's technical and organizational experience in the area put him in demand as a consultant, and his reputation and that of his Observatory spread rapidly."

76 The details of JSP's visit to U.S. observatories comes from *Report ... Ending March 31, 1907*, 47–73.

77 For an excellent description and simulation of good and bad seeing, see http://www.weatheroffice.gc.ca/astro/seeing_e.html.

78 The "Constitution of the International Union for Co-operation in Solar Research," *Trans. IUCSR* 1 (1906): 242–7.

79 Anthony Misch and William Sheehan, "Pioneering Telescope Turns 100," *S&T* (November 2008): 38; for a more detailed account, see Allan Sandage, *The Mount Wilson Observatory*, 159–69.

80 LOA, Copybook 6; in a letter to P. Lowell, September 1906, just after JSP's visit to the Lowell Observatory, V. Slipher says he hopes to obtain some new chemicals from Germany.

81 E.C. Pickering, "On the Spectrum of Zeta Ursae Majoris," *American Journal of Science* (January 1890): 46, a paper read at the Philadelphia meeting of the NAS, 13 November 1889. The pejorative term "harem" may give the wrong impression of Pickering, who was described by Annie Cannon as treating his women assistants as "equals in the astronomical world" and with "as full courtesy as if he were meeting them at a social gathering" (Jones and Boyd, *Harvard College Observatory*, 390).

82 JSP, "The Astronomical and Astrophysical Society of America," *JRASC* 2 (1908): 260.

83 UCSC, Directors' files, UA 36 Ser.03, W.W. Campbell to JSP, 10 October 1906.

84 UTA, A1974-0027/010, C.A. Chant's autobiography, chap. 4, "A Summer at the Lick Observatory," pp. 319–56; Heber D. Curtis had also spent summer vacations at Lick as a volunteer assistant.

85 UCSC, Directors' files, UA 36 Ser.03, JSP to W.W. Campbell, 18 June 1907.

86 UCSC, Directors' files, UA 36 Ser.03, JSP to W.W. Campbell, 2 August 1907.

87 UCSC, Directors' files, UA 36 Ser.03, W.W. Campbell to JSP, 9 August 1907, Chant to R.G. Aitken, 20 June 1930: "The summer I spent on the summit [of Mount Hamilton] is still a lively memory, and I hope Mrs. Chant and I may find ourselves there again some time in the future."

88 Martin L. Friedland, *The University of Toronto*, 218–19.

89 UCSC, Directors' files, UA 36 Ser.03, Chant to W.W. Campbell, 10 March 1910.

4 The Sun and the Stars, 1906–1911

1 Most of the technical details found in this chapter can be found in the annual *Reports of the Chief Astronomer* for the years 1906–11.

2 JSP lectured to the RASC Ottawa Centre on 28 May 1908; his talk was published as "The design of Spectrographs for Radial Velocity Determinations," *JRASC* 3 (June 1909): 190–209.

3 John Hearnshaw, *Astronomical Spectrographs and Their History*, 14–17.

4 JSP, "Instruments and Astrophysical Work," Appendix 2 to *Report … Ending June 30, 1906*, 60. JSP summarized his experiments and modifications in "Adapting a Universal Spectroscope for Radial Velocity Determinations," *JRASC* 1 (April 1907): 104–21.

5 JSP, "The Character of the Star Image in Spectrographic Work," *ApJ* 25 (April 1907): 195–217 and *ApJ* 27 (March 1908): 139–51. The first part was read at the RASC Ottawa Centre in January 1907 and was published with some minor changes in *JRASC* 1 (1907): 297–322 and as Appendix A to *Report … Ending March 31, 1907*, 187. JSP finished writing the second installment in December 1907.

6 JSP, *JRASC* 1 (1907): 322. Not everyone was thrilled with the work of Brashear and Company; W.W. Campbell had been frustrated in 1902 by their bumbling with a 94 cm mirror intended for the Lick Observatory in Chile: D. Osterbrock, *Pauper and Prince*, 87.

7 This was at a dispersion of 10 Angstroms per mm.

8 JSP, "The Star Image in Spectrographic Work II," *Ap J* 27 (1908): 146.

9 LOA, E.B. Frost to V. Slipher, 5 October 1908.
10 UCSC, Directors' files, UA 36 Ser.03, JSP to W.W. Campbell, 18 June 1907.
11 For the information in this paragraph, see JSP, "The Ottawa Spectrographs," *JRASC* 3 (1909): 287–305.
12 JSP, *Report … Ending March 31, 1909*, 148.
13 *Optical Society of America Centennial History Book*, chap. 1 draft, available at www.osa.org/osaorg/media/osa.media/History/Prehistory-of-the-Optical-Society-of-America.pdf, 7.
14 John Hearnshaw, *Astronomical Spectrographs and Their History*, 16–18.
15 JSP, "The Ottawa Spectrographs," *JRASC* 3 (1909): 287.
16 The probable error in radial velocities was 0.5 km/s at a dispersion of 10 Angstroms per mm [A/mm], and 0.7 km/s at 33 A/mm, W.F. King, "The Dominion Astronomical Observatory," *Proc. Trans. RSC* 4 (1910): Appendix D, lxxiii–lxxiv.
17 JSP first wrote about the effect of slit width in *Report … Ending March 31 1907*, 170 and *Report … Ending March 31 1908*, 86. Publications on this subject included *ApJ* 28 (November 1908): 259–65 and *Proc. Trans. RSC* 3 (1909): 209–14. They were all summarized and brought up to date in *JRASC* 4 (October 1910): 333–44.
18 JSP, *Report … Ending March 31, 1909*, 147.
19 JSP, "Camera Objectives for Spectrographs," *ApJ* 29 (May 1909): 290–300.
20 JSP, "The Collimation of the Correcting Lens," *ApJ* 32 (October 1910): 243–8.
21 LOA, correspondence between V. Slipher and J.B. McDowell, December 1912.
22 JSP, *Report … Ending March 31, 1907*, 46.
23 C.S. Beals, "Ralph Emerson DeLury," *MNRAS* 117 (1957): 251–2.
24 JSP, *Report … Ending March 31, 1908*, 67, 258.
25 JSP, "Astrophysical Work," *Report … Ending March 31, 1911*, section on solar research, esp. pp. 116–18, and R.E. DeLury, "Solar Physics," ibid., 254–6.
26 Walter Adams at Mount Wilson also experienced trouble with gratings from Michelson's lab (HL, W.S. Adams to JSP, 29 June 1911).
27 LAC, RG 48, vol. 48, file 5–12, JSP to C.E. St. John, 21 January 1919.
28 JSP and R.E. DeLury, "The Solar Rotation in 1911," *ApJ* 37 (1913): 73–104.
29 UCSC, Directors' files, UA 36 Ser.03, JSP to W.W. Campbell, 15 April 1907.
30 JSP, "The Spectrum of Mira Ceti," *JRASC* 1 (1907): 45–59.
31 UCSC, Directors' files, UA 36 Ser.03, W.W. Campbell to JSP, 16 March 1907.
32 JSP, "Astrophysical Work," *Report … Ending June 30, 1906*, 61; these figures are borne out in W.W. Campbell and Heber D. Curtis, "First Catalogue of Spectroscopic Binaries," *LicOB* 3 (1905): 136.

33 JSP, "Astrophysical Work," *Report … Ending March 31, 1909*, 177.
34 JSP, "Astrophysics" in "Report of the Dominion Observatory," *JRASC* 1 (1907): 372 and W.F. King, "Determination of the Orbits of Spectroscopic Binaries," *ApJ* 27 (1908): 125.
35 LAC, RG 48, vol. 49, file 6–13, E.B. Frost to JSP, 19 June 1906.
36 W.E. Harper, "The Spectroscopic Binary α Draconis," *JRASC* 1 (1907): 237-45.
37 JSP, "The Spectrograph," in "Notes from the Dominion Observatory," *JRASC* 1 (1907): 69.
38 JSP and W.E. Harper, "The Spectroscopic Binary Iota Orionis," *ApJ* 27 (1908): 272-79.
39 F. Schlesinger, "The Determination of the Orbit of a Spectroscopic Binary by the Method of Least Squares," *Publications of the Allegheny Observatory* 1 (1910): 33-44 (written 25 July 1908).
40 W.F. King, "Determination of the Orbits of Spectroscopic Binaries," *ApJ* 27 (1908): 125–38.
41 LAC, RG48, vol. 47, file 4–14, F. Schlesinger to JSP, 1 June, 19 June, and 17 August 1908.
42 UPA, 5/1 Allegheny Observatory, box 17g, folder 33, JSP to F. Schlesinger, 9 June 1908.
43 JSP, "The Orbit of ι Orionis," *ApJ* 28 (1908): 274–7; JSP and W.E. Harper, "Two Curiously Similar Spectroscopic Binaries," *ApJ* 30 (1909): 373–82. (The second one was BD −1° 1004 = HR 1952.)
44 To cite but two examples: J.M. Pittard and E.R. Parkin, "3D Models of Radiatively Driven Colliding Winds in Massive O + O Star Binaries – III. Thermal X-ray Emission," *MNRAS* 403 (2010): 1657–83 and S.V. Marchenko et al., "Coordinated Monitoring of the Eccentric O-star Binary Iota Orionis: Optical Spectroscopy and Photometry," *MNRAS* 317 (2000): 333–42.
45 J.R. Percy, "Joseph Miller Barr Revisited," *JRASC* 109 (2015): 270.
46 J. Miller Barr, "The Orbits and 'Velocity Curves' of Spectroscopic Binaries," *JRASC* 2 (1908): 70–5 is followed by notes on 75–7 by W.F. King, JSP, and a response by Barr on 78–81.
47 UPA, 5/1 Allegheny Observatory, box 17g, folder 33, JSP to F. Schlesinger, 9 June 1908: "[Barr] jumped severely with Dr. King and myself for daring to criticize his hypothesis."
48 H.A. Abt, "The Spectral-type Limits of the Barr Effect," *PASP* 121 (2009): 811–13.
49 Robert Fritzius has annotated Barr's paper, at www.datasync.com/~rsf1/barr1908.htm.
50 JSP, "Astrophysical Work," *Report … Ending March 31, 1911*, 98–116.

51 W.F. King, *Report ... Ending March 31, 1911*, 5.
52 W.F. King noted that JSP had photographed Comet Daniel on 20 July 1907, *AN* 175 (1907): 339. The photo was used as the frontispiece to *JRASC* 1 (August 1907).
53 J.H. Taylor, *Ottawa*, 133.
54 For a recent examination of the Science Academy of the RSC, see Trevor H. Levere, "The Most Select and the Most Democratic: A Century of Science in the Royal Society of Canada," *Scientia Canadensis* 20 (1996): 3–99. (The sections of the RSC were renamed academies in 1975.)
55 The *Proceedings* of the RSC, issued annually, included lists of fellows, of those attending the meeting, and of those elected to fellowship as well as general business of the society; the *Transactions* included the papers that were presented to each section of the society; both were published together.
56 LAC, MG30, vol. 3, Klotz diaries, 10 May 1910.
57 For an overview of the society and its centres, see Broughton, *Looking Up*.
58 "The Ottawa Section of the Royal Astronomical Society of Canada," *JRASC* 1 (1907): 60–3.
59 W.F. King, "Astronomy as a Science," *JRASC* 1 (1907): 22–37.
60 "The Optics of the Telescope" was summarized in *JRASC* 1 (1907): 131; "The Star Image in Spectrographic Work" was published in *JRASC* 1 (1907): 297–322, with a summary of part 2 in *JRASC* 1 (1907): 365–6.
61 JSP spoke to the RASC in Toronto on spectroscopic work at the DO on 5 March 1907 (*JRASC* 1 (1907): 66) and on the optics of the telescope on 16 June 1908 (*JRASC* 2 (1908): 211).
62 LAC, MG30, vol. 3, Klotz diaries, 8 February 1908.
63 JSP gave three successive evening lectures to the Ottawa Centre in 1910. On 24 February, 24 March, and 21 April his topics were the optics of the telescope and spectroscope, the constitution and radial motions of the stars, and stellar evolution and world building, respectively. The March lecture was identified as the second in a course. See *JRASC* 4 (1910): 135, 215, 388.
64 LAC, MG30, vol. 3, Klotz diaries, 21 April 1910.
65 That the RSC papers had little lasting impact is evident in bibliographies that were compiled for JSP's obituaries. Neither the one by H. Spencer Jones, *Obit. Not. Fell. R. Soc.* 4 (1942): 67, nor the one prepared by C.S. Beals, R.M. Petrie, and K.O. Wright, *JRASC* 35 (1941): 408–11, includes any RSC papers.
66 C.A. C[hant], "Introduction," *JRASC* 1 (1907): 1.
67 They, and their director successors, continued as associate editors for decades.

68 David H. DeVorkin, ed., *The American Astronomical Society's First Century*, includes chapters by Donald E. Osterbrock ("AAS Meetings before There Was an AAS") and Richard A. Jarrell ("Honorary American Astronomers: Canada and the American Astronomical Society").
69 *Publications of the Astronomical and Astrophysical Society of America* 1 (1910): xii–xiii includes a list of members and when they joined.
70 C.A. C[hant], "The Astronomical and Astrophysical Society of America," *JRASC* 1 (1907): 67–8.
71 RASC, Chronological Archives, JSP to J.R. Collins, 21 October 1908.
72 Abstracts are in *PAAS* 1 (1910): 303, 304, and 311. The latter two topics were the subject of published papers by JSP, "Camera Objectives for Spectrographs," *ApJ* 29 (1909): 290–300 and "Effect of Increasing the Slit-Width on the Accuracy of Radial Velocity Determination," *ApJ* 28 (1908): 259–65, respectively.
73 JSP, "The Astronomical and Astrophysical Society of America," *JRASC* 2 (1908): 255–60.
74 Abstracts are in *PAAS* 1 (1910): 322 and *Science* 30 (1909): 727.
75 JSP, "Notes from the Dominion Observatory," *JRASC* 3 (1909): 315–16.
76 For details of the AAS meeting, see *Science* 32 (1910): 149 and 874; for JSP's account of it and the subsequent meeting of the Solar Union, see *Report … Ending March 31, 1911*, 130–1, and *JRASC* 4 (1910): 373–8.
77 JSP, "The Astronomical and Astrophysical Society of America," *JRASC* 4 (1910): 373–8; C.A. Chant, "The Mount Wilson Conference of the Solar Union," *JRASC* 4 (1910): 356–72; and *Science* 32 (1910): 541–6 and 874–87. For an up-to-date assessment of this important conference and cooperation in stellar spectroscopy generally, see R.A. Jarrell, "The 1910 Solar Conference and Cooperation in Stellar Spectroscopy," *JAHH* 13 (2010): 127–38.
78 JSP, *Science* 34 (3 November 1911): 606.
79 The AAS Committee on Co-operation in the Measurement of Stellar Radial Velocities eventually disbanded in 1916 (see http://had.aas.org/aashistory/4committ.html).
80 Howard Plotkin, "Edward Charles Pickering's Diary of a Trip to Pasadena to Attend Meeting of Solar Union, August 1910," *Southern California Quarterly* 40 (spring 1978): 29–44, 31.
81 AIP, interview of Harry Plaskett by D. DeVorkin 29 March 1978.
82 Plotkin, "Edward Charles Pickering's Diary," 34.
83 UCSC, Directors' files, UA 36 Ser.03, C.A. Chant to W.W. Campbell, 8 June 1910.
84 Chant, "The Mount Wilson Conference." For a photograph (with key) of the whole group at Mount Wilson, see *PA* 18 (1910): 489.

85 *Trans. IUCSR* 4 (1914): 4 shows that the RASC was admitted to the union in 1910 and lists ten individuals nominated by the RASC to the national committee. See also Chant, "The Mount Wilson Conference."
86 Chant, "The Mount Wilson Conference," 362.
87 JSP agreed to concentrate on the spectral range 550–70 nm in the solar spectrum, *Science* 32 (1910): 544.
88 Chant, "The Mount Wilson Conference, 356.
89 David H. DeVorkin, "Community and Spectral Classification in Astrophysics: The Acceptance of E.C. Pickering's System in 1910," *Isis* 72 (1981): 29–49, summarizes the many classification schemes that had been in existence and explores the factors that influenced consensus.
90 JSP, *Report … Ending March 31, 1911*, 134.
91 René Racine, "The Historical Growth of Telescope Aperture," *PASP* 116 (2004): 77.
92 JSP, *Report … Ending March 31, 1911*, 134.
93 Ibid., 140.
94 Ibid., 105.
95 W.F. King, *Report … Ending March 31, 1911*, 6.

5 The Dream of an Upright Man, 1911–1913

1 W.F. King, "The Dominion Astronomical Observatory," *Proc. Trans. RSC* 5 (1911): li–liii.
2 For summaries of the meeting see C.A. Chant, "The Astronomical and Astrophysical Society of America: Ottawa Meeting," *JRASC* 5 (1911): 327 and R.H. Curtiss, "The Astronomical and Astrophysical Society of America," *Science* 34 (1911): 520. Expenses amounting to $292 that were incurred by King on behalf of the government are found in Canada, *Sessional Papers of Parliament*, vol. 47, Auditor General's Report for 1911–12, vol. 1, part 1, J-135.
3 UPA, 5/1 Allegheny Observatory, box 17g, folder 34, JSP to F. Schlesinger, 29 March 1911.
4 LAC, RG48, vol. 48, file 5-5.
5 PUL, C0045, JSP to H.N. Russell, 27 June 1927, quoted in David H. DeVorkin, *Henry Norris Russell*, 92.
6 AIP, interview of Harry Plaskett by D. DeVorkin, 29 March 1978.
7 AIP, interview of Margaret Russell Edmonston by D. DeVorkin, 21 April 1977. Mrs Edmonston was the daughter of H.N. Russell; she thought her father made paper animals for Harry Plaskett, but presumably he did it for Stuart.

8 UPA, 5/1 Allegheny Observatory, box 17g, folder 34, JSP to F. Schlesinger, 17 August 1911.
9 Ibid., JSP to F. Schlesinger, 25 September 1911.
10 LAC, RG48, vol. 47, file 4-14, item 27S, F. Schlesinger to JSP, 1 September 1911.
11 R.H. Curtiss, "The Astronomical and Astrophysical Society of America," *Science* 34 (1911): 521.
12 UPA, 5/1 Allegheny Observatory, box 17g, folder 34, JSP to F. Schlesinger, 25 September 1911.
13 "Meetings of the Society," *JRASC* 5 (1911): 429, and 6 (1912): 39 and 42.
14 C.A. Chant, "The Astronomical and Astrophysical Society of America: Washington Meeting," *JRASC* 6 (1912): 24.
15 UPA, 5/1 Allegheny Observatory, box 17g, folder 34, JSP to F. Schlesinger, 15 January 1912.
16 JSP, "Some Interesting Developments in Astronomy," *JRASC* 5 (1911): 245–65; *Ann. Rep. Smithsonian Inst. for 1911* (1912): 255–70; *Sci. Am. Supp.*, no. 1869 (28 October 1911): 278.
17 JSP, "Some Interesting Developments," 265.
18 Membership lists for the Ottawa Field-Naturalists were published annually in the April numbers of their publication, *Ottawa Naturalist*.
19 According to the diary of Shutt's great-nephew Alan W. Ford in SA, JSP and Shutt were friends. Information about Shutt's career comes from T.H. Anstey, *One Hundred Harvests* (Ottawa: Research Branch, Agriculture Canada, 1986), 19; for his other interests, see Helen Smith and Mary Bramley, *Ottawa's Farm: A History of the Central Experimental Farm* (Burnstown, ON: General Store Publishing House, 1996), 45
20 JSP, "The Evolution of the Worlds," *Ottawa Naturalist* 26, no. 2 (May 1912): 29–34 and 26, no. 3/4 (June/July 1912): 52–60.
21 Carl Berger, *Science, God, and Nature in Victorian Canada* (Toronto: University of Toronto Press, 1983).
22 JSP, "The Evolution of the Worlds," 33. This botanical analogy is certainly not original to JSP; it seems to have originated with William Herschel, "Catalogue of a Second Thousand of New Nebulae and Clusters of Stars; With a Few Introductory Remarks on the Construction of the Heavens," *Phil. Trans. Roy. Soc.* 79 (1789): 226.
23 JSP's statement, from *JRASC* 5 (1911): 245, that the great nebula of Andromeda "must be tens of thousands of light years distant and probably forms a universe by itself" is cited by Robert W. Smith, *The Expanding Universe*, 13, as an example of a revival of the island universe theory in the early twentieth century. He acknowledges, though, that there were many sceptics.

24 JSP, "The Evolution of the Worlds," 29.
25 William Whewell, *Astronomy and General Physics Considered with Reference to Natural Theology* (London: Pickering, 1833), 356.
26 JSP, "The Evolution of Worlds," 59–60.
27 LAC, MG30, series B13, vol. 4, file 26, Klotz diaries, 23 April 1913.
28 B.R. Pettersen, "Sigurd Enebo and Variable Star Research: Nova Geminorum 1912 and the RV Tauri Stars," *JAHH* 15 (2012): 246.
29 JSP, "The Spectrum of Nova Geminorum," *JRASC* 6 (1912): 27–36 (also *PDO* 1 (1913): 157–69).
30 Ibid., 36.
31 W.S. Adams and A. Kohlschutter, "Observations of the Spectrum of Nova Geminorum No. 2," *Contributions of the Mount Wilson Observatory* 62 (1912): 1–29 and *ApJ* 36 (1912): 293–321.
32 *Science* 35 (1912): 491
33 The memorial to "His Royal Highness, the Governor General in Council" was dated 20 June 1912 and was signed by the president of the RSC, W.D. LeSueur, *Proc. Trans. RSC* 7 (1913): xxviii–xxx.
34 Ibid.
35 UPA, 5/1 Allegheny Observatory, box 17g, folder 34, JSP to F. Schlesinger, 18 May 1912.
36 One such letter is JSP to Karl Schwarzschild, director of the Potsdam Observatory, 18 May 1912, Niedersächsische Staats- und Universitätsbibliothek, Göttingen K. Schwarzschild, letter 594.
37 LAC, RG48, vol. 44, file 1-4, letters from W.W. Campbell (director, Lick Observatory), E.C. Pickering (Harvard College Observatory), F. Schlesinger (Allegheny), H.F. Newall (director, University Observatory, Cambridge), K. Schwarzschild (director, Royal Astrophysical Observatory, Potsdam), F.W. Dyson (Astronomer Royal), May–June 1912 endorsing the value of a large reflector for Canada.
38 LAC, RG48, vol. 49, file 6-13, E.B. Frost to JSP, 11 May and 11 June 1912.
39 JSP, "Description of Building and Equipment" [1919], *PDAO* 1 (1922): 23.
40 UPA, 5/1 Allegheny Observatory, box 17g, folder 34, JSP to F. Schlesinger, 18 May 1912, and LAC, RG48, vol. 47, file 4-14, F. Schlesinger to JSP, 7 June 1912.
41 The speeches were published in *Science* 36 (1912): 341, 417.
42 UPA, 5/1 Allegheny Observatory, box 17g, folder 34, JSP to Mrs Schlesinger, 4 September 1912.
43 According to T.S. Plaskett's family history, supplemented by census and other data from www.ancestry.com, Fred and Harry were JSP's two brothers who worked together in the seed business in Rochester. Fred and

Allie Plaskett had two children; Harry and Minta had none. Harry died on 5 September 1912 in Kelowna, BC, so he likely was not in Rochester on this occasion when JSP and Reba visited.

44 LAC, RG48, vol. 44, file 1-4, note of 15 August 1912, written on memorandum to G.F. Buskard, private secretary, Department of Interior, asking King to estimate the cost of the proposed observatory.

45 Ibid., W.F. King to Buskard, 19 September 1912.

46 JSP, "The History of Astronomy in British Columbia," *BCHQ* 4 (1939): 75, reprinted in *JRASC* 77 (1983): 117.

47 LAC, RG48, vol. 44, file 1-4, McLennan to Osler, 6 November 1912.

48 *Science* 37 (1913): 639; C.A. Chant, "The Astronomical and Astrophysical Society of America, Cleveland Meeting," *JRASC* 7 (1913): 31–34.

49 Chant, ibid., 34.

50 CCL, F. Schlesinger diary for 4 January 1913.

51 JSP and R.E. DeLury, "Methods and Preliminary Results in the Spectroscopic Determination of the Solar Rotation," read at the meeting of 17 May 1911, *Proc. Trans. RSC* 5 (1912): section 3, 107–21.

52 JSP and R.E. DeLury, "The Solar Rotation in 1911," *ApJ* 37 (1913): 73–104.

53 Using an empirical law originally proposed by Hervé Faye, JSP found the angular daily rate of rotation to be $10.32 + 4.05 \cos^2(\phi)$, and DeLury got $10.04 + 4.00 \cos^2(\phi)$, where ϕ is the solar latitude. At the solar equator, $\phi = 0$ and $\cos \phi = 1$, yielding the values of 14.37 and 14.04 given in the text.

54 LAC, RG48, vol. 49, file 6-16, correspondence between J.A. Anderson and JSP, 1910–12.

55 JSP, "Experiments Regarding Efficiency of Spectrographs," *PDO* 1 (1914): 173–203; "The Plane Grating for Stellar Spectroscopy," *ApJ* 37 (1913): 373–9; and "Improvements in the Optical System of the Stellar Spectrograph," *ApJ* 40 (1914): 127–36.

56 JSP, "Improvements in the Optical System of the Stellar Spectrograph," 135 (also *PDO* 1 (1914): 199).

57 Frost's letter and replies to it are included in F. Schlesinger, "Spectroscopic Binaries under Investigation at Different Institutions," *ApJ* 39 (1914): 264.

58 Ibid., 265–6.

59 R.M. Petrie, "The Spectrographic Orbit of θ^2 Tauri," *PASP* 52 (1940): 286–7.

60 Robert G. Aitken, *The Adolfo Stahl Lectures in Astronomy*, 256.

61 LAC, RG48, vol. 44, file 1-3, "Some Reasons Why …," date stamped 27 June 1913.

62 "Resolution Regarding Ottawa Telescope," *JRASC* 7 (1913): 46–7.

63 JSP, "Description of Building and Equipment" [1919], *PDAO* 1 (1922): 1–103.

64 Ibid., 9.
65 LAC, RG48, vol. 44, file 1-3, W.W. Cory to W.F. King, 2 April 1913. The year 1913 was the first when a Canadian cost of living index was produced. The website http://data.bls.gov/cgi-bin/cpicalc.pf shows the current value of a 1913 dollar as $22.02, putting the estimated cost of the telescope at just over a million dollars.
66 UPA, 5/1 Allegheny Observatory, box 17g, folder 35, JSP to F. Schlesinger, 4 March 1913
67 Most of the information in this section comes from JSP, "Description of Building and Equipment," 1–103.
68 LAC, RG48, vol. 44, file 1-4, Provisional specifications and invitations to tender, 7 March 1913, sent to Brashear, Ritchie, Alvan Clark, Sir Howard Grubb, T. Cooke, Carl Zeiss, C.A. Steinheil.
69 JSP, "Description of Building and Equipment," 23.
70 LAC, RG48, vol. 44, file 1-4, W.F. King to T.W. Crothers, acting minister of the interior, 20 March 1913.
71 Some biographical information is found in T.S. Plaskett, *Plaskett-Stanley Family History*, 25, George J. Fraser, *The Story of Osoyoos, September 1811 to December 1952* ([n.d, n.p.), and http://osoyoosmuseum.ca/index.php/about-us/history-of-osoyoos/. The only actual evidence I have found of family meetings comes from JSP's brother Frank, who recalled visiting JSP in Victoria during one of his trips west about this time (BCA, CBC recording, T4022: 56/0004, Voices of B.C., part of Series GR-3378, Provincial Educational Media Centre, school broadcasts), and a copy of a letter to George J. Fraser, Penticton, paying him $25.35 for photographing the stars ($10 per month plus supplies), LAC, RG48, vol. 44, file 1-3.
72 LAC, RG48, vol. 48, file 5-5, A.J. Cannon to JSP, 18 January 1913.
73 JSP, "Description of Building and Equipment," 9–10.
74 LAC, RG48, vol. 44, file 1-4, 22 March 1913: Ritchey wrote to King regarding the form of mounting and the dangers of casting a mirror with a central hole, and expressing the hope that he would be considered for the design and construction of the Canadian telescope.
75 LAC, RG48, vol. 44, file 1-4, G.W. Ritchey to W.F. King, 24 August 1912. He was referring to the Ritchey-Chretien design that he and his assistant had invented at Mount Wilson in 1910. It eventually became widely used, most notably in the Hubble Telescope.
76 LAC, RG48, vol. 48, file 5-12, W.S. Adams to JSP, 5 May 1913.
77 BHL, Bimu C455, Michigan University, Observatory, box 1. Three letters from R.H. Curtiss to W.J. Hussey, 15 April, 12 May, and 26 June 1913, refer to JSP's visit to Ann Arbor on 10 April, his desire to obtain blueprints of their telescope, and the availability of these plans.

78 E.C. Pickering had purchased the 1.5 m telescope for Harvard in 1904. It had originally been built in 1888 by the English astronomer A.A. Common, for his own use. According to a letter from Shapley to JSP, Harvard had the old telescope refurbished in 1921. It was then used in Harvard's Boyden Station in South Africa.

79 *Cleveland Plain Dealer*, 11 April 1913, 7.

80 UPA, 5/1 Allegheny Observatory, box 17g, folder 35, F. Schlesinger to JSP, 15 April 1913.

81 *Peterborough Evening Examiner*, 24 April 1913, 11.

82 UPA, 5/1 Allegheny Observatory, box 17g, folder 35, F. Schlesinger to JSP, 15 April 1913.

83 "Notes and Queries," *JRASC* 7 (1913): 312.

84 UPA, 5/1, Allegheny Observatory, box 17g, folder 35, JSP to Mrs Schlesinger, 13 May 1913.

85 JSP, "Description of Building and Equipment," 10–18. The specifications were also published in W.F. King, "The New Reflecting Telescope of the Dominion Observatory," *JRASC* 7 (1913): 216–28.

86 H. Spencer Jones, *Obit. Not. Fell. R. Soc.* 4 (1942): 67–82 highlighted these additional qualifications.

87 See R.A. Jarrell, "J.S. Plaskett and the Modern Large Reflecting Telescope," *JHA* 30 (1999): 359–90 and Henry C. King, *The History of the Telescope* (High Wycombe, UK: Griffin, 1955; reprint, Mineola, NY: Dover, 2003), which gives W.H. Smyth's 15 cm "Hartwell" refractor in London and the 1.22 m Grubb reflector in Melbourne, Australia, as mid-nineteenth-century examples of the English cross-axis mounting.

88 JSP's travels in 1913 are briefly described in "Description of Building and Equipment" [1919], *PDAO* 1 (1922): 18–19; those for which he was reimbursed are listed in Canada, *Sessional Papers of Parliament*, vol. 50, Auditor General's Report for 1913–14, vol. 1, part 1, K151.

89 Jack and Reba were the only two Plasketts arriving at Quebec from Southampton aboard the *Ascania* on 7 September 1913 (passenger lists from www.ancestry.com). Stuart was not with them, so I assume he stayed home with Harry.

90 LAC, RG48, vol. 44, file 1-3, JSP (Shelbourne Hotel, Dublin) to W.F. King, 17 July 1913. The three commissions to which JSP was (re)appointed in 1913 are named in *Trans IUCSR* 4 (1914): 175–6: sunspot spectra, solar rotation by radial velocities, and classification of stellar spectra.

91 LAC, RG48, vol. 44, file 1-3, JSP (Shelbourne Hotel, Dublin) to King, 17 July 1913.

92 Ibid., JSP (Strand Palace Hotel, London) to W.F. King, 25 July 1913.

93 Ibid.

94 N. Bohr, "On the Constitution of Atoms and Molecules," *Philosophical Magazine* 26 (1913): 1–25, cited by J.B. Hearnshaw, *The Analysis of Starlight*.
95 LAC, RG48, vol. 44, file 1-3, JSP (Shelbourne Hotel, Dublin) to W.F. King, 17 July 1913.
96 JSP, "Description of Building and Equipment," 19, 36.
97 Jarrell, "J.S. Plaskett and the Modern Large Reflecting Telescope," 375 cites J. Delloye to Ambrose Swasey, 13 June 1914, LAC, RG48, vol. 45. Even the Mount Wilson blank was eventually used to produce a perfectly fine mirror.
98 JSP, "The Solar Union," *JRASC* 7 (1913): 420 gives an extensive account of the Plasketts' time in Germany. The official report is in *Trans. IUCSR* 4 (1914), including the report of the committee on the solar rotation using displacement of spectral lines. Summaries of the meetings are found in several other places, including C.H. Gingrich, *PA* 21 (1913): 457; F. Slocum, *ApJ* 38 (1913): 301; *Obs.* 36 (1913): 356 or 382.
99 JSP, "The Solar Union," 420–37, is the source for the rest of this chapter, unless otherwise cited.
100 Ibid. 434. W.W. Campbell from Lick Observatory, and H.H. Turner from Oxford, England, were similarly impressed with the extraordinary hospitality of the Germans; see Crelinsten, *Einstein's Jury*, 78.
101 JSP recalled this many years later, *Victoria Daily Colonist*, 25 February 1931, 2.
102 Passenger lists from www.ancestry.com.

6 Transition, 1913–1917

1 JSP, "Description of Building and Equipment" [1919], *PDAO* 1 (1922): 19.
2 I am using the term *cabinet*, but technically the correspondence is with "The Governor General in Council, The Clerk of the Privy Council and His Excellency the Administration." See LAC, RG48, vol. 44, file 1-5; order-in-council http://www.collectionscanada.gc.ca/databses/orders, approved 16 October 1913.
3 JSP, "The History of Astronomy in British Columbia," *BCHQ* 4 (1939): 63, reprinted in *JRASC* 77 (1983): 118.
4 JSP, "Description of Building and Equipment," 20.
5 W.E. Harper, "Tests Made to Ascertain Where Conditions Were Most Suitable for the 72-inch Reflector," *PDO* 2 (1915): 275–320. When C.A. Chant borrowed this 11 cm Cooke refractor and its equatorial mount for the solar eclipse of 1918, he said the equipment weighed 656 pounds and could be packed into four boxes. UTA, 1974-0027, C.A. Chant to E.B. Frost, 25 April and 9 May 1918.

6 LAC, RG48, vol. 44, file 1-3, W.E. Harper to W.F. King, 23 July 1913.
7 Ibid., file 1-5, W.E. Harper to W.F. King, 9 August 1913.
8 Harper, "Tests Made," 319.
9 Ibid., 290–1.
10 Ibid., 314.
11 The report is in LAC, RG48, vol. 44, file 1-5, but it was published in a somewhat abridged form in Harper, "Tests Made," 275–320.
12 JSP, "The 72-inch Reflecting Telescope," *JRASC* 8 (1914): 180,
13 JSP, "History of Astronomy," 63.
14 JSP, "Description of Building and Equipment," 25.
15 George E. Webb, "Victoria Welcomes the Dominion Astrophysical Observatory: Science and Society in the Pacific Northwest," *Pacific Northwest Quarterly* 94 (2003): 171 wrote about McBride's record. See also *Journals of the Legislative Assembly of British Columbia*, vol. 44, 2 March 1915, 61 re the purchase of the site, also vol. 45, 31 March 1916, 59 and vol. 46, 2 May 1917, 125 re the construction of the road.
16 The site of the observatory and the Saanich Interurban Electric line is shown on a map in the *Victoria Daily Colonist* (*VDC*), 17 May 1914, 12.
17 JSP, "History of Astronomy," 63.
18 JSP, "Description of Building and Equipment," 25.
19 *VDC*, 19 October 1913, 17 and 10 May 1914. The Gonzales observatory actually had a small equatorially mounted telescope too, made by local resident O.C. Hastings twenty years earlier.
20 Ibid., 3 March, 6; 5 March, 4; 6 March, 6, and 7 March, 7 (all 1914).
21 A.D. Watson, "Astronomy in Canada," *JRASC* 11 (1917): 76–7.
22 LAC, RG48, vol. 48, file 5-19, A.W. McCurdy to JSP, 12–24 March 1914. The *Daily Colonist*, in which the articles appeared, was a Tory newspaper. One clear piece of evidence is an editorial of 15 September 1916, 4, lamenting the defeat of the Conservatives in the provincial election.
23 Shepherd's role is discussed in an editorial in *VDC*, 13 May 1914.
24 Ibid., 13 May 1914, 1.
25 Ibid., 14 May 1914.
26 Ibid., 13 May 1914.
27 Webb, "Victoria Welcomes the Dominion Astrophysical Observatory," 171.
28 *VDC*, 13 May 1914, 11.
29 LAC, MG30, series B13, vol. 4, file 27, Klotz diaries, 2 June 1914, quoted by Hodgson, *The Heavens Above*, 45.
30 *VDC*, 16 May 1914, 6, reports Plaskett's departure.

31 UPA, 5/1 Allegheny Observatory, box 17g, folder 35, JSP to F. Schlesinger, 30 May 1914.
32 Ibid., JSP to F. Schlesinger, 25 June 1914.
33 *JRASC* 9 (1915): 63.
34 JSP, "Description of Building and Equipment," 21; JSP, "The 72-inch Reflecting Telescope," 401.
35 *Pittsburgh Press*, 25 August 1914, 14.
36 [C.A. Chant,] "The Allegheny and Ottawa Telescopes," *JRASC* 8 (1914): 303.
37 The problematic solar rotation has been summarized by Jarrell, *Cold Light*, 98–100 and analysed in detail by Klaus Hentschel, "A Breakdown of Intersubjective Measurement: The Case of Solar-Rotation Measurements in the Early 20th Century," *Studies in History and Philosophy of Science*, Part B, *Studies in History and Philosophy of Modern Physics* 29, no. 4 (1998): 473–507.
38 For reports of the meeting, see *PA* 22 (1914): 551 and *Chicago Tribune*, 28 August 1914, 9. Correspondence between Philip Fox and JSP, 1914, see AIP, AAS records, part II, series II, correspondence, box 3, folder 5.
39 Slipher showed lantern slides at an informal conversazione, *PA* 22 (1914): 551 and made a formal presentation, "Spectrographic Observations of Nebulae," *PA* 23 (1915): 21–4. He had also given a paper with the same title at the poorly attended Atlanta meeting eight months earlier; see *PA* 22 (1914): 146.
40 V.M. Slipher, "Nebulae," *Proc. Am. Phil. Soc.* 56 (1917): 403–9.
41 John S. Hall, "Vesto Melvin Slipher (1875–1969)," *American Philosophical Society Year Book 1970*, 164.
42 E.P. Hubble, in "A Relation between Distance and Radial Velocity among Extra-Galactic Nebulae," *Proc. NAS* 15 (1929): 168, used twenty-four nebulae for which ten had radial velocities identical to Slipher's published twelve years earlier. For a discussion of the development of the understanding of the expansion of the universe, see chapter 10 of Mario Livio, *Brilliant Blunders from Darwin to Einstein: Colossal Mistakes by Great Scientists that Changed Our Understanding of Life and the Universe* (New York: Simon & Schuster, 2013).
43 *Sci. Am. Supp.*, 30 October 1915, 274–5; 6 November 1915, 299.
44 JSP, "The Sidereal Universe," *JRASC* 9 (1915): 37.
45 LAC, RG48, vol. 44, file 1-22, H.H. Turner to JSP, 29 April 1915. Turner had written on this topic in "Stars, Motions in Space and the Two-Drift Theory: A Tentative Explanation of the 'Two Star Streams' in Terms of Gravitation," *MNRAS* 72 (1912): 387–407.

46 LAC, MG28-I382, Description of Photographic Art Club of Ottawa fonds.
47 F.A. Saunders, "Series in the Spectra of Calcium, Strontium, and Barium," *ApJ* 32 (1910): 153; see Saunders' biography online at http://www.nasonline.org/publications/biographical-memoirs/memoir-pdfs/saunders-frederick.pdf.
48 LAC, RG48, vol. 44, file 1-22, William Topley to JSP, 16 November 1918.
49 Ibid., file 1-23, JSP to W.J. Topley, 4 April 1923.
50 LAC, MG28-I382, Minute books of the Photographic Art Club of Ottawa (1908–21). Catalogues of exhibitions for 1909, 1911, 1913, and 1917, all containing lists of exhibited photographs by Plaskett, are in Arnprior, Macnamara fonds.
51 LAC, TR 646 C22085, The Photographic Art Club of Ottawa, *Special Exhibition ... October the 29th to November the 6th, 1915*.
52 *Christian Science Monitor*, 15 June 1917, 10.
53 Letter to the editor from C.E.K. Mees, *Science* 39 (1914): 464.
54 The Ottawa Naturalist 31, no. 1 (1917): 10 mentions JSP's talk on 16 December 1916 and the subsequent one delivered to the People's Forum. An announcement and report of this latter lecture is in the Ottawa Journal, 24 February 1917, 22 and in the Ottawa Citizen, 26 February 1917, 12 (the source of the passage quoted). The large size of the audience might be partly explained by the heightened public interest in the nearly complete observatory.
55 Heber D. Curtis "The Nebulae," *PASP* 29 (1917): 91 (the fifth Adolfo Stahl lecture, delivered in San Francisco, 9 March 1917).
56 UPA, 5/1 Allegheny Observatory, box 17h, folder 1, JSP to F. Schlesinger, 14 April 1915; CWRU, Warner & Swasey Collection, box 26, folder 7, JSP to the Warner & Swasey Co., 12 April 1915. JSP visited Pittsburgh in April 1915, presumably to check with Brashear on progress.
57 BHL, Bimu C455, Michigan University, Observatory, box 1, two letters from R.H. Curtiss to W.J. Hussey, 12 May and 26 June 1913.
58 JSP, Obituary of E.P. Burrell, *Science*, 25 June 1937, 597 and *PASP* 49 (1937): 141.
59 Approval of $35 000 for Warner & Swasey to construct the dome, dated 18 March 1915, http://www.collectionscanada.gc.ca/databases/orders.
60 *Proc. Trans. RSC* 9 (1915): xvii. Warner & Swasey had donated the model to the government of Canada. It apparently went to the DO in Ottawa, then was lent to the University of Toronto for display at the Canadian National Exhibition, then was borrowed back by Warner & Swasey for their fiftieth anniversary in 1929, and eventually got to the DAO, where it is still on display beneath the main observing floor in the dome. (UTA,

1974-0027/005, R.M. Stewart to C.A. Chant, 8 June 1929.) A photograph of the model appeared in the *Toronto Telegram*, 24 July 1922.
61 JSP, "Description of Building and Equipment," 55–76.
62 JSP, "Canada's 72-Inch Reflecting Telescope Installation," *Canadian Machinery* 16 (1916): 56.
63 *VDC*, 13 June 1915, 7.
64 JSP, "Description of Building and Equipment," 26; LAC, RG48, vol. 34, file 676.13, JSP to W.F. King, 19 June 1915.
65 LAC, RG48, vol. 44, file 1-6, Secretary to F.H. Shepherd, 3 March 1915.
66 Ibid., F.H. Shepherd to Gray Donald, 10 March 1915. According to C.A. Chant's autobiography, the Civil Service Commission was instituted in Ottawa in 1908, but not extended to affect all appointments until 1918 (UTA, A1974-0027, 156).
67 *Church Life* (Toronto), 22 July 1915, 354.
68 A full account of the meeting, including a group photo is in *PA* 23 (1915): 492.
69 Alan E. Leviton, Robert I. Bowman, and Michele L. Aldrich, *The Pacific Division of the American Association for the Advancement of Science: A Brief History* (San Francisco: AAAS, 2002), available online at www.archives.aaas.org/AAASPD_History.pdf.
70 LAC, MG30, series B12, vol. 4, file 28, Klotz diaries, 27 July 1915.
71 Ibid., 18 August 1915.
72 CWRU, Warner & Swasey Collection, box 37, folder 4, contracts and related materials.
73 LAC, RG48, vol. 34, file 676.13, several letters, especially from the clerk of the privy council, 20 March 1914 and 4 October 1916.
74 Ibid., JSP to W.F. King, 6 April 1915.
75 Ibid., JSP to W.F. King, 27 August 1915.
76 *Science* 40 (1914): 204 mentions JSP's hope that the telescope would be operational by the end of 1915. According to *VDC*, 24 April 1918, work on the mirror was suspended from September 1915 to August 1916, when the figuring began. The article gives many reasons for the delay in the optics.
77 R.A. Jarrell, "J.S. Plaskett and the Modern Large Reflecting Telescope," *JHA* 30 (1999): 375.
78 JSP, "Description of Building and Equipment," 27.
79 Jarrell, *Cold Light*, 102, cites copies of correspondence enclosed in a letter from R. McBride to JSP, 2 December 1915.
80 LAC, RG48, vol. 34, file 676.13, JSP (in Cleveland) to W.F. King, 24 February 1916.

81 Ibid.
82 JSP, "William Frederick King," *Obs.* 39 (1916): 338 and "W.F. King," *JRASC* 10 (1916): 267. Countering some of JSP's effulgent praise, Jarrell in his article on King in the *Dictionary of Canadian Biography,* vol. 14, online at http://www.biographi.ca/en/bio/king_william_frederick_14E.html, suggests that "his small empire fell into disarray" during the last year of his life and that his administrative abilities were "insufficient to control the disparate range of institutions he headed."
83 UTA, A89-0011/74, academic record cards. At the end of his second year, he was "below the line." *Toronto Globe,* 5 June 1914, 15. AIP, interview of Harry Plaskett by D. DeVorkin, 29 March 1978; UTA, A1974-0027/004, JSP to C.A. Chant, 12 February 1919, says that HHP did not shine at examinations.
84 H.H. Plaskett, "A Variation in the Solar Rotation," *ApJ* 43 (1916): 145.
85 V. Gaizauskas in an email to the author, 4 May 2010.
86 Harold Averill, Marnee Gamble, and Loryl MacDonald, *We Will Do Our Share: The University of Toronto and the Great War* (Toronto: Thomas Fisher Library 2014).
87 Harry's attestation papers stated he had a dark complexion, blue eyes, and dark brown hair, and gave his height as 5 feet 8½ inches: www.bac-lac.gc.ca/eng/discover/military-heritage/first-world-war/personnel-records/Pages/item.aspx?IdNumber=579948.
88 Ibid. He was taken on strength as lieutenant at Kingston with the 33rd Battery, Canadian Field Artillery, on 18 January 1917. He gave his profession as "astronomer" and his address as "Dominion Astrophysical Observatory, Victoria BC."
89 *JRASC* 11 (1917): 239; HL, Hale papers, JSP to W.S. Adams, 23 December 1916, also quoted by Klaus Hentschel, "Breakdown of Intersubjective Measurement."
90 LAC, MG30, series B13, vol. 4, file 29, Klotz diaries, 7 December 1916.
91 See Hentschel, "Breakdown of Intersubjective Measurement," n. 142.
92 Robert Howard, "Solar Rotation," *Annual Review of Astronomy and Astrophysics* 22 (1984): 131-55.
93 H.H. Plaskett, "Solar Granulation," *MNRAS* 96 (1936): 402–24.
94 *JRASC* 11 (1917): 239.
95 JSP, "Canada's 72-Inch Reflecting Telescope Installation," 55 and *Canadian Engineer* 31 (1916): 51, 74.
96 *Saturday Night,* 1 May 1915, 4; 10 February 1917; and 20 July 1918 – only the last one was identifiably written by JSP.
97 [H.H. Turner,] "Our Astronomical Column," *Nature* 93 (12 March 1914): 40; 93 (27 August 1914): 671–2; 95 (4 March 1915): 17; 97 (15 June 1916):

323–4; 102 (12 December 1918): 292–3; C.A. Chant, "A Great Telescope for Canada," *Nature* 93 (2 July 1914): 459; JSP, "The Dominion of Canada's 72-in. Telescope," *Nature* 103 (10 April 1919): 105–8.

98 LAC, RG48, vol. 34, file 676.13, JSP to W.F. King, 19 February 1916.

99 LAC, MG30, vol. 4, Klotz diaries, 20 May 1916.

100 HL, JSP to G.E. Hale, 15 May 1916; UPA, 5/1 Allegheny Observatory, box 17h, folder 1, JSP to F. Schlesinger, 15 May 1916.

101 UTA, A74-0027-004, JSP to C.A. Chant, 14 June 1916, thanking Chant for his strong letter of support. The letter from Chant to W. Roche, dated 1 June 1916, is in UTA, 1974-0027/005 RO. Chant has handwritten a PS: "This letter was written without any suggestion from Dr. Plaskett."

102 UPA, 5/1 Allegheny Observatory, box 17h, folder 1, JSP to F. Schlesinger, 15 May 1916, with attached circularized letter.

103 Ibid., F. Schlesinger to JSP, 18 May 1916.

104 Ibid., JSP to F. Schlesinger, 30 May 1916.

105 Jarrell, *Cold Light*, 104, citing HL, Hale Papers 95, W. Roche to G.E. Hale, 21 June 1916.

106 Hodgson, *Heavens Above*, 29–30.

107 Ibid., 82.

108 LAC, MG30, series B12, vol. 4, file 29, Klotz diaries, 27 May 1916.

109 LAC, RG48, vol. 34, file 676.13, June–July 1916 orders include Warner & Swasey Co., authorizing them to proceed with the spectrograph, at $3950 (optics by Brashear) and spectrograph attaching frame, at $2700, and E. Dent, 61 Strand, London, asking for two clocks to be supplied – mean time and Sidereal time, as advertised on the back cover of the *Observatory*.

110 *VDC*, 30 April 1916, 15, includes photos showing the house, "now awaiting the arrival of Dr. Plaskett," and the steel ribs of the observatory, then under construction. JSP may have been there in May too. There is a photo of the observatory under construction, taken 5 May 1916 in JSP, "The 72-Inch Reflecting Telescope," *JRASC* 10 (1916): 275.

111 *VDC*, 13 August 1916, 19, illustrated with five photos.

112 Ibid., 24 August 1916, 14.

113 AIP, AAS records, part II, series II, correspondence 1898–1935, box 3, folder 5, JSP 1912–17, JSP to Secretary Philip Fox, 17 August 1916.

114 LAC, RG48, vol. 44, file 1-12, Harper to JSP, 6 September 1916.

115 JSP, "Description of Building and Equipment," 27–8.

116 JSP's report to the Board of Trade as reported in *VDC*, 15 September 1916, 10.

117 JSP, "Description of Building and Equipment," 27.

118 Ibid., 64–8.

119 JSP, "Canada's 72-Inch Reflecting Telescope Installation," 55.
120 JSP, "Description of Building and Equipment," 31.
121 *Victoria Daily Times*, 21 October 1916, 11 and *VDC*, 22 October 1916.
122 *Thirty-Seventh Annual Report of the Victoria British Columbia Board of Trade* (Victoria, 1916), 65–6.
123 LAC, MG30 B13, vol. 12, the clippings, "Largest Telescope Is Near Victoria," *Toronto Globe*, 25 November 1916 and "What Canada Is Doing …," *Ottawa Journal*, 27 November 1916, are in Klotz's scrapbook.
124 *JRASC* 11 (1917): 106. Brashear and Swasey were on their way to China, see *PASP* 29 (1917): 149.
125 There was widespread resentment in Canada that Woodrow Wilson had campaigned in the fall of 1916 on keeping the United States out of the war.
126 *PA* 25 (1917): 228.
127 LAC, MG30, series B13, vol. 4, file 29, Klotz diaries, 26 December 1916.
128 Ibid., file 30, Klotz diaries, 27 March 1917 and 18 October 1917.
129 For JSP's appointment, see LAC, RG48, vol. 34, file 676.13, copy of minutes from clerk of the Privy Council, PC 31/992; *Ottawa Journal*, 27 March 1917, 14; salaries are found in Canada, *Sessional Papers of Parliament*, vol. 54, Auditor General's Report for 1917–18, vol. 1, part 1, K15.
130 LAC, RG48, vol. 34, file 676.13, 13 November 1916, memorandum from JSP to W.W. Cory.
131 For Klotz's appointment as chief astronomer, approved 4 October 1917, see http://www.collectionscanada.gc.ca/databases/orders/.
132 LAC, RG48, vol. 48, file 5-2, R.K. Young to JSP, 16 March 1912, 3 May 1913, 21 May 1913.
133 Ibid., R.K. Young to JSP, 5 July 1916, presents a table showing the number of plates taken by each observer and for whom during the previous three years. Young had obtained 661, JSP only 101.
134 Detroit border crossings for 4 July 1917, www.ancestry.com.
135 The U.S. Census for 1920 shows Wilhelmina as divorced and a secretary at the Alameda Sanatorium, www.ancestry.com.
136 Rod Millard, "The Crusade for Science: Science and Technology on the Home Front, 1914–1918," in *Canada and the First World War: Essays in Honour of Robert Craig Brown*, ed. David Mackenzie (Toronto: University of Toronto Press, 2005).
137 See, for example, M. "Bo" Sears Wheeler Jr., *Helium: The Disappearing Element* (Springer eBook 2015).
138 Mel W. Thistle, *The Inner Ring* and Wilfrid Eggleston, *National Research in Canada: The NRC, 1916–1966* (Toronto: Clarke, Irwin, 1978).

139 For the response of the RSC to the war, see T. Levere, "The Most Select and the Most Democratic: A Century of Science in the Royal Society of Canada," *Scientia Canadensis* 20 (1996): 28.
140 *VDC*, 2 June 1914, 4.
141 UPA, 5/1 Allegheny Observatory, box 17g, folder 35, JSP to F. Schlesinger, 16 November 1914.
142 AIP, AAS records, Part II, series II, correspondence, box 3, folder 5, JSP to Fox, 12 July 1915.
143 UTA, A1974-0027/004, two letters from JSP to C.A. Chant, 4 December 1916 and 20 February 1917.
144 PUL, C0045 – 55/40 and 21/36, correspondence between JSP and H.N. Russell, 10 March 1917 and 25 February 1919, respectively.
145 LAC, RG48, vol. 48, file 5-5, undated letter from Annie Cannon to JSP.
146 HUA, HUGFP 125.12, box 3, H.H. Plaskett to Annie Cannon, 14 October 1917.
147 UTA, A73-0026/365, where H.H. Plaskett's whereabouts, with dates, during the war can be found; activities of the brigade can be found in Canadian war diaries, available online at http://www.collectionscanada.gc.ca/archivianet/020152_e.html.

7 This Is the House That Jack Built, 1917–1921

1 John Lankford, *American Astronomy*, 383, shows that only one of seven telescopes with an aperture greater than 1.5 m during the period covered in this book was in Europe. On page 397 he offers an analysis of the *Handbuch der Astrophysik* (1928–36) as a test of Europe's theoretical dominance.
2 Reported by Augustin Symonds, acting secretary for the RASC Victoria Centre on 26 October 1922, *JRASC* 17 (1923): 45. JSP's very similar sentiments were recorded two years later in the *Victoria Daily Colonist* (*VDC*), 30 August 1925, 5.
3 LAC, RG48, vol. 49, file 6-18, miscellaneous "M," JSP to R.M. Motherwell, 27 July 1917. Hodgson, *The Heavens Above*, 83–4 summarizes Klotz's diaries for 26 April–3 May 1917 (the diaries are in LAC, MG30, series B 13, vol. 4, file 29.) The decline but not elimination of astrophysics at the DO after JSP's departure can also be traced through the annual reports of the DO in *MNRAS*.
4 Unless otherwise noted, all the details are found in JSP, "Description of Building and Equipment" [1919], *PDAO* 1 (1922): 1–103.
5 Ibid., 83.

6 UTA, A1974-0027/004, Plaskett file, JSP to Chant, 4 February 1919. JSP continued to use the borrowed prism until he obtained a new one from Hilger in 1919.
7 *VDC*, 4 July 1917.
8 Ibid., 2 August 1918, 8.
9 Ibid., 9 June 1918, 15.
10 C.A. C[hant], "Observatory Hill," *JRASC* 12 (1918): 128; LAC, RG48, vol. 45, file 2-14; *VDC*, 17 January 1918, 7.
11 LAC, RG48, vol. 49, file 6-27, letters from JSP to Victoria Public School Board and Saanich School Trustees, July–September 1917.
12 The basis for this statement as well as the following one is HUA, HUGFP 125.12, box 3, H.H. Plaskett to Annie Cannon, 16 February 1918.
13 LAC, RG48, vol. 45, file 2-3, JSP to J. Brashear, 8 March 1918.
14 Robert C. Brown and Ramsay Cook, *Canada, 1896-1921*, chap. 12 "The War Economy," especially 231–2.
15 LAC, RG48, vol. 45, series 4, file 2-15, JSP to P. Marchand, 18 March 1918. The director's salary range at the time was $3700–$4000, LAC, RG48, vol. 12, file 1179.
16 LAC, RG48, vol. 44, file 1-12, correspondence between W.E. Harper and JSP, July–September 1917.
17 *VDC*, 20 November 1917, 1, gives the itinerary of the governor general, including a drive to the observatory. Notice of the visit was sent five days earlier, LAC, RG48, vol. 49, file 6-27.
18 UTA, 1974-0027, R.K. Young to C.A. Chant, 10 April 1918.
19 UPA, 5/1 Allegheny Observatory, box 17h, folder 2, JSP to Schlesinger, 23 August 1917, and his reply, 17 September 1917; also JSP to Schlesinger, 17 April 1919, re his earlier consultations with Adams at Mount Wilson and Campbell at Lick.
20 LAC, RG48, vol. 48, file 5-12, G.E. Hale to JSP, 21 November 1916.
21 Ibid., file 5-5, E.C. Pickering to JSP, 30 June 1917. JSP subsequently tried using different gratings a few times, but Pickering's death in 1919 brought the collaboration to an end. (See JSP, "Report …," *:S* 80 (1920), 392).
22 S. Chapman and P.J. Melotte, "On the Application of Parallel Wire Diffraction Gratings to Photographic Photometry," *MNRAS* 74 (1913): 50. For an easy to read, good description of various techniques used in the past to find the magnitudes (brightness) of stars, including the wire grating, see Reginald L. Waterfield, *A Hundred Years of Astronomy* (London: Duckworth, 1938), 90–105, or J.B. Hearnshaw, *The Measurement of Starlight: Two Centuries of Astronomical Photometry* (New York: Cambridge University Press, 1996) for a more thorough account.

23 LAC, RG48, vol. 47, file 4-16, J. C. Kapteyn to JSP, 16 March 1918. Kapteyn's advice to JSP was not unique. Unable to generate much data at the small Groningen observatory in the poor Dutch climate, Kapteyn had to appeal to colleagues elsewhere. The day after he wrote JSP, Kapteyn sent an extensive memorandum to Hale at Mount Wilson, suggesting suitable radial velocity work with the large telescopes there. (See O. Gingerich, "Kapteyn, Shapley, and Their Universes," in *The Legacy of J.C. Kapteyn*, ed. P.C. Van der Kruit and K. van Berkel, 199.

24 *VDC*, 23 March 1918, 7.

25 Ibid., 24 April 1918, 4; JSP; "The Optical Parts of the 72-inch Telescope," *JRASC* 14 (1920): 177. Following McDowell's sudden death in 1923, JSP wrote an obituary (*JRASC* 18 (1924): 185) in which he quoted McDowell's widow in refuting the newspaper reports of suicide (e.g., *New York Times*, 1 December 1923). In subsequent correspondence, LAC, RG48, vol. 49, file 6-12, September 1924, Schlesinger explained to JSP why he was sure that McDowell had taken his own life.

26 *VDC*, 5 April 1918, 7 and 28 April 1918, 18.

27 Ibid., 1 and 2 May 1918, 7; *Vancouver Daily Sun*, 3 May 1918, 7.

28 Jarrell, *Cold Light*, 103 cites a letter from JSP to Pickering giving the date of first light as 3 May, presumably for visual observation. The first spectrogram was taken three days later.

29 JSP, "Description of Building and Equipment," 35.

30 Ibid., 76.

31 JSP, "The Stellar Spectrograph of the 72-Inch Reflecting Telescope," *ApJ* 49 (1919): 212–13.

32 *VDC*, 14 May 1918, 7.

33 JSP would not have read Albert Einstein's original papers in German, but would certainly have known that the prediction was going to be tested during the 1918 eclipse.

34 See Crelinsten, *Einstein's Jury*, 116–17, and a reference he cites, Katherine Bracher, "The Famous Eclipse of June 8, 1918," *S&T* 58 (1979): 411–13.

35 W.W. Campbell, *Science* 48 (1918): 34 and "The Crocker Eclipse Expedition from the Lick Observatory, University of California, June 8, 1918," *PASP* 30 (1918): 219–40.

36 JSP, "Notes on the Spectrum of Nova Aquilae, No. 3," *JRASC* 12 (1918): 350.

37 *VDC*, 12 June 1918, 15.

38 J.A. Brashear, *John A. Brashear*, 236–7 quotes part of a letter he wrote to his friend Edward D. Adams, 5 November 1918.

39 *VDC*, 12 June 1918, 1, 4.

40 "The Dominion Astrophysical Observatory," *PASP* 30 (1918): 261; very similar comments were reported in *VDC*, 12 June 1918, 4.
41 Brashear, *John A. Brashear*, 237. The current distance is given as 21 000 ly.
42 LAC, RG48, vol. 44, file 1-6, correspondence between JSP and Meighen, Shepherd, and Alfred Thompson.
43 JSP wrote to Meighen, not only during his tenure as minister of the interior, but also three times during his first short term as prime minister, on 9 July 1920, 25 August 1921, and 15 December 1921. The letters and Meighen's replies are in LAC, RG48, vol. 44, file 1-6.
44 George E. Webb, "Victoria Welcomes the Dominion Astrophysical Observatory: Science and Society in the Pacific Northwest," *Pacific Northwest Quarterly* 94 (2003): 171.
45 LAC, RG48, vol. 48, file 5-5, 1 September 1918, A. Cannon to JSP.
46 *VDC*, 23 June 1918, 5, with a photo of Harry Plaskett in uniform. However, in his Roll of Service (UTA, A73-0026/365 (25)) he stated that he was in Passchendaele in November 1917 and in Lens-Arras from December to April 1918, and was assigned to Amiens on 8 August, to Drocourt-Queant on 2 September, to Cambrai on 27 September, and to Valenciennes on 1 November (all 1918).
47 LAC, RG48, vol. 49, Herbert C. Wilson to JSP, 12 January 1918, asking about employment opportunities for his son.
48 LAC, RG48, vol. 48, file 5-5, JSP to E.C. Pickering, 3 December 1918.
49 Bessie Z. Jones and Lyle G. Boyd, *The Harvard College Observatory*, 399.
50 Hodgson, *The Heavens Above*, 92, quotes Klotz as saying that the first time a woman ever typed at the observatory was 11 February 1919, and that was just a temporary measure for one day. John H. Taylor, *Ottawa*, 120 gives the information about women filling the ranks.
51 LAC, RG48, vol. 45, series 4, file 2-9, JSP to the secretary of the Civil Service Commission, 20 June 1918.
52 Keay remained at the observatory until she moved with her parents to Vancouver in 1927, at which time JSP spoke strongly in her favour, urging that a suitable position be found for her (LAC, RG48, vol. 47, JSP to W.W. Cory, 29 April 1927); she was replaced by Lora M. Blake from the Deputy Minister's Office in Ottawa (LAC, RG48, vol. 47, Cory to JSP, 1 February 1927).
53 "The Great Telescope of Birr Castle, Ireland," *The Hitchhiker's Guide to the Galaxy: Earth Edition*, available at http://h2g2.com/edited_entry/A87835477.
54 Allan Sandage, *The Mount Wilson Observatory*, 171–6; Mike Simmons, "Building the 100-inch Telescope," https://www.mtwilson.edu/Simmons4.html.

55 Gale E. Christianson, *Edwin Hubble*, 122, gives 11 September 1919 as the date the telescope was cleared for general use.
56 Donald E. Osterbrock, *Pauper and Prince*, 139–56.
57 JSP, "The Great Canadian Telescope," *Canadian Magazine* 52 (1919): 861. Plaskett may have overlooked the historic 1.2 m reflector that was still operating in its urban setting in Melbourne, Australia, though in terms of light-gathering power, he was still correct since $2(1.2)^2 < (1.8)^2$.
58 Ibid.
59 JSP gave the cost of the building and dome as $57 000 and the telescope and spectrograph as $97 000 in "The Dominion Astrophysical Observatory," *PASP* 39 (1927): 90. For 1913–18 government expenditures, which include many other items, see the following volumes of Canada, *Sessional Papers*, Auditor General's Reports: vol. 50, K151; vol. 51, V179; vol. 52, K170 and V45; vol. 53, K156–K157 and V39–V40; vol. 54, V123.
60 Langton, *Sir John Cunningham McLennan*, 71.
61 A.H. Joy, "The Mount Wilson Observatory of the Carnegie Institution of Washington," *PASP* 39 (1927): 9 and http://www.mtwilson.edu/Simmons4.php. Assuming the cost is proportional to volume, the ratio of Mount Wilson compared to DAO would be $(2.54/1.83)^3 = 2.7$. In fact the ratio of costs was 3.7.
62 LAC, RG48, vol. 49, file 6-1, JSP to F.W. Dyson, 23 February 1928. Inflation, especially during the war, when it ran at of over 10 per cent annually, would have accounted for a doubling in price.
63 See http://www.thecanadianencyclopedia.ca/en/article/khaki-university/ for a summary and further references.
64 LAC, RG48, vol. 44, file 1-22, correspondence between JSP and H.M. Tory, 18 January and 19 February 1919.
65 AIP, interview of Harry Plaskett by David DeVorkin, 29 March 1978. The extent to which Harry absorbed the work that Fowler had done is evident in papers he wrote within three years of 1919. See H.H. Plaskett, "The Pickering Series and Bohr's Atom," *JRASC* 16 (1922): 137–49 and "The Spectra of Three O-Type Stars," *PDAO* 1 (1922): 325–74.
66 Margaret Wilson, *Ninth Astronomer Royal; the Life of Frank Watson Dyson* (Cambridge: W. Heffer, 1951), 185; see also LAC, RG48, vol. 49, file 6-1, correspondence between JSP and F. Dyson, 1919.
67 HUA, HUGFP 125.12, cox 3, Reba Plaskett to Annie Cannon, 6 March 1919; LAC, RG48, vol. 44, file 1-23, H.H. Turner to JSP, 23 May 1920. Harry was elected a FRAS at a meeting of the society on 9 May 1919, when Fowler was president (*MNRAS* 79 (1919): 467). JSP was nominated

by Dyson on 13 June (*MNRAS* 79 (1919): 541) and elected on 12 December (*MNRAS* 80 (1919): 95).

68 Christianson, *Edwin Hubble*, 72–3; 109–10.

69 Alan Bowker, *A Time Such as There Never Was Before: Canada after the Great War* (Toronto: Dundurn, 2014).

70 Surprisingly, Jarrell, *Cold Light*, 87, wrote of the harmonious interaction of the institutions, noting "One rarely comes across evidence of the kinds of dissension that occasionally surfaced in other great observatories and departments of astronomy in the United States and Europe."

71 UTA, A1974-0027/004, memorandum to the members of the RASC Ottawa Centre, 18 December 1917.

72 UTA, A1974-0027/004, JSP to C.A. Chant, 29 November 1918.

73 Ibid., JSP to C.A. Chant, 21 January 1919.

74 The names of officers of the RASC and its centres were published in the *Journal* in annual reports. JSP became the society's honorary president once again for 1928–30.

75 LAC, MG30, series B13, Klotz diaries, vol. 3, 20 December 1906.

76 LAC, RG48, vol. 49, file 6-1, correspondence between JSP and F. Dyson, February–April 1919; the *Photographic Journal* 62 (October 1922): 19 and 437–8 describes the four photos, and includes a comment from Charles R. Davidson of the Royal Greenwich Observatory that "the smallest star discs do not exceed one second of arc in diameter." Plaskett had earlier sent the Royal Photographic Society thirteen photographs showing stages in the construction of the DAO, *Photographic Journal* (1916): 23.

77 "Meeting of the RAS, Friday 1919, April 11," *Obs* 42 (1919): 185, 189; *MNRAS* 80 (1920): 439. The RAS catalogue numbers of the slides were 288 and 289.

78 HL, Hale papers, JSP to G.E. Hale, 31 July 1918. For a thorough discussion of Plaskett's persistent efforts to promote cooperation in radial velocity work from 1910 to 1920, see R.A. Jarrell, "The 1910 Solar Conference and Cooperation in Stellar Spectroscopy," *JAHH* 13 (2010): 127–38. Astronomer Ray Carlberg pointed out in "The Plaskett Telescope at 90: The Creation of Canadian Observatories," *JRASC* 102 (2008): 178 that, "the DAO adopted the collaborative research style … partly based on the view that a national observatory could and should tackle scientific problems of greater complexity than those that were within the grasp of individual scientists."

79 LAC, RG48, vol. 47, file 4-16. JSP to Kapteyn, 31 July 1918. The letter referred to is in HL, Hale papers, G.E. Hale to JSP, 15 June 1918.

80 LAC, RG48, vol. 48, file 5-12, correspondence between JSP and W.S. Adams, 25 May–3 September 1918 and between G.E. Hale and JSP, 18 June–3 September 1918.
81 *JRASC* 13 (1919): 244.
82 JSP, "Dominion Astrophysical Observatory," *MNRAS* 80 (1920): 393.
83 JSP, "The Spectroscopic Binary H.R. 8170," *JRASC* 13 (1919): 174–8. An unfortunate mistake was discovered and all the velocities were reduced by 1.28 km/s in the revised paper published in *PDAO* 1 (1922): 113-17.
84 UTA, A1974-0027/004, JSP to C.A. Chant, 24 February 1919.
85 JSP, W.E. Harper, and R.K. Young, "Fourth List of Spectroscopic Binaries," *JRASC* 13 (1919): 372–8.
86 W.S. Adams, "Seven Spectroscopic Binaries," *PASP* 27 (1915): 132 and "Ten Spectroscopic Binaries," *PASP* 29 (1917): 259.
87 JSP, "Notes from the DAO," *JRASC* 12 (1918): 460–3 was the first DAO list of stars with variable radial velocities, including Boss 5918. E.B. Frost called it "a star with disappearing bright lines" in *ApJ* 49 (1919): 61. Subsequently the star was found to be a variable (EW Lacertae) of the γ Cas type, where rapid rotation causes material to spin off into a ring or disk around the star. Plaskett measured the radial velocity of the visual binary, Struve 3050, five times in the fall of 1918: "An Interesting Double Star," *PASP* 31 (1919): 39. Eric Doolittle was the first to investigate it further, "The Stellar System Σ 3050," *PASP* 31 (1919): 158–61.
88 JSP, W.E. Harper, R.K. Young, and H.H. Plaskett, "One Hundred Spectroscopic Binaries" [1920], *PDAO* 1 (1922): 163–85.
89 Ibid., 164.
90 JSP, "Catalogue of Radial Velocities of O and B Type Stars," *PDAO* 5 (1930): 109.
91 JSP, "Notes on the Spectrum of Nova Aquilae No. 3," *JRASC* 12 (1918): 350–6.
92 Edward C. Pickering, "Nova Aquilae," *Bulletin of the Harvard College Observatory*, No. 662 (1918): 1.
93 JSP, "Notes on the Spectrum of Nova Aquilae" (abstract), *PA* 27 (1919): 93.
94 J. Evershed, note added 28 April 1919 to his paper, "The Spectrum of Nova Aquilae," *MNRAS* 79 (1919): 483.
95 F.E. Baxandall, "On the Identity of a Pair of Strong Lines of Peculiar Behavior in the Spectrum of Nova Aquilae III," *PASP* 31 (1919): 297–9 and "On the Presence of Absorption Lines of Nitrogen and Oxygen in the Spectra of Nova Aquilae III," *MNRAS* 81 (1920): 66–74.
96 JSP, Harper, and Young, "Fourth List of Spectroscopic Binaries," 376.

97 LAC, RG48, vol. 48, file 5-12, JSP to W.S. Adams, 10 October 1918; also, vol. 49, file 6-14, JSP to E.B. Frost, 10 October 1918.
98 LAC, RG48, vol. 49, file 6-14, JSP to E.B. Frost, 31 October 1918.
99 J.A. Pearce, "Some Recollections of the Observatory (1924–1935)," *JRASC* 62 (1968): 287.
100 Hodgson, *The Heavens Above*, 90. Harper was eventually appointed assistant director in 1924 (*Canada Gazette*, 6 December 1924).
101 LAC, MG30, series B13, vol. 4, file 31, Klotz diaries, 26 November 1918.
102 LAC, RG48, vol. 45, file 2-9, correspondence between JSP and W.J. Roche, 1918–19; for comparison, the "average annual earnings for supervisory office employees in manufacturing industries rose from $1317 in 1917 to $1810 in 1920," Brown and Cook, *Canada*, 232.
103 UPA, 5/1 Allegheny Observatory, box 17h, folder 2, JSP to Schlesinger, 18 February 1919; UTA, A74-0027/004, correspondence between JSP and C.A. Chant, 12 February and 7 March 1919.
104 UPA, 5/1 Allegheny Observatory, box 17h, folder 2, JSP to Schlesinger, 17 April 1919.
105 Ottawa city directories show John J. Smith, barrister and solicitor, living at 320 Fairmont Avenue from 1914 to 1923. After his death in 1924, his widow, Alice, continued living there.
106 LAC, RG48, vol. 45, file 2-15, item 230M.
107 Ibid., vol. 49, file 6-9, correspondence with S.L. Boothroyd, 1915–19.
108 Boothroyd, "The Orbit and Spectrum of H.R. 8803," *PASP* 33 (1921): 103–6.
109 JSP, "Description of Building and Equipment," 101.
110 JSP, "The Optical Parts of the 72-inch Telescope," 196–7.
111 Klotz described the details in his diary entry for 21 July 1921, LAC, MG30, series B13, vol. 4, file 32.
112 PUL, C0045, 55/40 and 21/36, correspondence between JSP and H.N. Russell, 18 and 26 February 1919.
113 For Russell's method of analysing light curves of eclipsing binaries, see R.G. Aitken, *Binary Stars*, 183–201.
114 Decades later, Russell would still recognize that "eclipsing variables really do provide the Royal Road of remarkably easy access to the solution of other inaccessible problems." H.N. Russell to B.J. Bok, 17 October 1946, cited by D.H. DeVorkin, *Henry Norris Russell*, 141.
115 A.S. Eddington, "The Internal Constitution of the Stars," *Obs* 43 (1920): 349.
116 JSP, "The Spectroscopic Orbits of the Eclipsing Variables U Ophiuchi, RS Vulpeculae, and TW Draconis" [1920], *PDAO* 1 (1922): 137–51, and *JRASC* 14 (1920): 1–15.

117 For modern data, see D.E. Holmgren, G. Hill, and W. Fisher, "Absolute Dimensions of Early-Type Eclipsing Binary Stars. III—U Ophiuchi," *Astronomy and Astrophysics* 248 (1991): 129–38; for RS Vul from D. Holmgren, "Absolute Dimensions and Evolutionary State of RS Vulpeculae," *Space Science Reviews* 50 (1989): 347; and for TW Dra from H. Bozic, J. Nemravova, and P Harmanec, "Standard UBV Photometry and Improved Physical Properties of TW Dra," *Information Bulletin on Variable Stars* (2013), No. 6086, #1.

118 Aitken, *Binary Stars*, 200.

119 A.S. Eddington, "On the Relation between the Masses and Luminosities of the Stars," *MNRAS* 84 (1924): 308–32.

120 JSP, "The Spectroscopic Orbits and Dimensions of the Eclipsing Variables U Coronae, TX Herculis, and Y Cygni," *JRASC* 14 (1920): 409–24.

121 JSP, W.E. Harper, R.K. Young, and H.H. Plaskett, "The Radial Velocities of 594 Stars" [1921], *PDAO* 2 (1924): 3–127.

122 Annual reports of the DAO were published in *MNRAS* starting in 1920 and in *PA* starting in 1923. The figures for 1920 appear in *MNRAS* 81 (1921): 290. This impressive total was way ahead of the 1344 spectra on 202 nights that would later become the norm (*MNRAS* 95 (1935): 367), but the apparent decline happened as fainter stars came under study, requiring longer exposures.

123 Most sources say that the umbrella organization was the International Research Council and that the International Council of Scientific Unions was not established until 1931. However the latter organization held its constitutive assembly in 1919; see International Council of Scientific Unions, *Report of the Constitutive Assembly Held at Brussels, July 18–28, 1919*, 1 (1920), 227.

124 G.E. Hale had called an organizational meeting of the proposed IAU on 8 March 1919; it was there that Plaskett's committee membership was proposed; see *Science* 49 (1919): 508. See also W.W. Campbell and J. Stebbins, "Report on the Organization of the International Astronomical Union," *Proc. NAS* 6 (1920): 349.

125 International Council of Scientific Unions, *Report*, 1 (1920), 227.

126 CITA, Hale papers, box 33, G.E. Hale to JSP, 9 July 1919.

127 *JRASC* 14 (1920): 219 and *Trans. IAU* 1 (1922) lists the national committee for Canada as Klotz (chair), DeLury (secretary), F. Henroteau, C.C. Smith, and R.M. Stewart (for the DO), Harper, Plaskett, and Young (for the DAO), Chant and Miller (RASC), and L.V. King (from McGill).

128 LAC, RG48, vol. 49, file 6-9, JSP to S. Boothroyd, 13 June 1919.

129 CITA, Hale papers, box 33, JSP to G.E. Hale, 3 July 1919.

130 Crelinsten, *Einstein's Jury*, 151 and 184–5. At the concurrent meeting of the American Physical Society, E.P. Lewis and E.T. Bell spoke about the physical and mathematical aspects of relativity.
131 LAC, RG48, vol. 49, file 6-9, Boothroyd correspondence; "Titles and Abstracts of Papers for the Seattle Meeting of the Society, June 17–19, 1920," *PASP* 32 (1920): 189, 201, and 251.
132 S.L. Boothroyd, "The Orbits of the Spectroscopic Components of Boss 4602," *PASP* 32 (1920): 329–31.
133 HL, Hale papers, JSP to G.E. Hale, 28 August 1920.
134 JSP, "The Dimensions of the Stars," *PASP* 34 (1922): 79–93.
135 LAC, RG48, vol. 48, file 5-12, correspondence between JSP and G.E. Hale, 4 November, 28 November, and 5 December 1921.

8 Challenges and Rewards, 1921–1923

1 Hodgson, *The Heavens Above*, 88–90.
2 LAC, MG30, series B13, vol. 4, file 32, Klotz diaries, 25 and 30 May 1921.
3 Ibid., 1 June 1921.
4 *Victoria Daily Times*, 21 July 1921.
5 LAC, MG30, series B13, vol. 4, file 32, Klotz diaries, 21 July 1921. Plaskett was quoted in the *Victoria Colonist* of 14 May 1918, 7, that the public "was admitted to the observatory every day from 9 to 5, excluding Sundays and holidays." If Klotz's experience reflected the norm, this arrangement continued in force well after once-a-week evening visits were instituted.
6 LAC, RG48, vol. 14, file 1516-1, O. Klotz to S.A. Mitchell, 18 October 1921. The origin of the microphotometer is uncertain. One was certainly ordered from Topley and Company in Ottawa in 1921 and arrived at the DAO in December 1922 (LAC, RG48, vol. 44, file 1-23, JSP to Topley, 19 July 1921 and 28 November 1922), yet Harry Plaskett says that the final mechanical design of the microphotometer he was using about this time was due to JSP "and it was constructed by a machine shop in Victoria" ("The Wedge Method and Its Application to Astronomical Spectroscopy" [1923], *PDAO* 2 (1924): 221).
7 LAC, RG48, vol. 12, file 1082.
8 H.H. Plaskett, "The Intensity Distribution in the Continuous Spectrum and the Intensities of the Hydrogen Lines in γ Cassiopeiae," *MNRAS* 80 (1920): 771–82; M.G. Adam and Roger Hutchins, "Plaskett, Harry Hemley," *Oxford Dictionary of National Biography* (Oxford 2004).
9 J.H. Oort, "Some Peculiarities in the Motion of Stars of High Velocity," *BAN* 1 (1922): 133–7.

10 JSP, W.E. Harper, R.K. Young, and H.H. Plaskett, "The Radial Velocities of 594 Stars" [1921], *PDAO* 2 (1924): 3–127. I have figured that 800 – 594 = 206, though Plaskett treats the numbers in greater detail on page 4.
11 JSP, "The Dominion Astrophysical Observatory [Report]," *MNRAS* 83 (1923): 275–7.
12 LAC, RG48, vol. 49, file 6-28, correspondence with Travelers Insurance Company 1921–3.
13 J.A. Pearce, "Some Recollections of the Observatory (1924–1935)," *JRASC* 62 (1968): 293. Plaskett's second accident is confirmed in "A Remarkable Variable Spectrum," *PASP* 35 (1923): 145, where he says he was prevented from observing in December 1922 and January 1923 by personal injury.
14 The full title is *The Dominion Astrophysical Observatory, Victoria, B.C.: A sketch of the development of astronomy in Canada and of the founding of the Observatory. A Description of the Building and of the Mechanical and Optical Details of the Telescope; an Account of the Principal Work of the Institution.*
15 BCA, death records, B13121, 1922-09-306642. It may be of interest that the doctor who certified Annie's death came from Oroville, Washington.
16 *Penticton Herald*, 29 March 1922.
17 BCA, BC Supreme Court Probate/Estate Files, GR-1578 B09565, 680/1922.
18 The marriage record is in AO, Reg. No. 9417, MS 934, reel 19. According to the *Ottawa Citizen*, 6 January 1921, 8, the wedding took place at 320 Fairmont Avenue (the Smiths' home), with the rector of St Matthias Anglican Church officiating. Judging by a letter dated 26 January 1921 to JSP from Alfred Thompson (LAC, RG48 vol. 44, file 106), Harry's parents did not attend. Even during the run-up to his wedding and during the honeymoon in Toronto, Harry kept astronomy near the top of his agenda. Over the Christmas holidays in 1920, he presented his father's paper on Z Vulpeculae to the AAS/AAAS in Chicago and during his honeymoon attended a series of lectures on relativity by Ludwik Silberstein. This series of lectures may possibly have partly inspired C.A. Chant's interest in looking for the relativistic deflection in star positions at the 1922 eclipse. For the lectures, see Ludwik Silberstein, *The Theory of General Relativity and Gravitation: Based on a Course of Lectures Delivered at the Conference on Recent Advances in Physics held at the University of Toronto, in January, 1921* (Toronto: University of Toronto Press, 1922).
19 Baptismal record, Christ Church Cathedral, Victoria.
20 According to Chant's autobiography (UTA, 1974-0027/010, p. 460), he had planned to have only his wife and daughter assisting him at the eclipse, but suddenly realized a month before departing that he needed

a competent assistant. He sent a desperate telegram to Young, who, to Chant's great relief and joy, was willing and able to accept.

21 Crelinsten, *Einstein's Jury*, 200–12 describes the 1922 eclipse. On page 201, he quotes a letter from Heber D. Curtis to W.W. Campbell, "I suppose you have your hands full advising some of the weak sisters who are going" (UCSC, Directors' files, UA 36 Ser.03, 1 June 1922).

22 UCSC, Directors' files, UA 36 Ser.03, C.A. Chant to W.W. Campbell, 28 April 1920; it seems to be more than coincidence that, in this two-paragraph letter, Chant enquires about the eclipses of 1922 and 1923 in the first paragraph and speaks of his "agitation" and "propaganda" for an observatory in the second.

23 UTA, A 1974-0027, E.B. Frost to C.A. Chant, 1 June 1922, replying to a letter that Chant had written on 29 March

24 UTA, A1974-0027/004, JSP to C.A. Chant, 6 May 1922. The Department of the Interior agreed not only to Young's paid leave but also covered his steamer fare. UTA A 1974-0027, R.K. Young to C.A. Chant, 20 May 1922. The auditor general's reports for 1921–22 (vol. 59, K18) and for 1922–23 (vol. 60, K18) show that RKY was kept on full salary.

25 UTA, A1974-0027/010, , C.A. Chant autobiography, p. 475.

26 C.A. Chant, "An Astronomical Trip to Australia," *JRASC* 16 (1922): 251 and 267.

27 Ibid., 249–54, "The Eclipse Expedition to Australia," *JRASC* 16 (1922): 281–3 and 319–21, and "The Eclipse Camp at Wallal," *JRASC* 17 (1923): 1–9.

28 "Mrs. J.S. Plaskett" is mentioned by the *Victoria Daily Colonist* in connection with the YWCA on at least three occasions in 1920: 17 July, 8; 26 September, 28; and 12 October, 13.

29 The dates and titles of JSP's talks and citations to them are listed in Appendix B. In addition, we know that JSP often attended other meetings of the RASC Victoria Centre because he is on record as thanking the speaker or proposing a motion, for example. For his two contributions to the syndicated newspaper columns on astronomy in 1929–31, see W.E. Harper, "Articles on Astronomy for the Newspapers," *JRASC* 23 (1929): 357 and "Popular Articles on Astronomy for the Press," 25 (1931): 452.

30 JSP's audiences certainly enjoyed his well-illustrated lectures. One recorded instance was a talk he gave to the Victoria Teachers' Association on 29 February 1918. The teachers followed him "with the closest attention and interest" and at the end "very heartily applauded" his "wonderfully lucid" account (*Victoria Daily Colonist*, 1 March 1918, 10). His ability as a lecturer was acclaimed by Helen Hogg after she and her husband had moved to Victoria and started attending JSP's talks.

Plaskett, she said, "did a fine job. Got a big hand too. He is very well thought of around here, and he should be." (UTA, B94-0002/054 (02), H. Hogg to her family in Massachusetts, 11 March 1933.)

31 *San Francisco Chronicle*, 17 December 1919, 2; *Bellingham [WA] Herald*, 16 December 1919, 1.

32 Two newspaper clippings in a Rotary Club of Vancouver scrapbook in the City of Victoria Archives summarize JSP's address to the club on 30 August 1923. One is from *Victoria Daily Times*, 30 August 1923, 2. The articles in several U.S. papers can be found using the historic newspapers section of www.ancestry.com. The earliest is the *Lincoln [NB] Star*, 31 August 1923, 1 whose headline I have used. The latest is *New Castle [PA] News*, 27 November 1923, 3.

33 JSP, *The Dominion Astrophysical Observatory*, 35–6.

34 A.S. Eddington, "The Internal Constitution of the Stars," *Obs* 43 (1920): 354 proposed that the loss in mass when four hydrogen atoms are synthesized into helium (as found by Francis Aston) is the source of stellar energy. F. Wesemael, "Harkins, Perrin and the Alternative Paths to the Solution of the Stellar-Energy Problem, 1915–1923," *JHA* 40 (2009): 277–96, discusses similar ideas of others about the same time.

35 *JRASC* 15 (1921): 257.

36 LAC, RG48, vol. 49, file 6-9, JSP to S. Boothroyd, 26 April 1921.

37 The dates and topics of JSP's five speeches to the Vancouver Institute are listed in Appendix B and can be found at www.library.ubc.ca/archives/vaninsti.html.

38 UTA, A74-0027/004, JSP to Chant, 28 December 1918.

39 LAC, RG48, vol. 45, file 2-10, JSP to W. Roche, 26 July 1922.

40 *JRASC* 16 (1922): 338 and 17 (1923): 39–40.

41 LAC, RG48, vol. 48, file 5-5, JSP to H. Shapley, 2 January 1923.

42 Ibid., vol. 12, file 1082, correspondence from Klotz, Stewart, and Cory, 1922.

43 Ibid., vol. 44, file 1-6, correspondence 22 December 1921–19 April 1923.

44 UTA, A1974-0027/004, JSP to C.A. Chant, 9 January 1923.

45 Ibid., JSP to C.A. Chant, letters of 29 June 1920, 18 January 1921, 1 February 1922, 9 and 19 January 1923. Chant and R.K. Young were elected in 1923, andHarper joined them the following year (*Proc. Trans. RSC* 17 (1923): xvii, and 18 (1924): xxiii).

46 LAC, RG48, vo. 44, file 1-6, JSP to W.L.M. King, 27 March 1923.

47 JSP, "Work of Observatory to Observe Stars," *Victoria Daily Colonist*, 18 January 1920, 29.

48 JSP et al., "The Radial Velocities of 594 Stars" [1921], *PDAO* 2 (1924): 7.

49 Inventories of all the equipment on hand were required to be submitted annually to Ottawa. The Property Returns for 1935–7 list this equipment and the dates acquired, LAC, RG48, vol. 50. Another machine was desperately needed, as five men with a great deal of orbital work were competing for time. A second hand-operated Monroe was purchased for $300 in 1926.
50 JSP, "Description of Building and Equipment" [1919], *PDAO* 1 (1922): 46–8.
51 David Crampton, "From 1.8 to 8 m Telescopes," *JRASC* 87 (1993): 302.
52 JSP, "Temperature Regulation of Spectrograph and Mirror," *JRASC* 16 (1922): 91–100.
53 [Blank note]
54 JSP, "The Optical Parts of the Victoria Spectrograph," *ApJ* 59 (1924): 65–75. The procedures and examples he gave were still being cited years later, for example, by Ralph A. Sawyer, *Experimental Spectroscopy* (New York: Prentice Hall, 1951), 83, 106.
55 Ira S. Bowen, "Instrumentation at the Mount Wilson and Palomar Observatories," *PASP* 69 (1957): 378.
56 Joseph A. Pearce, "Some Recollections of the Observatory, 1924–1935," *JRASC* 62 (1968): 289–90. Re-silvering three or four times a year seems excessive. As far as I could determine, only the Helwan Observatory, which was in a dusty desert location, re-silvered its mirrors more often than twice a year. S.H. Trimes, "Note on and Attempt to Protect a Silvered Mirror from Tarnishing," *Helwan Institute of Astronomy and Geophysics Bulletin* 10 (1913): 79. An aluminum coating did not tarnish as fast as silver, but the process of aluminizing was not used until after Plaskett's retirement.
57 C.S. Beals, "Early Days at the Dominion Astrophysical Observatory," *JRASC* 62 (1968): 301.
58 Online Journals of the Legislative Assembly of BC, 9 November 1921, p. 58, at http://archives.leg.bc.ca/civix/document/id/leg_archives/legarchives/757376896; Correspondence with Travelers Insurance Company, LAC, RG48, vol. 49, file 6–28, correspondence with Travelers Insurance Company; also LAC, RG48, vol. 47, file 4-3, JSP to R. M. Stewart, 23 April 1924. For the original description of Brashear's silvering process, see H. [*sic*] Brashear, "Hints on Silvering Specula," *English Mechanic* 31 (1880): 327. Several techniques, including Brashear's, are discussed in H.D. Curtis, "Methods of Silvering Mirrors," *PASP* 23 (1911): 13.
59 JSP, "The O-Type Stars," *PDAO* 2 (1924): 287–358.

60 HUA, HUGFP 125.12, box 3, JSP to Annie Cannon, 7 February 1921.
61 The three orbits are all found in *PDAO* 2 (1924): for BD 6°1309, see p. 147; for BD 44°3639, see p. 183; and for BD 56°2617, p. 269. JSP later corrected this last orbit in *PDAO* 4 (1931): 103.
62 The identification of Plaskett's Star by Henry Draper number, HD 47129, is more commonly used now than the Bonner Durchmusterung number, BD 6°1309.
63 JSP, "The Spectrographic Orbit of BD 6°1309" [1922], *PDAO* 2 (1924): 147–57; "The Star of Greatest Known Mass," *JRASC* 16 (1922): 284-93. These papers are the source of the detailed information in this paragraph and the following one.
64 Since a^3/P^2 is proportional to mass, it is easy to see that a larger value of a and a smaller value of P, implies a larger value of M.
65 LAC, RG48, vol. 49, file 6-1, JSP to F. Dyson, 11 October 1926. Plaskett's feelings were undoubtedly the same in 1922.
66 *Obs* 45 (1922): 237.
67 "Prof. Plaskett's Massive Star," *Nature* 110 (1922): 53; JSP, "A Very Massive Star," *MNRAS* 82 (1922): 447–57.
68 JSP, "The Spectrographic Orbit of BD 6°1309" and "The Star of Greatest Known Mass."
69 LAC, RG48, vol. 49, files 6-18 and 6-27, Select Pictures Corporation to JSP, 20 and 22 October 1922; in the Toronto papers alone, between 14 July and 11 August 1922, there were at least ten articles about the discovery.
70 W.J. Loudon, *Studies of Student Life*, 5: 6–7, 221–2.
71 The National Library of Australia has recently digitized archival copies of the country's newspapers, which are available at trove.nla.gov.au/newspaper/.
72 *Sydney Morning Herald*, 26 July 1922, 6.
73 See, for example, *New York Times*, 24 July 1922 and *Toronto Globe*, 14 July 1922.
74 For a while, it looked as if Joseph Pearce, also at the DAO, might have equaled his boss's record with his finding that another star system, HD 698, had a mass at least 158 times the Sun's: *MNRAS* 92 (1932): 877. Otto Struve and M. Rudkjøbing, "A Note on the Spectrum of HD 698 (J.A. Pearce's Star of Large Mass)," *ApJ* 108 (1948): 537–40 cast serious doubt on Pearce's finding. A new interpretation of the spectrum of Plaskett's Star as showing gas streaming from one component to the other meant that the orbital velocity was reduced, and the mass of the binary was somewhat less than Plaskett had found. A. Underhill, "Plaskett's Star (HD 47129): More and More Curious," *ApJ* 410 (1993): 365–9. For a recent

popular article on the most massive stars, see Yael Nazé, "The Quest for the Most Massive Star," *S&T* 119 (May 2010): 22–7.

75 For the optimistic outlook for 1922 and the reasonably good news of 1922 see *Toronto Globe*, 31 December 1921, 6, and 30 December 1922, 6. For an account of the AAS meeting, see *PA* 30 (1922): 479, which includes a group photo showing JSP and Reba in front of Yerkes Observatory.

76 "The Ultra-violet Spectrograph of the 72-inch Telescope" was summarized in *PA* 31 (1923): 20. Like all items of DAO equipment, its cost and the year of its purchase was listed in LAC, RG48, vol. 50, Property Returns for DAO 1935–8. For further information about this spectrograph, see LOA, W.E. Harper to V. Slipher, 18 June 1928.

77 LOA, JSP to V.M. Slipher, 9 September and 9 October 1922; *Coconino Sun* (Flagstaff, AZ), 29 September 1922, 6.

78 UTA, A1974-0027/004, JSP to C.A. Chant, 22 January 1922. In the end, Klotz was the only Canadian to attend in Rome, as Young was in Australia with Chant.

79 LAC, RG48, vol. 49, file 6-1, JSP to F.W. Dyson, 20 December 1921; LAC, RG48, vol. 48, file 5-12, JSP to G.E. Hale, 21 January 1922.

80 *Trans. IAU* 1 (1922): 95–109. The report of Commission 30 on stellar radial velocities, whose "American" members were Campbell, Adams, and JSP, was originally submitted by the American section of the IAU and subsequently endorsed by the European members, Deslandres, Hamy, and Newall.

81 JSP's election as FRS was noted in the *Times* (London), 24 February 1923, 12; JSP, "A Great Astronomical Meeting," *JRASC* 17 (1923): 219 and *Nature* 111 (3 March 1923): 298. One should not confuse fellowship in the elite Royal Society with fellowship in the RAS, which is open to almost anyone with a serious interest in astronomy or geophysics.

82 LAC, RG48, vol. 47, file 4-17, JSP to F.J.M. Stratton, 20 March 1923.

83 For a list of all fellows since inception of the Royal Society, see: https://royalsociety.org/~/media/Royal_Society_Content/about-us/fellowship/Fellows1660-2007.pdf?la=en-GB. On the same day that JSP's election was made known (5 March 1923), the Royal Society announced J.J.R. Macleod's fellowship. Macleod, though technically not a Canadian, was working at the University of Toronto and won the 1923 Nobel Prize, along with Frederick Banting, for his discovery of insulin.

84 LAC, RG48, vol. 44, file 1-6, W.L.M. King to JSP, 19 April 1923.

85 JSP, "A Great Astronomical Meeting," 219.

86 LAC, RG48, vol. 49, file 6-17, correspondence between JSP and R.W. Boyle. It may be of interest to note that Boyle was the first McGill PhD in physics (1909).

87 See *Edmonton Journal*, 17 May 1923, 3, for an account of the convocation ceremonies and a brief summary of JSP's address. It seems to have been similar to the speech, already cited, that he later gave to the Rotary Club in August.

88 LAC, RG48, vol. 49, file 6-17, JSP to R.W. Boyle, 25 April 1923.

89 *Toronto Star*, 7 June 1923, 1.

90 UTA, A1974-0027/004, JSP to C.A. Chant, letters of 29 June 1920, 18 January 1921, 1 February 1922, and 9 and 19 January 1923.

91 *Toronto Star*, 8 June 1923, 2.

92 Ibid., 3.

93 LAC, RG48, vol. 48, file 5-5, correspondence between A. Cannon and JSP, 18 June–8 August 1923.

94 *PA* 31 (1923): 555.

95 LAC, MG30, series B13, vol. 4, file 32, Klotz diaries, 10 September 1923.

96 JSP, "A Great Astronomical Meeting," 325. Annie Cannon planned to go and hoped to meet up with JSP, Harry, and their wives; see her letter to JSP, 23 February 1921, in LAC, RG48, vol. 48, file 5-5.

97 *Science* 58 (1923): 334–6; abstract in *PA* 31 (1923): 658.

98 JSP, "A Great Astronomical Meeting," 331.

99 Ibid. See also R.A. Jarrell, "The Reception of Einstein's Theory of Relativity in Canada," *JRASC* 73 (1979): 358-69.

100 Thomas S. Kuhn, *Structure of Scientific Revolutions* (Chicago: University of Chicago Press, 1962), 151, citing a 1949 quote from Max Planck's *Scientific Autobiography*: "A new scientific truth does not triumph by convincing its opponents and making them see the light, but rather because its opponents eventually die, and a new generation grows up that is familiar with it."

101 E.R. Paul, *The Milky Way Galaxy*, 101–6, provides a succinct summary of the understanding of general absorption during this period.

102 V.M. Slipher, "Peculiar Star Spectra Suggestive of Selective Absorption of Light in Space," *Lowell Observatory Bulletin* 2 (1909): 1–2; *AN* 189 (1911): 5–8.

103 LOA, JSP to V. Slipher, 18 September 1911.

104 W.G. Hoyt, "Vesto Melvin Slipher" *NAS Biog. Mem.* 52 (1980): 411–49.

105 R.K. Young, "The Calcium Lines H and K in early Type Stars" [1920], *PDAO* 1 (1922): 219–31.

106 LAC, RG48, vol. 48, file 5-18, JSP to R.S. Dugan, 23 November 1928.

107 JSP, "The O-Type Stars," *PDAO* 2 (1924): 335.

108 Osterbrock, *Yerkes Observatory*, 94, is incorrect in saying "J.S. Plaskett … believed that each B star had its own cloud … Struve's measurements … disproved this simple model (and aroused the Plasketts' criticisms)."

109 *PA* 31 (1923): 660.
110 This analogy was used by H.N. Russell, R.S. Dugan and J.Q. Stewart, *Astronomy: A Revision of Young's Manual of Astronomy*, vol. 2, *Astrophysics and Stellar Astronomy* (Boston: Ginn, 1938), 657.
111 The current estimate of the Sun's speed relative to the local standard of rest is 13.4 km/s. See James Binney and Michael Merrifield, *Galactic Astronomy* (Princeton, NJ: Princeton University Press, 1998), 628.
112 The discussion appears in *Obs.* 47 (1924): 4–8, and the paper itself as JSP, "The H and K Lines of Calcium in O-Type Stars," *MNRAS* 84 (1923): 80–93.
113 "Fixed Calcium Clouds in Interstellar Space," *Nature* 112 (22 Dec. 1923): 912.
114 H.H. Turner, letter to the editor, *Times* (London), 18 December 1923, 8.
115 "Fixed Calcium Clouds in Interstellar Space," 912. See also Turner's comments in "From an Oxford Note-book," *Obs.* 47 (1924): 32, and "Stars of the Month," *Times* (London), 4 February 1924, 20.
116 R.J. Trumpler, "Absorption of Light in the Galactic System," *PASP* 42 (1930): 214–27.
117 LAC, MG30, series B13, vol. 4, file 34, Klotz diaries, 28 June 1923.
118 JSP, "The Star of Greatest Known Mass," 284.
119 LAC, RG48, vol. 49, file 6-17, correspondence between JSP and C. Bowman, January 1924; Bowman mentioned his friendship with Klotz, Plaskett, Harper, and others in *Ottawa Editor: The Memoirs of Charles A. Bowman* (Sidney, BC: Gray, 1966), 7.
120 LAC, RG48, vol. 47, file 4-3, JSP to R.M. Stewart, 17 January 1924.

9 The Farthest Stars, 1924–1926

1 JSP's lecture to the Vancouver Institute, "The Evolution of Stars," 7 February 1924, is included in the list of lectures available at www.library.ubc.ca/archives/institute/vaninst2.html.
2 N.J. Crichlow and C.D. Turney, *Stained Glass Windows*, 27–8.
3 Joseph Plaskett, *A Speaking Likeness* (Vancouver: Ronsdale Press, 1999), 40–1.
4 *Islander*, 17 August 1975, 1; Victoria *Daily Colonist*, 21 June 1924, 1 and 3, 5 July 1924, 9.
5 LAC, RG48, vol. 49, file 6-1, JSP to F.W. Dyson, 26 November 1923.
6 Ibid., file 6-12, F. Schlesinger to JSP, 25 March 1927, and JSP's reply, 1 April 1927.
7 Ibid., vol. 47, file 4-3, JSP to R.M. Stewart, February 1924.

8 Ibid.
9 Alan Batten, in a private communication, provided this assessment of the state of affairs when he edited the *Publications* of the DAO in the 1960s.
10 LAC, RG48, vol. 50. The annual returns document what scientific equipment was acquired when, and at what cost.
11 LAC, RG48, vol. 47, file 4-3, JSP to R.M. Stewart, 7 June 1924.
12 Canada, *Official Reports of the Debates in the House of Commons*, 10 June 1925, 4073–6.
13 LAC, RG48, vol. 47, file 4-3, JSP to R.M. Stewart, 26 June 1924; Jarrell, *Cold Light*, 114, says that JSP wrote to King directly in March 1923 and that this led to expenses for the office building being approved. The features of the office building were briefly described in *PA* 33 (1925): 25.
14 On at least two occasions, money for residences seemed assured; see *Victoria Colonist*, 16 November 1915 and 6 May 1919, 7.
15 LAC, RG48, vol. 12, file 1082, JSP to W.W. Cory, 23 November 1923; also vol. 46, file 3-1, correspondence between JSP and J.H. King, Minister of Public Works, 1925.
16 Jarrell, *Cold Light*, 114 cites JSP to W.L.M. King, 3 January 1927 (LAC RG48, vol. 44, file 1-6) and correspondence between W.W. Cory and JSP, 20 and 27 January 1927 (RG48, vol. 47).
17 LAC, RG48, vol. 47, file 4-3, JSP to R.M. Stewart, 23 April 1924.
18 JSP admitted as much in a letter to W.W. Cory, 28 November 1927 (LAC, RG48, vol. 47), though he assured the deputy minister that three of the astronomers were keen to live on the grounds.
19 *MNRAS* 84 (1924): 250. Correspondence and his "Record of Employment Card" giving many biographical details are in LAC, RG48, vol. 47, file 4-4.
20 Ibid., vol. 48, file 5-12, JSP to Sanford, 14 January 1930: "Your letter and the M.S. arrived on the 8th and I have made the very modest corrections and additions asked by you. I still think that you have been exceedingly generous to Mr. Hill and I appreciate very highly your self-sacrifice in the matter." S.N. Hill was superannuated because of ill health, 31 March 1934. See Harper, *MNRAS* 95 (1935): 367.
21 LAC, RG48, vol. 47, JSP to W.W. Cory, 20 June 1927.
22 Crelinsten, *Einstein's Jury*, 210–12.
23 See chapter 8, note 6.
24 C.A. Chant and R.K. Young, "Evidence of the Bending of the Rays of Light on Passing the Sun, Obtained by the Canadian Expedition to Observe the Australian Eclipse" [1923], *PDAO* 2 (1924): 275–85; R.K. Young, "The Canadian Eclipse Expedition," *JRASC* 17 (1923): 129–37. For

three years after he moved to Toronto, Young continued to publish work based on observations at Victoria (with Plaskett's consent), both in *PDAO* and *JRASC*.

25 There was a lot of behind-the-scenes negotiating at the University of Toronto related to Young's desire to do research with a large telescope at a good site and to be appointed associate (rather than assistant) professor. UTA, A1974-0027, correspondence between C.A. Chant and R.K. Young between December 1922 and June 1924.

26 UTA, A74-0027/004, JSP to C.A. Chant, 6 May 1922 and 14 December 1923.

27 UTA, A1974-0027, R.K. Young to C.A. Chant, 8 June 1924.

28 The thesis was published: R.J. McDiarmid, "The Elements of the Eclipsing Systems TV, TW, TX Cassiopeiae, and T Leonis Minoris," *ApJ* 42 (1915): 412–33.

29 JSP, "The Spectroscopic Orbit of TV Cassiopeiae" [1922], *PDAO* 2 (1924): 141.

30 LAC, RG48, vol. 47, file 4-3, JSP to R.M. Stewart, 7 May 1924.

31 Ibid., vol. 48, file 5-5, H. Shapley to JSP, 29 October 1924. Menzel's early career path is discussed by John Lankford, *American Astronomy*, 162–3.

32 LAC, RG48, vol. 48, file 5-5, JSP to H. Shapley, 14 Nov. 1924.

33 PUL, C0045 21/36, Russell's letter to JSP recommending Menzel, Edwin F. Carpenter, and Payne, 4 November 1924. For an extensive discussion of the role of women in astronomy, see chapter 9, "Science and Gender," in John Lankford, *American Astronomy*.

34 C.S. Beals, "Early Days at the Dominion Astrophysical Observatory," *JRASC* 62 (1968): 304.

35 LAC, RG48, vol. 48, file 5-5, H. Shapley to JSP, 9 December 1924. William Francis Herschel Waterfield joined the Harvard College Observatory staff in 1926. His obituary, by Leon Campbell, is in *PA* 41 (1933): 415; another (unsigned) in *MNRAS* 94 (1934): 285 corrects several mistakes in the former.

36 Jarrell, *Cold Light*, quotes correspondence from R.M. Stewart at the DO to Aitken at Lick, 14 May 1929, saying, "according to the general regulations it's not possible to appoint anyone other than a Canadian unless we are able to establish the fact that no properly qualified Canadian is available." The regulation, as stated originally in 1918, required that an applicant be "a natural born or naturalized British subject, and also has been a resident of Canada for at least three years" (1918, c. 12, s. 41). However, a revised statute (1927, c. 22, s. 59) states, "In any case where the [Civil Service] Commission decides that it is not practicable nor in the public interest to apply this Act to any position or positions, the Commission

may, with the approval of the Governor in Council, exclude such position or positions in whole or in part from the operation of the Act, and make such regulations as are deemed advisable prescribing how such position or positions are to be dealt with."

37 LAC, RG48, vol. 47, file 4-3, memo to W.W. Cory from R.M. Stewart, 15 May 1924.

38 Lankford, *American Astronomy*, 136–8.

39 *PA* 31 (1923): 661.

40 *PDAO* 5 (1935). Table 1 on p. 3 shows that Pearce contributed 40 per cent of the observations to Plaskett's 29 per cent; in measurements, Pearce did 37 per cent and Plaskett 28 per cent.

41 LAC, RG48, vol. 48, file 5-1, JSP to B. Boss, 31 August 1923.

42 JSP, "The O-Type Stars," *PDAO* 2 (1924): 287–358.

43 E.A. Milne, *Obs* 48 (1925): 63.

44 Hans Kienle, ed., *Probleme der Astronomie: Festschrift für Hugo v. Seeliger, dem Forscher und Lehrer, zum fünfundsiebzigsten Geburtstage* (Berlin 1924). See 328–37 for JSP's chapter, "Problems of the O-Type Stars."

45 F.J.M. Stratton, "The O-Type Stars," *Obs* 47 (1924): 297-99.

46 JSP, "The O-Type Stars," *PDAO* 2 (1924): 341.

47 JSP, "The O-Type Stars and Their Relation to the Stellar Evolutionary Sequence," *JRASC* 18 (1924): 321 and 327.

48 Ibid.; see also HL, Dunham mss, 1/57, H.H. Plaskett to A. Unsold, 31 March 1929.

49 JSP, "O-Type Stars," 345.

50 A.S. Eddington, "Bakerian Lecture: Diffuse Matter in Interstellar Space," *Proceedings of the Royal Society* 111(A) (1926): 424–56.

51 Ibid.

52 Statements by Eddington crediting JSP appear, for example, in "Bakerian Lecture," 444, in a report of a radio lecture by Eddington in the *Times* (London), 16 April 1929, 21, and in Eddington's *New Pathways in Science* (Cambridge: Cambridge University Press, 1935), 188–9. In "Interstellar Matter," *Obs* 60 (1937): 100, Eddington still credits JSP with "giving decisive evidence for a calcium cloud detached from the stars and nearly at rest in space." However, he goes on say that "it was recognized that adequate evidence of the same kind had been given by V.M. Slipher in 1909, but his work escaped attention."

53 JSP, "Diffuse Matter in Interstellar Space," *Proc. Am. Acad. Arts. Sci.* 64 (1930): 337.

54 "Eddington: The Most Distinguished Astrophysicist of His Time," was the title that Subrahmanyan Chandrasekhar used for his Eddington

Centenary Lecture, published in chapter 6 in his book *Truth and Beauty* (Chicago: University of Chicago Press, 1986). Of course, Eddington was not the only astrophysicist of the time to enjoy a large public following: Owen Gingerich, in a letter to *S&T*, June 1982, 563, quotes *Time* magazine of 16 April 1934 as stating that Sir James Jeans' *The Mysterious Universe* had sold 161,000 copies while Eddington's "fatter and costlier tome," *The Nature of the Physical World*, had sold 53,000.

55 This was becoming clear at the DAO even as early as 1938; see C.S. Beals, "Evidence for Complex Structure in Lines of Interstellar Sodium," *ApJ* 87 (1938): 568–72.

56 Gerrit L. Verschuur, *Interstellar Matters* (New York: Springer-Verlag, 1989), 92–7.

57 JSP, "The O-Type Stars and Their Relation to the Stellar Evolutionary Sequence" is an amplification of JSP's address as retiring president of section III, RSC.

58 The complete, official account is BAAS, *Report of the Ninety-Second Meeting*, available at http://www.biodiversitylibrary.org/item/96033#page/7/mode/1up.

59 The McLennan party was described in the *Toronto Globe*, 12 August 1924. Eddington is quoted by A. Vibert Douglas, *The Life of Arthur Stanley Eddington* (London: Thomas Nelson and Sons, 1956), 96; the British Astronomical Association meetings and excursions are summarized in *Obs.* 47 (1924): 325.

60 Non-mathematicians JSP, Chant, and McLennan were among the thirteen members of the organizing committee of the International Mathematical Congress. See John C. Fields, ed., *Proceedings of the International Mathematical Congress held in Toronto, August 11-16, 1924* (Toronto: University of Toronto Press, 1928), 13. Although the BA meetings influenced the time and location of the congress, there were apparently other considerations as well. See, for example, Guillermo P. Curbera, *Mathematicians of the World, Unite!* (Wellesley, MA: A.K. Peters, 2009), 75–81, who discusses the influence of postwar politics on the decision to hold the congress in Toronto.

61 LAC, RG48, vol. 49, file 6-10, J. Stebbins to JSP, 20 October 1923: "Chant tells me that the British Association will probably meet from September 3–10, so that we can probably arrange for the [AAS] meeting in Dartmouth [College in Hanover, NH] to fit in well with the gathering in Toronto."

62 *Report of the BAAS for 1924*. The description of the official journeys begins on p. 470.

63 LAC, RG48, vol. 49, file 6-27, JSP to C.P. Schwengers (president of the Chamber of Commerce), 12 May 1924.
64 *Vancouver Sun*, 21 August 1924, 18 and 22 August 1924, 1 and 9.
65 *Manitoba Free Press* (Winnipeg), 22 August 1924, 6.
66 *Lethbridge Daily Herald*, 23 August 1924, 1. This same article reported on dubious signals detected at Dulwich, England, at 1 a.m. on 23 August.
67 LAC, RG48, vol. 47, file 4-1, JSP to W.W. Cory, 21 January 1925. This letter is the source for many of the following details of the ways that JSP was honoured.
68 Ibid., correspondence between JSP and W.W. Cory 29 January–25 March 1925, for all quotes in this paragraph.
69 Victor Gaizauskas, "The Grand Schism in Canadian Astronomy," *JRASC* 106 (2012): 198.
70 F. Henroteau, "The International Astronomical Union at Cambridge," *JRASC* 19 (1925): 197–200 gives a good brief account of the meeting.
71 *Proc. Trans. RSC* 19 (1925): xxxvii. The actual paper by McLennan, "On the Origin of the Auroral Green Line" was not published here, just listed among those presented at the meeting.
72 Passenger lists from www.ancestry.com.
73 *Obs.* 48 (1925): 206.
74 Quoted in Margaret Wilson, *Ninth Astronomer Royal*, 214–15.
75 LAC, RG48, vol. 49, file 6-1, F.W. Dyson to JSP, 30 May 1925. The original invitation apparently did not reach Plaskett, as he sailed for England on 29 May.
76 The source of this information and much of what follows concerning JSP's time in England is LAC, RG 48, vol. 47, file 4-1, JSP [at Regent Palace Hotel] to W.W. Cory, 25 July 1925; "Luncheon to Dr. and Mrs. J.S. Plaskett," *JRASC* 19 (1925): 247.
77 *JRASC* 19 (1925): 253.
78 JSP, *Trans. IAU* 2 (1925): 117, 121–3, 207–9.
79 Cecilia Payne-Gaposhkin, *An Autobiography*, 169. Her objection was that the spectra of Wolf-Rayet stars sometimes showed absorption lines, but JSP was vindicated at the next IAU meeting, three years later, when his plan was approved, with partial support from Payne.
80 LAC, RG48, vol. 48, file 6-12, F. Schlesinger to JSP, 4 November 1926.
81 CCL, Frank Schlesinger diary.
82 UTA, A1974-0027/010, Chant's autobiography, p. 719.
83 LAC, RG48, vol. 49, file 6-1, F.W. Dyson to JSP, 3 July 1925.
84 S.A. Mitchell, "The Two Hundred and Fiftieth Anniversary at Greenwich," *JRASC* 19 (1925): 137 includes a photograph of King George V at the observatory. A short video clip of the event includes

their majesties and Sir Frank Dyson: http://www.britishpathe.com/video/g-h-q-time/.

85 R.J. Tayler, *History of the Royal Astronomical Society*, 27.

86 *Victoria Daily Times*, 1 September 1925, 1, also states that the Plasketts were guests in Perthshire of relatives of Mrs R. Green of Victoria.

87 Canadian Passenger lists at www.ancestry.com.

88 LAC, RG 48, vol. 47, file 4-1, correspondence between JSP and W.W. Cory, 25 July–14 September 1925.

89 Ibid. JSP to W.W. Cory, 27 January 1925. The articles are: "Canada Prominent at Astronomical Union," Natural Resources Canada (Department of the Interior) newsletter 4, no. 10 (1925); "Interesting Data on Remarkable Class of Stars," 5, no. 7 (1926); "Astronomers Seeking New Source of Energy," 6, no. 4 (1927); "Study of Atom Aids Modern Development," 7, no. 2 (1928); "How the Universe Revolves in Space," 8, no. 5 (1929).

90 "Luncheon to Dr. and Mrs. J.S. Plaskett," 247, 303, gives an account of his speeches of 25 September to the Victoria branch of the Canadian Club and on 2 October to the RASC Victoria Centre. Plaskett subsequently spoke to the Victoria Canadian Club in September 1926.

91 *Victoria Daily Times*, 1 September 1925, 9.

92 JSP told Cory in a letter of 8 September 1925, that he had already been asked to give three talks about his trip, LAC, MG48, vol. 47, file 4-1. The purposes of the Canadian Club are taken from their website. On 11 January 1935, the club presented him with a complimentary dinner and a beautiful desk set (*JRASC* 29 (1935): 73). His long-standing membership in the Canadian Club was noted in the *Victoria Colonist*, 6 April 1943.

93 *JRASC* 19 (1925): 256; LAC, RG48, vol. 47, file 4-1, JSP to W.W. Cory, 28 September 1925.

94 The citations for all UBC honorary degree recipients can be found at http://www.library.ubc.ca/archives/hdcites/hdcites1.html.

95 J.A. Pearce, "The Portland Meeting of the American Association for the Advancement of Science," *JRASC* 19 (1925): 207–16. Presumably the "two interesting B-type binaries" that Pearce reported on were those that JSP described in "The Orbits of Two Double-Lined Binaries" [1925], *PDAO* 3 (1927): 179–88.

96 JSP, "Three Spectroscopic Binary Orbits (21 Cas, Boss 3354 and HD 191201)" [1926], *PDAO* 3 (1927): 247–64.

97 Ibid., 248–55.

98 LAC, RG48, vol. 47, file 4-15, April 1925, correspondence between J. Stebbins and JSP.

99 G. Shajn and O. Struve, "On the Rotation of Stars," *MNRAS* 89 (1929): 222–39, include a brief history of the subject up to that time. Their paper

was included in Owen Gingerich and Kenneth Lang, eds., *A Source Book in Astronomy and Astrophysics, 1900–1975* (Cambridge: Harvard University Press, 1979).

100 JSP, "Note on Three Peculiar Spectra," *MNRAS* 87 (1926): 31–4, "Three Peculiar Spectra" [1926], *PDAO* 4 (1931): 1–26, and "Two Spectroscopic Orbits and Notes on υ Sagitarii" [1928], *PDAO* 4 (1931): 103–18.

101 *Obs.* 49 (1926): 354–5; LAC, RG48, vol. 49, file 6-1, JSP to F. Dyson, 11 October 1926.

102 JSP, "Notes on υ Sagittarii," 111.

103 JSP and J.A. Pearce, "A Catalogue of the Radial Velocities of O and B Type Stars" [1930], *PDAO* 5 (1935): 108 along with the notes starting on p. 160, show that there were many doubtful cases among the stars that were announced as spectroscopic binaries.

104 Lists of DAO spectroscopic binaries were published in the *JRASC* in 1918 and 1919 and a cumulative list of the first hundred in *PDAO* 1 (1922): 163. A supplementary list of eighty-eight more came out in *PDAO* 1 (1922): 287 with a few corrections in *JRASC* 18 (1924): 367. The Lick tradition that held Plaskett back from producing a comprehensive catalogue is discussed by R.A. Jarrell, "The 1910 Solar Conference and Cooperation in Stellar Spectroscopy," *JAHH* 13 (2010): 127–38. However, the DAO did eventually take over (by mutual agreement) the task of updating the catalogues of spectroscopic binaries. Alan Batten produced three editions, in 1968, 1978, and 1989. The job has now been assumed by an international group, with the ninth catalogue appearing online.

105 JSP and J.A. Pearce, "The Radial Velocities of 523 O and B Type Stars Obtained at Victoria, 1923–1929" [1930], *PDAO* 5 (1935): 8, credits JSP with 36 spectroscopic binary discoveries, which, added to 58 in earlier lists referred to in the previous note, gives a total of 94. The number of Harper's orbits is mentioned in his obituaries and biographical articles.

106 LAC, RG48, vol. 47, file 4-7, JSP to W.W. Cory, 11 February 1926. JSP's involvement with the NRC dated back at least to May 1925, when he wrote to Cory (file 4-1, 6 May 1925) that the Associate Committee of the Research Council would pay his transcontinental travel expenses to attend its meetings in Ottawa.

107 Correspondence of 1926 quoted in Mel W. Thistle, *The Inner Ring*, 196–7.

10 Beyond the Stars, 1927–1930

1 JSP's address to section III: "Sixty Years Progress in Astronomy," adapted in *JRASC* 21 (1927): 295–310.

2 *Proc. Trans. RSC* 21 (1927): lxxxvi.
3 *JRASC* 22 (1928): 69, 70.
4 LAC, RG48, vol. 47, JSP to R.A. Gibson, 9 July 1927. For Young's description of the 48 cm telescope, see his paper in *JRASC* 24 (1930): 17.
5 LAC, RG48, vol. 47, JSP to W.W. Cory, 29 April 1927.
6 JSP, "The Orbit of the B-type Spectroscopic Binary, HD 193536," *PASP* 39 (1927): 258.
7 JSP, "Two Spectroscopic Orbits and Notes on υ Sagitarii" [1928], *PDAO* 4 (1931): 103-18. Y Cygni was another case in which the period, 2.9963 days, was almost exactly a whole number.
8 AAAS *Proceedings* 77 (1925): 757, shows Plaskett became a member in 1922, a fellow in 1923, and was allied to the astronomy section. JSP's certificate of election as fellow of the AAAS, dated 26 April 1923, is in LAC, RG48, vol. 44, series 1. JSP was elected vice-president of Section D of AAAS (*Nature* 121 (28 January 1928): 142).
9 LAC, RG48, vol. 47, file 4-10, JSP to R.M. Stewart, 15 November 1928.
10 *PA* 35 (1927): 481.
11 Alan H. Batten, *Resolute and Undertaking Characters: The Lives of Wilhelm and Otto Struve* (Dordrecht: Reidel, 1988).
12 LAC, RG48, vol. 49, file 6-14, correspondence between E.B. Frost and JSP, June 1922. For articles about the plight of Russian astronomers and relief efforts, see "An Astronomer's Life in Russia," *PA* 29 (1921): 524–5; "Astronomical News from Russia," *PA* 30 (1922): 1–3; and "Relief for Russian Astronomers," *PA* 30 (1922): 255–6 and 323–4.
13 O. Struve, "On the Calcium Clouds," *PA* 33 (1925): 639–53 and 34 (1926): 1–14.
14 Donald E. Osterbrock, *Yerkes Observatory*, 90–1.
15 O. Stuve, "Interstellar Calcium," *ApJ* 65 (1927): 163–99, quotes from 163 and 165.
16 JSP, "Diffuse Matter in Interstellar Space," *Proc. Am. Acad. Arts Sci.* 64 (1930): 339.
17 See DeVorkin, *Henry Norris Russell*, 150–2 for a discussion of this. For an almost opposite interpretation, see T.G. Cowling, "Otto Struve," *Biogr. Mems. Fell. R. Soc.* 10 (1964): 283–304.
18 H.N. Russell, "Some Problems of Sidereal Astronomy," *Proc. NAS* 5 (1919): 391–416 reprinted *PA* 28 (1920): 212–24, 264–75.
19 A.S. Eddington, "The Internal Constitution of the Stars," *Obs* 43 (1920): 341–58.
20 *Obs.* 52 (1929): 345; JSP and J.A. Pearce, "The Motions and Distribution of Interstellar Matter," *MNRAS* 90 (1930): 243–68.
21 JSP, "Two Spectroscopic Orbits and Notes on Upsilon Sagitarii" [1928], *PDAO* 4 (1931): 108.

22 LAC, RG48, vol. 47, file 4-7, correspondence between JSP and R.M. Stewart, 1926.

23 Annual inventories of all the equipment on hand were required to be submitted annually to Ottawa. For the Property Returns for 1935–7 that list this equipment and the dates acquired, see LAC, RG48, vol. 50.

24 A.R. Hassard, "Visit of Mr. W.E. Harper to Toronto," *JRASC* 23 (1929): 233–5.

25 C.S. Beals, "Early Days at the Dominion Astrophysical Observatory," *JRASC* 62 (1968): 304.

26 *JRASC* 14 (1920): 428; the seminars were held weekly, starting in December 1926 (see *JRASC* 22 (1928): 49). They sometimes appeared in the *JRASC* under the heading of "Papers Prepared for the DAO Club."

27 The number of visitors was reported for each year from 1927 to 1930 in the reports of the DAO published in *MNRAS*. The year 1929 seems to have set the record, with 39,027 recorded visitors.

28 J.A. Pearce, "Some Recollections of the Observatory (1924–1935)," *JRASC* 62 (1968): 290.

29 *Victoria Daily Colonist*, 18 August 1918, 17.

30 LAC, RG48, vol. 47, file 4-7, JSP to R.M. Stewart, 18 November 1926.

31 P. Broughton, "Fifty Years of Canadian Ph.D.s in Astronomy," *JRASC* 97 (2003): 21–3.

32 The names of personnel, including summer assistants, are found in the reports of the DAO published each year in *MNRAS*. For a brief obituary of Christie, see *PASP* 67 (1955): 366. Plaskett mentions that Christie was "a distant relative of the late Astronomer Royal" of the same name in a letter to V. Slipher, 13 October 1927, LOA.

33 John Lankford, *American Astronomy*, 342.

34 Broughton, "Fifty Years of Canadian Ph.D.s," 21.

35 J.S. Foster and A.V. Douglas, "Analysis of Profiles of Helium Lines in Spectra of B Stars," *Nature* 134 (1934): 417, thank JSP for the opportunity to work at the DAO.

36 H.H. Plaskett, "The Wedge Method and Its Application to Astronomical Spectrophotometry," *PDAO* 2 (1924): 213–59, and correspondence between JSP and H.N. Russell, 11 March, 26 April, and 28 August 1921, PUL, C0045, 55/40 (for the first two) and 21/36 for the last. Russell may also have visited the DAO in the spring of 1928: see his letter to JSP, 28 November 1927, PUL, C0045 21/36.

37 LAC, RG48, vol. 48, file 5-5, H. Shapley to JSP, 14 May 1925, re Lassovsky, and the following references in *MNRAS* 88(1928): 279 for Zanstra; 90 (1930): 407 for Pannekoek; 91 (1931): 358 for Bruggencate; and *JRASC* 25 (1931): 268 for Petersson. JSP recalled that astronomers from abroad also included

Eddington and Jackson from England, Deslandres from France, Bourgeois from Belgium, Duffield from Australia, and "several eminent astronomers" from China and Japan, see *Victoria Daily Colonist*, 25 February 1931, 2.

38 LAC, RG48, vol. 49, file 6-4, Pannekoek correspondence.

39 A. Pannekoek, "Line Intensities in Spectra of Advanced Type," *PDAO* 8 (1946): 141-223.

40 Lord Rayleigh, "Some Recent Work on the Light of the Night Sky," *Nature* 122 (1 Sept 1928): 315–17; LAC, RG48, vol. 48, file 5-8, E. Hertzsprung to JSP, 27 January 1931, and file 5-18, K. Walter to JSP, 20 June 1933.

41 PUL, C0045, correspondence between JSP and H.N. Russell, January–February 1923; LAC, RG48, vol. 48, file 5-5, correspondence between JSP and H. Shapley, January 1923.

42 LAC, RG48, vol. 48, file 5-5, H.H. Plaskett to H. Shapley, 19 July 1927.

43 Ibid., vol. 47, file 4-17, JSP to Stratton, 2 December 1927.

44 Cecilia Payne-Gaposhkin, *An Autobiography*, 156–7, 222.

45 M.G. Adam and Roger Hutchins, "Plaskett, Harry Hemley," *Oxford Dictionary of National Biography* (Oxford 2004).

46 ADBCA, Baptismal record, Christ Church Cathedral, Victoria.

47 LAC, RG48, vol. 49, file 6-1, JSP to F.W. Dyson, 14 December 1927.

48 Beals apparently began work in the fall of 1927, after it was known that Harry Plaskett would be leaving but before he actually left.

49 LAC, RG48, vol. 47, file 4-7, JSP to R.M. Stewart, 14 May 1926.

50 K.O. Wright, "Obituary – Beals, Carlyle S.," *QJRAS* 21 (1980): 212; for correspondence from Harper wanting authority to contract for a grating spectrograph with Sidney Girling, see LAC, RG48, vol. 45, file 2-1, Harper to R.M. Stewart, 4 July 1928. Beals remained at the DAO until he went to Ottawa as Dominion astronomer in 1947.

51 LAC, RG48, vol. 49, file 6-1, JSP to F.W. Dyson, 9 February 1926.

52 Ibid., vol. 47, file 4-17, JSP to Stratton, 22 February 1927.

53 R.F. Griffin and Richard Woolley, "Roderick Oliver Redman," *Biogr. Mems. Fell. R. Soc.* 22 (1976): 335. The title of Redman's thesis is "An Investigation into the Parallaxes, Absolute Magnitudes and Motions of Certain Stars." He published the results of his main research at the DAO in "The Galactic Rotation Effect in Late Type Stars," *MNRAS* 90 (1930): 690.

54 LOA, JSP to V. Slipher, 3 February 1931.

55 JSP and J.A. Pearce, "The Motions of the O and B Type Stars and the Scale of the Galaxy" [1934], *PDAO* 5 (1935): 305. JSP returned to this idea of three stages in "Modern Conceptions of the Stellar System," *PA* 47 (1939): 239.

56 This is an oversimplification of a long and fascinating story told by Marcia Bartusiak, *The Day We Found the Universe.*

57 J. Kapteyn, "First Attempt at a Theory of the Arrangement and Motion of the Sidereal System," *ApJ* 55 (1922): 302.

58 H. Shapley, "Remarks on the Arrangement of the Sidereal Universe," *ApJ* 49 (1919): 311 (the twelfth paper in his series of Mount Wilson Contributions).

59 JSP, "The O-Type Stars," *PDAO* 2 (1924): 320.

60 E. Freundlich and E. von der Pahlen, "Untersuchung des K-Effektes auf Grund des Katalogs von Radialgeschwindigkeiten von J. Voûte," *AN* 218 (1923): 369.

61 J.A.H. Gyldén, *Öfversigt af K. Vetenskapsakademiens Förhandinger* 28 (1871): 947. This reference is cited by A.S. Eddington, *Rotation of the Galaxy*, 13, who credits B. Lindblad for providing it. Even now Gyldén's great insight is rarely remembered. His biography in the *Biographical Encyclopedia of Astronomers* does not mention it. Strictly speaking, rotation refers to a body rotating on an axis, as Earth does daily, while revolution refers to motion about a central point, as Earth does yearly in its journey around the Sun. Plaskett and his contemporaries thought the stars were revolving about the centre of the Galaxy, but nevertheless often used the term "rotation" to describe their motion, terminology that has stuck.

62 Lindblad's papers (in Swedish) are cited in Jan H. Oort, "Observational Evidence Confirming Lindblad's Hypothesis of a Rotation of the Galactic System," *BAN* 3 (1927): 275–82.

63 Oort, ibid., and subsequent papers in *BAN* 4 (1928): 79–89, 91–3. For a first-hand account by Oort's student Bart Bok, see "The First Five Years of Jan Oort at Leiden 1924–1929," in Hugo van Woerden, W.N. Brouw, and H.C. van de Hulst, eds., *Oort and the Universe: A Sketch of Oort's Research and Person* (Dordrecht: Reidel, 1980), 55–8.

64 For a straightforward mathematical proof of Oort's equations, see M. Longair, *The Cosmic Century*, 97–9.

65 J.H. Oort, "Investigations Concerning the Rotational Motion of the Galactic System Together with New Determinations of Secular Parallaxes, Precession and Motion of the Equinoxes," *BAN* 4 (1927): 87.

66 For an easily accessible and clear discussion of these points, see https://en.wikipedia.org/wiki/Oort_constants.

67 LAC, RG48, vol. 49, file 6-12, correspondence between JSP and F. Schlesinger, 15–27 July 1927.

68 AIP, selected correspondence of J.H. Oort (microfilm), correspondence between J.H. Oort and JSP, 28 November and 22 December 1927.

69 *Obs.* 51 (1928): 33; JSP, "The Rotation of the Galaxy," *MNRAS* 88 (1928): 395–403.
70 JSP, "The Motion of Stars," *JRASC* 22 (1928): 111–34.
71 Ibid., 134.
72 David Baneke, "Teach and Travel: Leiden Observatory and the Renaissance of Dutch Astronomy in the Interwar Years," *JHA* 41 (2010): 167–98.
73 LAC, RG48, vol. 47, file 4-9, JSP to R.M. Stewart, 31 March 1928.
74 Ibid., JSP to R.M. Stewart, 3 May 1928.
75 I heard Bart Bok tell the story at the sixtieth anniversary of the London Centre of the RASC. Bok, in 1928, was a young graduate student, designated to meet IAU guests at the train station. Bok went on to be "the man who sold the Milky Way," to use the title of his biography by David Levy. Bok and his wife, Priscilla, wrote an influential book, *The Milky Way*, which was issued in several editions over the course of forty years.
76 LAC, RG48, vol. 47, file 4-9, copy of a letter from R.M. Stewart to the deputy minister, along with JSP to Stewart, 31 March 1928.
77 R.E. DeLury, "News and Comments," *JRASC* 22 (1928): 199, 247.
78 Ibid. Passenger lists from www.ancestry.com confirm that the Plasketts travelled from Montreal on *Ascania* and arrived at London on 17 June 1928. The Chants got to Liverpool on 28 June, and the Buchanans arrived in London on 1 July.
79 "Meeting of the British Astronomical Association," *Obs.* 51 (1928): 261–3.
80 C.A. Chant, "Two European Astronomical Meetings," *JRASC* 22 (1928): 383–92, gives a full account of the receptions and excursions with photos.
81 *JRASC* 23 (1929): 116.
82 J.H. Oort to the author, *JRASC* 82 (1988): 31.
83 Adriaan Blaauw, *History of the IAU; Birth and First Half-Century of the International Astronomical Union* (Dordrecht: Kluwer, 1994).
84 The first Zeiss planetarium was unveiled at the Deutsches Museum, Munich, in October 1923, according to O. Gingerich, *The Physical Sciences*, 274.
85 C.A. Chant, "Two European Astronomical Meetings," 391.
86 Passenger lists from www.ancestry.com.
87 LAC, RG48, vol. 47, file 4-10, JSP to R.M. Stewart, 21 November 1928.
88 *JRASC* 23 (1929): 115.
89 Canada, *Report of the Auditor General for the Year Ended March 31 1929*, vol. 2, *Public Accounts* (Ottawa 1929), p. K16 for salary and K32 for travel expenses.
90 LAC, RG48, vol. 48, file 5-8, correspondence between E. Hertzsprung and JSP, 12–13 December 1928 and 7 January 1929; this resulted in E.

Hertzsprung, "A New Probable Member of the Ursa Major Group," *BAN* 6 (1930): 60.

91 LAC, RG48, vol. 48, file 5-8, JSP to Jan Oort, 5 April 1929; reply, 11 May 1929.

92 For abstracts of the Ottawa papers, see *PA* 38 (1930): 90–1, or *PAAS* 6 (1931): 277.

93 UTA, A1974–0027/004, JSP to C.A. Chant, 22 May 1929.

94 The report of the Berkeley meeting is in *PASP* 41 (1929): 244 and includes an abstract of Plaskett and Pearce's paper, p. 251.

95 JSP, "The Expansion of the Universe," *JRASC* 27 (1933): 243; cf. Robert W. Smith, *The Expanding Universe*, 165.

96 Bartusiak, *The Day We Found the Universe*, 236, cites *Los Angeles Times*, 10 November 1929, F4.

97 Summaries of these papers appear in *PA* 38 (1930): 90–2; and shorter summaries in *Science News-Letter* 16 (14 September 1929): 159.

98 The idea that certain B-type stars apparently moved through the Galaxy in streams was promulgated by Kapteyn in 1913 and incorporated by Shapley in 1918. See O. Gingerich "Kapteyn, Shapley, and Their Universes," in P.C. Van der Kruit and K. van Berkel, eds., *The Legacy of J.C. Kapteyn*, 198.

99 Plaskett and Pearce give a historical account of the K effect up to their time in *PDAO* 5 (1935): 244. For two later perspectives written at two different times, see H. Weaver, "The K-effect in Stellar Motions," *Vistas in Astronomy* 1 (1955): 228–38 and H. Arp, "Redshifts of High-luminosity Stars: The K Effect, the Trumpler Effect and Mass-loss Corrections," *MNRAS* 258 (1992): 800–10.

100 For an abstract, see *PA* 38 (1930): 90 or *PAAS* 6 (1931): 277.

101 Pearce gives credit to Pannekoek for the suggestion in "Some Recollections of the Observatory," *JRASC* 62 (1968): 294, but Oort had already done this using the DAO data available to him in 1927.

102 JSP and J.A. Pearce, "The Motions and Distribution of Interstellar Matter," *MNRAS* 90 (1930): 243–68; Eddington, *Rotation of the Galaxy*, 11 is the source of the quote.

103 *Times* (London), 2 June 1930, 8.

104 Eddington, *Rotation of the Galaxy*, 13.

105 B.P. Gerasimovič and O. Struve, "Physical Properties of a Gaseous Substratum in the Galaxy," *ApJ* 69 (1929): 7-33, who also cite J. Oort *BAN* 3 (1927): 275.

106 JSP and J.A. Pearce, "Motions and Distribution of Interstellar Matter," 243–68.

107 LAC, RG48, vol. 47, file 4-11, W.W. Cory to JSP, 5 October 1929.

108 The text of JSP's Toronto lecture, "The Rotation of the Galaxy," given 4 November 1929, is available at https://www.canadianclub.org/Events/EventDetails.aspx?id=786.

109 The story was recalled by J.A. Pearce, *JRASC* 62 (1968): 296. The transcript, referred to in the preceding note, suggests that JSP was able to show his slides only after his formal presentation was finished.

110 *New York Times*, 12 November 1929, 31.

111 CITA, Hale papers, box 33, correspondence between JSP and G.E. Hale, December 1929–January 1930.

112 LAC, RG48, vol. 47, file 4-11, JSP to R.M. Stewart; LOA, a letter from JSP to V. Slipher, 15 November 1929, confirms JSP's one-day visit to the Lowell Observatory on 24 November.

113 JSP, "The Motions of the B Stars," *Science* 71 (1930): 225–30; for a popular account, see *Science News-Letter* 17 (4 January 1930): 2; another with many photographs, including one of JSP, is "New York Meeting of the AAAS," *Science Monthly* 28 (1929): 5–40; for a brief report, see *New York Times*, 1 January 1930, 19.

114 *Science* 71 (1930): 152.

115 JSP and J.A. Pearce, "A Catalogue of the Radial Velocities of O and B Type Stars" [1930], *PDAO* 5 (1935): 99–165. It is stated on p. 100 that the preparation was mainly due to Pearce.

116 To give a couple of examples: for the star 60 Cygni, see Petr Harmanec et al., *Information Bulletin on Variable Stars*, no. 2912 (July 1986): 1; for HD 28446, HD 42087, and HD 195592, see Pavel Mayer, Magdy A. Hanna, Marek Wolf, and Drahos Chochol, "Radial Velocities of Six Early-type Evolved Stars," *Astrophysics and Space Science* 262 (1999): 163, available at http://link.springer.com/article/10.1023%2FA%3A1001840718226#page-1.

117 J.H. Moore, "A General Catalogue of the Radial Velocities of Stars," *PLO* 18 (1932).

118 LAC, RG48, vol. 47, file 4-11, JSP to R.M. Stewart, 7 March 1930.

119 Ibid., vol. 48, file 5-12, JSP to W.S. Adams, 27 March 1930.

120 See table 1 in JSP and J.A. Pearce, "The Radial Velocities of 523 O and B Type Stars Obtained at Victoria, 1923-01929" [1930], *PDAO* 5 (1935): 3.

121 J.A. Pearce, "Some Recollections of the Observatory," *JRASC* 62 (1968): 287.

122 Ibid. The Esquimalt location of the golf course suggests that JSP had become a member there after he took up residence on Armit Road.

123 Beals, "Early Days at the Dominion Astrophysical Observatory," 310.

124 H.H. Plaskett, "The Wedge Method and Its Application to Astronomical Spectrophotometry," *PDAO* 2 (1924): 213–59.
125 LAC, RG48, vol. 47, file 4-3, JSP to R.M. Stewart, 5 March 1924.
126 Ibid., vol. 49, file 6-12, JSP to F. Schlesinger, 13 February 1922.
127 Ibid., vol. 47, file 4-11, JSP to R.M. Stewart, 4 May 1931.
128 Ibid., file 4-15, JSP to J. Stebbins, 13 August 1926.
129 A.H. Batten to the author, private communication, April 2012.
130 J.A. Pearce, "Some Recollections," 293, and Beals, "Early Days," 305.
131 I am indebted to Karen James, a resident of Plaskett Place, for pointing out that, in the Esquimalt Municipal Archives, the Esquimalt building permit registry shows that JSP obtained the permit for 318 Armit Road in March 1926; the proposed two-storey, six-room house of frame and stucco construction on a concrete foundation was valued at $4600, above average in comparison to similar ones in the registry. *Wrigley's British Columbia Directory* for 1929 lists J.S. Plaskett and Stuart Plaskett on Armit Road. Reeve Alfred Wurtele had Armit Road renamed Plaskett Place during his time in office (1952–63), according to e-mails from his daughters. Will/probate files for Rebecca Hemley Plaskett (1967) includes an appraisal of the property with a thorough description of the land and dwelling (BCA, 23160-40/B).
132 LAC, RG48, vol. 49, file 6-18, "Miscellaneous 'G'," 20 July 1919.
133 "Imperialism was one form of Canadian nationalism": Carl Berger, *The Sense of Power: Studies in the Ideas of Canadian Imperialism, 1867–1914*, 2nd ed. (Toronto: University of Toronto Press, 2013), 259.
134 BCA, MS 2736, British Columbia Historical Association fonds, vol. 1, file 7, lists the officers and their terms. Ibid., MSS 2779, box 1, file 2, is an association scrapbook that contains clippings, some mentioning the Plasketts.
135 For a history of the cathedral, see Sel Caradus, *A Temple Not Made with Hands: A History of Christ Church Cathedral* (Victoria: Printorium Bookworks, 2004). Information about JSP's involvement with the cathedral comes from correspondence with Jacquie Nevins, archivist, ADBCA, and items in the archives as follows: Christ Church Cathedral: Cathedral Buildings Ltd. Minutes T223, 1; Correspondence with R. Meldrum Stewart, June 1926, b.1; circular to Synod Journal, 1928; clippings from *Victoria Daily Colonist*, 26 and 27 September 1929; "Civic Dinner" program; and vestry minutes, 15 January 1929 and 15 January 1930, T208. Detailed reports of the consecration and banquet were also front-page news in the *Victoria Daily News* of 28 and 29 September 1929.
136 ADBCA, T.208, minutes of the vestry meeting of 15 January 1930, p. 396.

11 The Big Picture, 1930–1934

1 Roger J. Tayler, ed., *History of the Royal Astronomical Society*, 2: 22.

2 Crommelin's citation of JSP's accomplishments and career are summarized in *Obs.* 53 (1930): 65–9, and published in full in A.C.D. Crommelin, "Address ..., " *MNRAS* 90 (1930): 471–7 and also in Crommelin, "The Astronomical Work of John S. Plaskett," *JRASC* 24 (1930): 217–32. The report of the meeting at which the presentation was made is in *Obs.* 53 (1930): 161–4. A summary of JSP's lecture is in JSP, "The George Darwin Lecture 1930," *Obs.* 53 (1930): 223–9; for the full lecture, see JSP, "The High Temperature Stars," *MNRAS* 90 (1930): 616–35.

3 *Obs.* 53 (1930): 65.

4 LAC, MG26-J1, microfilm C-2323, vol. 183, pp. 155980–2, Prime Minister King's correspondence, February 1930.

5 Ibid.

6 Ibid., vol. 182, pp. 155475–7, Prime Minister King's correspondence, February 1930. Contrast this letter of Charles Stewart's with his much more supportive comments made in 1925 in the House, noted in chapter 9. Permission had originally been given in 1926 for Stuart Plaskett to do occasional clerical chores at the DAO for 50 cents an hour (LAC, RG48, vol. 47, file 4-8, JSP to R.M. Stewart, 6 December 1927).

7 JSP, "Diffuse Matter in Interstellar Space," *Proc. Am. Acad. Arts Sci.* 64 (1930): 335–46; "The Structure and Rotation of the Galaxy," *Proc. Am. Phil. Soc.* 69 (1930): 401–17.

8 LAC, RG48, vol. 48, file 5-5, telegrams. I have been unable to ascertain if Harry played any role in nominating his father.

9 Ibid. Canadians were not honoured again with the Rumford Premium until 1971, when a team of radio astronomers shared the prize with two other teams, *JRASC* 65 (1971): 188.

10 JSP, "Diffuse Matter in Interstellar Space," 335.

11 For JSP's lectureship at Harvard, along with that of Paul ten Bruggencate, see *Harvard Crimson*, 15 April 1930, www.thecrimson.com/article/1930/4/15/prominent-astronomers-come-to-lecture-at/.

12 C.A. Chant, "The R.A.S. Gold Medal," *JRASC* 24 (1930): 90; "Canadian Discusses Spiral Form in Astronomical Session of Philosophical Society," *New York Times*, 27 April 1930, 21. For details of the symposium and Plaskett's speech, see *Proc. Am. Phil. Soc.* 69 (1930): xv and 401.

13 Incoming passenger lists, www.ancestry.com. This was the third ship to be named *Arabic* in the White Star Line.

14 JSP, "The High Temperature Stars," *MNRAS* 90 (1930): 616, but see also Crommelin, "The Astronomical Work of John S. Plaskett," 217–32; JSP, *PA* 38 (1930): 90; *MNRAS* 91 (1931): 358; *Obs.* 53 (1930): 161.
15 JSP, "High Temperature Stars," 635.
16 The two meetings are outlined in *Obs.* 53 (1930): 201 and 222. The Plasketts arrived at Montreal from Liverpool aboard the *Doric* on 6 July 1930, according to passenger lists on www.ancestry.com.
17 The meeting in Winnipeg is described in *JRASC* 25 (1931): 106 and in the *Winnipeg Tribune*, 18 July 1930, 3.
18 LAC, RG48, vol. 44, file 1-6, 27 September 1920, JSP to W.L.M. King, and reply of 1 October 1920; King had also planned to visit in 1924 but sent a telegram to JSP, 18 October 1924, expressing his great disappointment at his inability to do so.
19 H. Blair Neatby, "King, William Lyon Mackenzie," *Dictionary of Canadian Biography*, vol. 17, online at http://www.biographi.ca/en/bio/king_william_lyon_mackenzie_17E.html; Allan Levine, *King: William Lyon Mackenzie King: A Life Guided by the Hand of Destiny* (Vancouver: Douglas & McIntyre, 2011).
20 Diaries of William Lyon Mackenzie King, 7 July 1930, http://www.bac-lac.gc.ca/eng/discover/politics-government/prime-ministers/william-lyon-mackenzie-king/Pages/item.aspx?IdNumber=12350&. The date being the seventh day of the seventh month was significant for King and he recalled his memorable visit in his diary on 8 July 1937, 7 July 1939 and 7 July 1942.
21 *Times* (London), 4 December 1931, 4 and 2 October 1933, 3. For a broader picture, see Jack Morrell, *Science at Oxford, 1914-1939; Transforming an Arts University* (Oxford: Clarendon, 1997).
22 Madge G. Adam and Roger Hutchins, "Plaskett, Harry Hemley," *Oxford Dictionary of National Biography* (Oxford 2004).
23 David Levy, *The Man Who Sold the Milky Way*, 28.
24 After Harry obtained further observations at Oxford and wrote about his results ("Solar Granulation," *MNRAS* 96 (1936): 402), Harlow Shapley drew attention to his research as one of ten "conspicuous astronomical advances of the year" *Science* 84 (1936): 430–1.
25 Though the younger Plasketts did not move to Oxford until June, Harry spent January there. *Montreal Gazette*, 17 December 1931, 1.
26 It is often claimed that the telescope of the Perkins Observatory was the first large telescope to be made entirely in the United States. However, the 94 cm reflector at Steward Observatory, which was completed in 1922, also claims that honour.

27 LAC, RG48, vol. 46, H. Stetson to JSP, 12 October 1931, cited by R. Jarrell, "J.S. Plaskett and the Modern Large Reflecting Telescope," *JHA* 30 (1999): 378.
28 LAC, RG48, vol. 48, file 5-16, JSP to R.M. Stewart, 6 November 1931.
29 *Ohio Wesleyan Transcript* (Delaware, OH), 4 December 1939, 1.
30 The telescope was moved to a much better site at Lowell Observatory in 1961. Three years later, the mirror was replaced with a slightly larger one, 1.83 m in diameter, equal to that at the DAO. The original mirror is now on display at Perkins Observatory, its original site in Ohio.
31 JSP, "Ambrose Swasey," *JRASC* 31 (1937): 413.
32 M.C. Urquhart and K.A.H. Buckley, eds., *Historical Statistics of Canada*, 2nd ed. (Ottawa 1983), H50.
33 Hodgson, *The Heavens Above*, 126–31, discusses the impact of the Bennett government's austerity measures on the civil service as well as the DO and DAO.
34 LAC, RG48, vol. 47, files 4-11, 4–12, correspondence between JSP and R.M. Stewart, 1932–3.
35 Ibid., R.M. Stewart to JSP, 25 February 1932.
36 Ibid., file 4-13, JSP to R.M. Stewart, 15 May 1933.
37 Redman left in February and the Hoggs arrived in August 1931 (*MNRAS* 92 (1932): 288), though they got to know one another when Redman returned to the DAO for two months in the summer of 1934 (UTA, B94-0002, box 54, H.S. Hogg to her family, 28 July 1934). Technically, the Hoggs' departure for Toronto, and Sherwood Hill's retirement, at the end of 1934, also took effect during JSP's directorship, but the subsequent reorganization and appointment of McKellar and Petrie occurred after Harper was appointed acting director.
38 AIP, interview of Helen Hogg by David DeVorkin, 17 August 1979, available at https://www.aip.org/history-programs/niels-bohr-library/oral-histories/4679.
39 UTA, B94-0002/054 (02), H.S. Hogg to family, 1933.
40 Ibid., H.S. Hogg to "Ellie," 31 December 1931.
41 UTA, B94-0002/54 (01), H.S. Hogg letters to her family in Massachusetts, 1932.
42 According to Canadian Passenger Lists found at www.ancestry.com, when the Plasketts sailed for England on 3 March 1935, Stuart gave his occupation as "fisherman." Pearce also commented on fishing with Stuart, *JRASC* 62 (1968): 311.
43 UTA, B94 0002/054 (01), H.S. Hogg to her family in Massachusetts, 3 August 1932; SA, diary of Alan Ford, 29 July 1934.

44 BCA, T4022: 56/0004, Voices of B.C., part of Series GR-3378, Provincial Educational Media Centre, school broadcasts, as recalled in 1956 by Frank Plaskett in a CBC recording.
45 LAC, RG48, vol. 47, file 4-17, correspondence between JSP and Arthur Beer, 1934–5.
46 Ibid. Plaskett tried to interest Chant in hiring Beer on a Carnegie fellowship when the David Dunlap Observatory opened in 1935, but that never happened. (UTA, A1874-0027/008, JSP to C.A. Chant, 6 March [1935]). A recent article on Beer's career is Hilmar W. Duerbeck and Peter Beer, "Arthur Beer and His Relations with Einstein and the Warburg Institute," *JAHH* 9 (2006): 93.
47 D. DeVorkin, "History Is Too Important to Be Left to the Historians," *Organizations, People, and Strategies in Astronomy* 2 (2013): 431, online at http://venngeist.org/opsa2_devorkin.pdf.
48 *Toronto Globe*, 22 May 1931, 13. JSP used similar words in his speech to the Canadian Club, see *Toronto Globe*, 5 November 1929, 19.
49 BCA, T4022:56/0004, Voices of B.C., recalled by JSP's brother Frank.
50 Quoted by J. Lawler, *Canadian Churchman*, 30 October 1941, 614.
51 See, for example, Ronald L. Numbers, *Science and Christianity in Pulpit and Pew* (Oxford: Oxford University Press, 2007).
52 JSP, "Science and Religion," *Canadian Churchman*, 15 March 1934, 1.
53 "An Earnest Lay-Reader," *Canadian Churchman*, 5 April 1934, 215.
54 Alan H. Batten, "A Most Rare Vision: Eddington's Thinking on the Relation between Science and Religion," *QJRAS* 35 (1994): 249–70; Matthew Stanley, *Practical Mystic: Religion, Science and A.S. Eddington* (Chicago: University of Chicago Press, 2007); Meg Weston Smith, *Beating the Odds; the Life and Times of E.A. Milne* (London: Imperial College Press, 2013); David DeVorkin, *Henry Norris Russell: Dean of American Astronomers* (Princeton, NJ: Princeton University Press, 2000).
55 *Toronto Daily Star*, 12 November 1926, 29.
56 Interview by R.E. Knowles, *Toronto Daily Star*, 3 June 1935, 3.
57 I am indebted to Elizabeth Griffin, who looked through all the observing logs for 1930–4 and found that Plaskett did not observe during those years.
58 JSP and J.A. Pearce, "A Catalogue of the Radial Velocities of O and B Type Stars" [1930], *PDAO* 5 (1935): 99; R.O. Redman, *Obs.* 56 (1933): 27. wrote an extensive review of this work, his only quibble being with the authors' handling of statistical errors.
59 JSP and J.A. Pearce, "The Motions of the O and B Type Stars and the Scale of the Galaxy" [1934], *PDAO* 5 (1935), table 1 on p. 249, table 14 on p. 281, and table 24 on p. 299.

60 V.M. Slipher, "Peculiar Star Spectra Suggestive of Selective Absorption of Light in Space," *Lowell Observatory Bulletin* 2 (1909): 2; see JSP and J.A. Pearce, "The Problems of the Diffuse Matter in the Galaxy" [1931], *PDAO* 5 (1935): 167–237.

61 Alan Batten has pointed out that a careful calibration of the intensity of the K line as a distance marker was made by C.S. Beals and J.B. Oke, "On the Relation between Distance and Intensity for Interstellar Calcium and Sodium Lines," *MNRAS* 113 (1953): 530–52. Although this paper was published after Beals left Victoria, it was, of course, strongly based on DAO data.

62 O. Struve, "Matter in Interstellar Space," *PA* 41 (1933): 423–31, a later paper than those specifically criticized by Plaskett.

63 *Ann. Rep. Smithsonian Inst. for the Year Ending June 30, 1933* (1935): 147. I assume that this "advertisement" at the beginning of the appendix as well as the brief note on page ix at the beginning of the report itself were written by the Smithsonian's secretary, C.G. Abbot. He was the named author of "The Contents of Interstellar Space," 211–18. JSP's original paper "The Structure and Rotation of the Galaxy," is reprinted on pp. 189–210.

64 *JRASC* 26 (1932); 118; *Manitoba Free Press* (Winnipeg), 16 May 1931, 18.

65 *Proc. Trans. RSC* 25 (1931): c–ci.

66 *Woodstock Daily Sentinel Review*, 22 May 1931, 2.

67 LAC, RG48, vol. 14, file 1523-2, R.M. Stewart to JSP, 8 May 1931.

68 Ibid., vol. 48, file 5-2, JSP to Mrs D.A. Dunlap, 23 May 1931.

69 LOA, letters from JSP to V. Slipher 28 May, 2 June, and 13 June 1931.

70 J.H. Moore, "The Meeting of the Astronomical Society of the Pacific Held in Conjunction with the Pasadena Meeting of the American Association for the Advancement of Science (with Abstracts of Papers)," *PASP* 43 (1931): 291.

71 Recently, all the Bruce medalists have been the subject of a series of articles written by Joe Tenn for the ASP. For the article on JSP, see J. Tenn, *Mercury* 24 (1995): 34–5. The only other Canadian so far to receive the Bruce Medal was Sydney van den Bergh, also a DAO director, in 2008.

72 Alfred H. Joy, "Address of the Retiring President of the Society in Awarding the Bruce Gold Medal to John Stanley Plaskett," *PASP* 44 (1932): 5–20. See also *PASP* 44 (1932): 66, 80 and 192; R.E. DeLury, "News and Comments," *JRASC* 26 (1932): 91; *PA* 40 (1932): 313.

73 Joy, *PASP* 44 (1932): 6–7.

74 JSP, "The Structure and Rotation of the Galaxy," *PASP* 44 (1932): 141–66; LAC, RG48, vol. 47, file 4-12, JSP to R.M. Stewart, 26 March 1932, re expenses.

75 JSP, "The Structure and Rotation of the Galaxy," *Ann. Rep. Smithsonian Inst. for the Year Ending June 30, 1933* (1935): 189–209.
76 "Notes Added by Author, April, 1934," *Ann. Rep. Smithsonian Inst. for the Year Ending June 30, 1933* (1935): 209–10.
77 *New York Times*, 2 July 1935, 14.
78 *Washington Post*, 2 July 1935, 20.
79 *Chicago Daily Tribune*, 1 July 1935, 13.
80 *Toronto Globe*, 1 July 1935, 1.
81 Frederic M. Lee, "The Scale of the Universe," *ASP Leaflet* 3, no. 122 (April 1939): 171–6.
82 See *JRASC* 21 (1927): 254 for a brief summary of his talk and of the stereoscopic photos of the Moon, planets, and comets that he showed. Fifteen of "Plaskett's slides" illustrating the powers of ten from electrons and nuclei to spiral galaxies are at UTA, Department of Astronomy lantern slide collection.
83 LAC, RG48, vol. 47, file 4-12, JSP to R.M. Stewart, 26 March 1932.
84 Ibid., vol.,48, file 5-12, J.A. Pearce to P.W. Merrill, 21 May 1932; *Proc. Trans. RSC*, 26 (1932): xxiii–iv.
85 W.E. Harper, "News and Comments," *JRASC* 26 (1932): 179; *Montreal Gazette*, 27 May 1932, 2.
86 H. Boyd Brydon, "Visit of Eclipse Party to Victoria," *JRASC* 26 (1932): 360–3.
87 LAC, RG48, vol. 46, file 3-2 JSP to R.M. Stewart, 21 October 1929, and J.G. Brown, District Resident Architect, to W.E. Harper, Acting Director, DAO, 17 June 1930. Since the Plasketts were living in Esquimalt by this time, I conclude that they occupied both houses. Helen Hogg, in "Memories of the Plaskett Era," *JRASC* 82 (1988): 328, referred to visits "both at the Director's residence and at their seaside home for retirement at Esquimalt."
88 "The National Council of Education … made arrangements with the Astronomer-Royal for a series of lectures to be given at various cities across Canada," Margaret Wilson, *Ninth Astronomer Royal*, 242. His address to the RASC Winnipeg Centre on 20 August was "Eclipse of the Sun," *JRASC* 27 (1933): 114.
89 Bamfield Historical Society, newspaper clipping dated 8 August 1932.
90 Brydon, "Visit of the Eclipse Party to Victoria."
91 *Trans. IAU* 4 (1932) is the official report.
92 JSP, "A Review of the Progress of Astronomy," *PASP* 44 (1932): 215–29.
93 J. Murray Luck, "The Sixteenth Annual Meeting of the Pacific Division," *Science* 76 (7 October 1932): 303–16; Ferdinand J. Neubauer, "The Pullman

Meeting of the Astronomical Society of the Pacific (with Abstracts of Papers Presented)," *PASP* 44 (1932): 253–60.

94 JSP, "The Expansion of the Universe," *JRASC* 27 (1933): 235–52.

95 LAC, RG 48, vol. 48, file 5-5, correspondence between H. Shapley and JSP, July–August 1933.

96 Shapley and Curtis put forth their positions in "The Scale of the Universe," which appeared in the *Bulletin of the National Research Council* 2 (May 1921): 171–217, online, https://apod.nasa.gov/htmltest/gifcity/cs_nrc.html. The debate became the source of much subsequent discussion. See, e.g., Michael A. Hoskin, "The Great Debate: What Really Happened" *JHA* 7 (1976): 169–82.

97 Helge Kragh, *Matter and Spirit in the Universe: Scientific and Religious Preludes to Modern Cosmology* (London: Imperial College Press, 2004), ch. 4, 124–60.

98 JSP, "A Review of the Progress of Astronomy," 219.

99 Roy Macleod and Philip F. Rehbock, *Pacific Science* 54 (2000): 209–25, provide insight into the American origins of the Pacific Science Congress. The organization was named the Pacific Science Association in 1926.

100 The objectives of the Pacific Science Association are quoted in *Proceedings of the Fifth Pacific Science Congress Held under the Auspices of the National Research Council of Canada and through the Generosity of the Government of Canada* (Toronto 1934), 2.

101 LAC, RG48, vol. 48, file 5-16, JSP to H.T. Stetson, 8 January 1931.

102 The official account of the congress, including two group photos, is a five-volume, 4227-page compendium, *Proceedings of the Fifth Pacific Science Congress*. For a short summary, see W.E. Harper, "The Pacific Science Congress," *JRASC* 27 (1933): 293–7.

103 University of Alberta, Peel Library, newspaper collection, *Western Globe*, 6 July 1933, 2. A couple of years earlier, in an interview, Plaskett was quoted as saying, "Scientists are generally disposed to believe that there is some type of life on Mars ... but it would be different in every way from anything known here" (*Toronto Daily Star*, 20 May 1931, 4).

104 JSP and J.A. Pearce, "Determination of the K-term, the Solar Motions, and the Galactic Rotation from the Radial Velocities of 849 O to B7 Class Stars," *Proceedings of the Fifth Pacific Science Congress* 2 (1934): 1101–4, and JSP, "The Work of the Dominion Astrophysical Observatory," ibid., 1125–30.

105 JSP and J.A. Pearce, "The Distance and Direction to the Gravitational Centre of the Galaxy," *MNRAS* 94 (1934): 679, and "The Motions of the O- and B-type Stars," 241.

106 Ferdinand J. Neubauer, "Radial Velocities of 351 Stars Observed at the Chile Station of the Lick Observatory in the Period 1924 to 1929," *LicOB* 15 (1930): 47–85; the observatory was sold to the Catholic University of Chile in 1929: see W.W. Campbell, "Sale of the Chile Station of the Lick Observatory," *PASP* 40 (1928): 249–52.
107 JSP and Pearce, "Motions of the O and B Type Stars."
108 A.S. Eddington, "Interstellar Matter," *Obs.* 60 (1937): 99–103.
109 Abstracts at *PAAS* 6 (1931): 277, and 7 (1933): 225; *PA* 39 (1931): 18; *PASP* 43 (1931): 278, and 45 (1933): 207–8.
110 Gerald Prevost, "Dr Plaskett on 'The Expanding Universe for V.I.,'" *Ubyssey* 15 (21 February 1933): 1.
111 JSP, "Observational Confirmation of the Rotation of the Galaxy," *Obs.* 56 (1933): 328–32.
112 Ibid., 330. JSP would have learned about Joy's work in Salt Lake City that June. It was published in Alfred H. Joy, "Rotation Effects, Interstellar Absorption, and Certain Dynamical Constants of the Galaxy Determined from Cepheid Variables," *ApJ* 89 (1939): 356–76.
113 For an account of this meeting and JSP's speech, see *Science* 79 (1934): 100.
114 *Time*, 8 January 1934.
115 A possible basis for this statement is the finding by F.H. Seares and P.J. van Rhijn that the number of stars in the Galaxy is between 30 and 38 billion: "Distribution of the Stars with Respect to Brightness and Distance from the Milky Way," *Proc. NAS* 11 (1925): 358.
116 *Science News-Letter* 26: 388.
117 *Obs.* 57 (1934): 207–9.
118 *JRASC* 28 (1934): 328.
119 LAC, RG48, vol. 47, file 5-12, JSP to R.M. Stewart, 17 April 1934; JSP's consulting work for the 2.0 m telescope for the McDonald Observatory will be discussed in chapter 12.
120 UTA, B94-0002/54, H.S. Hogg to her Massachusetts family, 28 May 1934.
121 An abstract was published in *Proc. Trans. RSC* 29 (1935): xcix.
122 JSP and Pearce, "Motions of the O and B Type Stars."
123 An object in a circular gravitational orbit about an attracting central mass M is subject to a centripetal acceleration $v^2/r = GM/r^2$ where v is the object's velocity, r is its distance from the centre, and G is the universal gravitational constant. This equation yields a rough estimate of the mass of the Galaxy interior to the Sun's distance from the centre.
124 Allan Sandage, *The Mount Wilson Observatory*, 452.
125 Virginia Trimble and Markus J. Aschwanden, "Astrophysics in 1999," *PASP* 112 (2000): 481.

126 Owen Gingerich, *The Physical Sciences in the Twentieth Century*, 65. This album of historic illustrations and detailed captions included only thirty figures in the chapter "The Stars and Beyond," and JSP's diagram was one of them.
127 JSP and Pearce, "Motions of the O and B Type Stars," 327.
128 *Obs.* 57 (1934): 208.
129 Jarrell, *Cold Light*, 123.
130 LAC, RG48, vol. 45, file 2-15, [W.E. Harper] to P. Marchand, controller of expenditure, Department of the Interior, 14 May 1919, gives details of the staff, their individual salaries, and itemized expenditure. Harper gives the annual expenditure as $11 500.
131 For a summary of research funding available to American observatories in 1918, see Charles R. Cross, "American Association for the Advancement of Science," *Science* 48 (1918): 75. Here Plaskett gives the figure $13 000. Although I have given comparative figures for the major U.S. observatories, the DAO budget was very close to that for the Allegheny Observatory and a great deal more than all the smaller observatories, most of which had no research funding at all.
132 UTA, B94-0002/54 (03), Helen Hogg cites an article in "tonight's paper," presumably the *Victoria Times*, 19 June 1934.
133 LAC, RG48, vol. 48, file 5-2, W.E. Harper to R.K. Young, 9 October 1935.

12 Retirement, 1934–1941

1 *Victoria Daily Times*, 24 August 1934, 1.
2 UTA, B94-0002 (box 6), Helen S. Hogg to Harlow Shapley, 14 September 1934.
3 LAC, RG48, vol. 12, file 1511-3, order-in-council, Privy Council 1/1561, 31 July 1933.
4 W.E. Harper, "Notes from the Dominion Astrophysical Observatory," *JRASC* 29 (1935): 73. Harper said that JSP's retirement was effective 31 January 1934, but he surely meant 1935, a mistake commonly made early in a calendar year.
5 Roger Hutchins, *British University Observatories, 1772–1939* (Aldershot: Ashgate, 2008), 209, bases his opinion on Harry Plaskett's introductory note to *Five Halley Lectures*.
6 LAC, RG48, vol. 48, file 5-12, JSP to Edwin Hubble, 26 November 1934.
7 HL, box 13.274, JSP to F.H. Seares, 2 January 1935, and box 3.63, correspondence between JSP and A.H. Joy, December 1934.
8 LAC, RG48, vol. 48, file 5-12, E. Hubble to JSP, 4 December 1934.

9 Jeff Hughes, "'Divine Right' or Democracy? The Royal Society 'Revolt' of 1935," *Notes and Records of the Royal Society of London* 64, Supplement 1 (2010): S101.
10 LAC, RG48, vol. 47, JSP to W.W. Cory, 27 January 1927.
11 LAC, M-1079, pp. 251556–62, correspondence between JSP and R.B. Bennett.
12 *Victoria Daily Colonist*, 12 January 1935, 1, 6, 8. Unsworth followed with a long column, "Local Astronomer Makes Public His Christian Belief," *Victoria Daily Colonist*, 13 January 1935, 12, in which he reported further on the dinner and drew heavily on JSP's article in *Canadian Churchman*, 15 March 1934, 1.
13 LAC, M-1079, 251561–2, JSP to R.B. Bennett, 29 January 1935.
14 See www.nasonline.org/search.html?q=Plaskett&submit.x=8&submit.y=8.
15 V.M. Slipher, "Presentation of the Henry Draper Medal for 1934 to John Stanley Plaskett," *Science* 81 (1935): 415.
16 F.E. Wright, "Medalists of the National Academy of Sciences," *Scientific Monthly* 41 (1935): 183.
17 UK incoming passenger lists, at www.ancestry.com.
18 UTA, B94-0002, box 6, JSP to H.S. Hogg, 9 April 1935.
19 Correspondence between the author and J.S. Plaskett (JSP's grandson), 2008–12.
20 UTA, A1874-0027/008, file "P," JSP to C.A. Chant, 6 March [1935].
21 *Obs.* 58 (1935): 104; much is made of the rivalry among the English theoretical astrophysicists by Arthur I. Miller, *Empire of the Stars: Friendship, Obsession and Betrayal in the Quest for Black Holes* (London: Little, Brown, 2005), esp. 98–104.
22 *Obs.* 58 (1935): 165.
23 Ibid., 188–9.
24 RASCA, "Plaskett" file, JSP to C.A. Chant, 6 January 1936.
25 *Five Halley Lectures* was reviewed by R. Woolley in *Obs.* 60 (1937): 26–7, and by H. Spencer Jones in *Nature* 139 (13 February 1937): 266–7. Both reviewers credit Harry Plaskett with the idea of publishing them. JSP wrote a condensed version of his lecture in "The Dimensions and Structure of the Galactic System," *JRASC* 30 (1936): 153–64.
26 Roger Hutchins, "Astronomical Measurements at the Radcliffe Observatory," in *A History of the Radcliffe Observatory: The Biography of a Building*, edited by Jeffery Burley and Kristina Plenderleith (Oxford: Oxford Green College, 2005), 92.
27 *Trans. IAU* 4 (1932): 251; A.D. Thackeray, "H. Knox-Shaw," *QJRAS* 12 (1971): 197–201.

28 A.H. Batten made this point in an unpublished paper he presented to the ASP in 1998. Michael Feast, who was present on that occasion and who worked at the Radcliffe Observatory from 1952 to 1974, confirmed that a condition of his appointment was that he should spend some specified fraction of his time working on that extension.
29 UTA, B94-0002, box 6, JSP to H.S. Hogg, 24 July 1935.
30 C.A. Chant, "The International Astronomical Union: Fifth Meeting," *JRASC* 29 (1935): 316–20.
31 "Science: Organizer of Heaven," *Time* 26 (29 July 1935).
32 CCL, Schlesinger diary; Harry Plaskett's address comes from *Canadian Who's Who* 2 (1936–7), 878.
33 *JRASC* 30 (1936): 261; *Victoria Daily Colonist*, 7 February 1936, 3.
34 LAC, RG48, vol. 13, file 1511-4-1, W.E. Harper to R.M. Stewart, 2 February 1937.
35 R.M. Stewart, in his obituary of Harper (*JRASC* 34 (1940): 233), referred to his "entire absence of any spirit of self-seeking," and gave an apt illustration of his acquiescence to a temporary pay cut. In another obituary, J.A. Pearce (*PASP* 52 (1940): 231) spoke of Harper's "habits of punctuality, of diligence, and industry" as well as his loyalty to his superiors.
36 R.M. Petrie, "John Stanley Plaskett, 1865–1941," *PA* 50 (1942): 2.
37 J.A. Pearce, "Some Recollections of the Observatory," *JRASC* 62 (1968): 297.
38 This assessment was expressed by Batten after reading an earlier draft of this book. Similarly, another DAO astronomer, Robert McClure, in "Scientific Highlights from the D.A.O.," *JRASC* 87 (1993): 282, noted that "the directorship of K.O. Wright [1966–76] was a time of very significant transition for the D.A.O."
39 W.M. Smart, "On the K Term of the Radial Velocities of B-Type Stars," *MNRAS* 96 (1936): 568–73; "The Scorpio-Centaurus Cluster (The Southern Stream)," *MNRAS* 100 (1939): 60–85; "The K Term and the Galactic Rotational Constants," *MNRAS* 100 (1940): 370–7.
40 See Aaron Rizzuto, M.J. Ireland, and J.G. Robertson, "Multidimensional Bayesian Membership Analysis of the Sco OB2 Moving Group," *MNRAS* 416 (2011): 3108–17.
41 JSP, "An Analysis of the K Term in the B-Type Stars," *MNRAS* 98 (1938): 518–27.
42 JSP, "Modern Conceptions of the Stellar System," *PA* 47 (1939): 239–55; "The Galaxy," *Scientia* 60 (1936): 309–17.
43 *PASP* 48 (1936): 201; T.S. Jacobson, "Seventy-five Years of Astronomy at the University of Washington," *PASP* 78 (1966): 388.

44 Steven Soter and Neil deGrasse Tyson, eds., *Cosmic Horizons: Astronomy at the Cutting Edge* (New York: New Press, 2001), "Fritz Zwicky's Extraordinary Vision," 135–7, online at http://www.amnh.org/explore/resource-collections/cosmic-horizons/profile-fritz-zwicky-s-extraordinary-vision.
45 Chapter 3 of David S. Evans and J. Derral Mulholland, *Big and Bright* deals with the legal contests.
46 LAC, RG48, vol. 49, file 6-17, correspondence between JSP and H.Y. Benedict, 1926.
47 Donald E. Osterbrock, *Pauper and Prince*, 264–5.
48 Ibid., 264–5.
49 UCL, Yerkes Observatory, Office of the Director, box 247, JSP to O. Struve, 29 October 1932.
50 Ibid., 29 June 1933.
51 LAC, RG48, vol. 47, file 5-12, JSP to R.M. Stewart, 17 April 1934; UCL, Office of the Director, box 247, JSP to Otto Struve, 5 June 1934; JSP's formal employment as optical consultant by Warner & Swasey seems not to have started until 1936.
52 Donald E. Osterbrock, *Yerkes Observatory*, 200–2.
53 UCL, Yerkes Observatory, Office of the Director, box 168, folder 7, JSP to O. Struve, 9 August 1937 and 15 November 1937.
54 Ibid., O. Struve to JSP, 19 November 1937.
55 Ibid., box 174, folder 7, O. Struve to JSP, 5 January 1938.
56 HL, Dunham papers, box 1, folder 54, JSP to Theodore Dunham, 30 March 1937.
57 *New York Times* 1 May 1937, 21.
58 C.A. Chant, "Completion of the Tube of the 200-inch Telescope," *JRASC* 31 (1937): 241.
59 For a full (though, according to some reviewers, not entirely reliable) story of the Palomar observatory, see Ronald Florence, *The Perfect Machine: Building the Palomar Telescope* (New York: Harper Collins, 1994).
60 *PA* 45 (1937): 360, 363.
61 JSP, "Obituary of E.P. Burrell," *Science*, 25 June 1937, 597 and "Edward P. Burrell," *PASP* 49 (1937): 141. According to the *New York Times*, 22 March 1937, 23, Burrell died that day, at a time when JSP was working closely with him.
62 According to JSP, "Ambrose Swasey, Engineer, Scientist, Philanthropist," *JRASC* 31 (1937): 409, Swasey died 15 June 1937.
63 JSP, "Preliminary Report on 82-inch Texas Mirror," 13 October 1938, CWRU, box 25, folder 5.

64 Osterbrock, *Yerkes Observatory*, 201, citing letters from Struve to Bart J. Bok, 3 April 1938 and to Henry G. Gale, 4 April 1938.
65 Helen S. Hogg, "The Centennial Celebration of the Loomis Observatory," *JRASC* 32 (1938): 301.
66 "Our Galaxy Biggest Known," *Science News Letter* 34 (1938): 3; JSP, "Modern Conceptions of the Stellar System," *PA* 47 (1939): 239. This latter paper was referred to earlier as one of Plaskett's "overviews" published after his retirement.
67 See HL, Dunham papers, box 1, folder 14, Theodore Dunham Jr. to Harry Plaskett, 2 August 1938; "The San Diego Meeting of the Astronomical Society of the Pacific," *PASP* 50 (1938): 210.
68 Passenger lists, www.ancestry.com. According to T.S. Plaskett, *Plaskett-Stanley Family History*, 27, "Jack and Reba were Warner and Swazey's [*sic*] guests at the I.A.U. meeting at Stockholm."
69 Frank S. Hogg, "The Sixth Meeting of the International Astronomical Union," *JRASC* 32 (1938): 376.
70 *Victoria Daily Colonist*, 22 November 1938, 6.
71 R.M. Stewart, "William Edmund Harper, 1878–1940," *JRASC* 34 (1940): 233.
72 J.F. Heard, "The Sixtieth Meeting of the American Astronomical Society," *JRASC* 32 (1938): 414; Leo J. Klosterman et al., *100 Years of Science and Technology in Texas: A Sigma Xi Centennial Volume* (Houston, TX: Rice University Press, 1986), 111–13.
73 JSP, "Preliminary Report on 82-inch Texas Mirror," 13 October 1938, CWRU, box 25, folder 5.
74 Evans and Mulholland, *Big and Bright*, 77.
75 JSP, "The 82-Inch Mirror of the McDonald Observatory," *ApJ* 89 (1939): 84. Struve was the editor of *ApJ* at this time.
76 A.G. Ingalls, "Telescoptics," *Sci. Am.* 166 (1942): 158.
77 *Victoria Daily Colonist*, 22 November 1938, 6.
78 "Report on Installation and Testing of Optical Parts of McDonald Telescope," CWRU, box 25, folder 5; letter from JSP to Mrs Burrell, 18 March 1939, CWRU, box 25, folder 4.
79 G. Van Biesbroeck, "The Nebulosity Near S Monocerotis," *ApJ* 89 (1939): 554, and adjoining plate XXVII.
80 Evans and Mulholland, *Big and Bright*, 88–9.
81 Helen S. Hogg, "The Dedication of the McDonald Observatory," *JRASC* 33 (1939): 201.
82 AIP, AV 714.

83 Otto Struve, "Acceptance of Completed Observatory," in a booklet entitled *Addresses Made at the Dedication of the McDonald Observatory* (1939), 30; also in *Science* 89 (1939): 493.
84 Struve, "Acceptance of Completed Observatory" and *Science* 89 (1939): 495–7.
85 Evans and Mulholland, *Big and Bright*, 83.
86 UCL, Yerkes Observatory, Office of the Director, box 247, folder 8, JSP to O. Struve, 29 October 1932.
87 Evans and Mulholland, *Big and Bright*, 79.
88 UCL, box 188, folder 7, O. Struve to JSP, 29 February 1940.
89 CWRU, box 25, folder 5, correspondence between JSP and C.J. Stilwell, October–November 1940.
90 I have not been able to substantiate these claims by Plaskett. In O. Struve, "The Organization of the Observatory," *Publications of the McDonald Observatory* 1 (1940): 11, Struve did credit the Warner & Swasey Company and its new president C.J. Stilwell for exceeding former standards of excellence in telescope building. And in his "Biographical Memoir of Edwin Brant Frost, 1866–1935," *Biographical Memoirs of National Academy of Sciences* 19 (1937): 30, online at http://www.nasonline.org/publications/biographical-memoirs/memoir-pdfs/frost-edwin.pdf, Struve said that "probably no other scientist has ever been admired and loved so much by those who knew him or knew of him." This is hard to reconcile with Plaskett's understanding of Struve's treatment of Frost but, as Osterbrock frequently points out in *Yerkes Observatory*, Struve certainly disparaged those who did not pull their weight as researchers.
91 O. Struve, "The Birth of McDonald Observatory," *S&T* 24 (December 1962): 320.
92 This is discussed by Evans and Mulholland, *Big and Bright*, 77–8.
93 BCA, MSS 2779, box 6, file 1, scrapbook of the British Columbia Historical Association includes newspaper clippings about the activities of the association, some mentioning the Plasketts.
94 JSP, "The History of Astronomy in British Columbia," *BCHQ* 4, no. 2 (1940): 63–78, and *JRASC* 77 (1983): 108–20.
95 *JRASC* 34 (1940): 164.
96 Clippings of obituaries of T.S. Plaskett are in UTA, A73-0026/365 (26). These include ones from the *Montreal Gazette*, 23 December 1939; the *Toronto Globe and Mail*, 25 December 1939, which includes the names of his four surviving brothers and one sister; and the *Toronto Telegram*, 26 December 1939, which names the six pallbearers (JSP was not among them).

97 UCL, Yerkes Observatory, Office of the Director, box 188, JSP to O. Struve, 12 April 1940.
98 Though the Warner & Swasey Observatory is an abandoned hulk in Cleveland, the Schmidt telescope lives on at the Kitt Peak National Observatory in Arizona.
99 JSP was quoted on his return trip to Victoria by the *Winnipeg Free Press*, 27 May 1940, 7.
100 *London [ON] Free Press*, 21–2 May 1940.
101 C.S. Beals, "John Stanley Plaskett," *JRASC* 35 (1941): 401–7, or a slightly expanded version in *Proc. Trans. RSC* 36 (1942): 107–13.
102 Photo of RASC banquet, *Victoria Colonist*, 15 December 1940; ADBCA, items relating to Christ Church Cathedral: Cathedral Buildings Ltd.
103 BCA, Death records, B13172, 1941-09-595219, states that he had lived for seven years in Esquimalt.
104 It would have been a strange coincidence indeed if he happened to be related to George Biddell Airy, Britain's famous Astronomer Royal from 1835 to 1881. I do not know which members of the family attended the funeral, but likely JSP's two brothers in BC, Frank (who died in 1962) and Bert (who died in 1957), were among the mourners. For death dates, see Neville J. Crichlow and C. Digby Turney, *Stained Glass Windows*, 27.
105 HUA, HUG 4773.10, papers of Harlow Shapley, box 22A, "Plaskett, Harry H."
106 In a letter to the author dated 8 December 2008, J.S. Plaskett, JSP's grandson, states that his father (Harry) did not visit the United States or Canada between 1932 and 1955. An e-mail to the author from JSP's great granddaughter, Deborah Tilley, dated 20 April 2012, gives the death dates of Deborah's parents: Barbara died on 24 March 1998, and Arthur on 6 September 1999, both in Halifax, NS.
107 *Montreal Gazette*, 8 July 1946.
108 B.J. Bok, "John Stanley Plaskett, 1865–1941," *Science* 94 (1941): 455.
109 BCA, BC Supreme Court probate/Estate Files GR-2083, B09153, 54/1942.
110 The medals presented to JSP by the RAS, ASP, American Academy of Arts and Sciences, and RSC were melted but replicas were presented to the BC Archives in 1943, along with the CBE, which is likely the original (email to the author from Katy Hughes). See also *Victoria Colonist*, 6 and 7 April 1943.
111 Black and white photos of the original windows can be found in *S&T*, July 1943, 10 and the *Victoria Colonist*, 7 April 1943. The details of the service are in the *St. John's Church Parish Leaflet*, 4 April 1943, Anglican Diocese of BC, and there are write-ups in the newspapers, in *S&T*, July

1943, 10, and in F.S. H[ogg], "Memorial to Dr. J.S. Plaskett," *JRASC* 37 (1943): 167–8.

112 Randall Rosenfeld has analysed this prayer and found almost all of it in various passages of the *Canadian Book of Common Prayer* (1922) (private communication).

113 Esquimalt Municipal Archives, newspaper clipping, "Mrs. J.S. Plaskett Celebrates 90 Years," *Esquimalt Sentinel*, clipping dated 18 November 1964. Either the handwritten date or the headline is incorrect, as Reba would have turned ninety-seven in 1964.

114 The obituary in the *Victoria Daily Times*, 17 June 1967 says that Reba was born in London, England, that she was cremated, and that she left two grandchildren. Canon George Biddle officiated at the service. I am grateful to neighbours and former neighbours for sharing their reminiscences with me.

115 BCA, 23160–40/B, will/probate files for Rebecca Hemley Plaskett.

116 British Columbia Directories, microfilmed for UBC Library, 2002; E. John Love, "A Day at the Hotel Yates," http://ejohnlovebooks.com/true-life/true-life/1972-1974-2/a-day-at-the-hotel-yates/.

117 BCA, Registration of Death, available at search-collections.royalbcmuseum.bc.ca/Genealogy/DisplayGenealogyImage?k=b2d2ea15-34ab-4e0d-a487-6bb51a4e6bb1; also, BCA, 23160-40/B, will/probate files for Stuart Stanley Plaskett.

118 J. Lawler, *Canadian Churchman*, 30 October 1941, 614; *Times* (London), 30 October 1941, 7; *New York Times*, 18 October 1941, 19.

119 These brief quotes are taken from obituaries by B.J. Bok, *Science* 94 (1941): 453, R.G. Aitken, *PASP* 53 (1941): 323, and R.O. Redman, *Obs.* 64 (1942): 208 respectively.

120 R.M. Petrie, "John Stanley Plaskett, 1865–1941," *PA* 50 (1942): 2.

121 C.S. Beals, "John Stanley Plaskett," *JRASC* 35 (1941): 401.

122 C.S. Beals, "Early Days at the Dominion Astrophysical Observatory," *JRASC* 62 (1968): 304.

123 F.S. Hogg, "Plaskett, John Stanley," *MNRAS* 102 (1942): 73.

124 R.O. Redman, "Obituary, J.S. Plaskett," *Obs.* 64 (1942): 211; Redman also wrote the article on JSP in the *Oxford Dictionary of National Biography*, a recognition by the editors that Plaskett, even though he never lived in Britain, "shaped an aspect of national life."

125 H. Spencer Jones, "John Stanley Plaskett, 1865–1941," *Obit. Not. Fell. R. Soc.* 4 (1942): 67; F.J M. Stratton, "Dr. J.S. Plaskett, C.B.E., F.R.S." *Nature* 148 (29 November 1941): 654.

126 R.F. Sanford, "John Stanley Plaskett, 1865–1941," *ApJ* 98 (1943): 141.

127 R.G. Aitken, "John Stanley Plaskett, 1865–1941," *PASP* 53 (1941): 323.
128 Bok, "John Stanley Plaskett," 455.

13 Regeneration, 1942–

1 LAC, Historic Sites and Monuments Board, microfilms T 14241-2, pp. 1775–1802; there were two articles in the *Victoria Daily Colonist* of 27 June 1952: "Plaque Honors Astronomer" and "Plaque Commemorates Work of Dr. Plaskett." As well, "Distinguished Scientists Honor Late Dr. Plaskett," *Victoria Daily Times*, 27 June 1952, includes a photo of Reba Plaskett with Dr W. Sage. See also R.M. Petrie, "The Joint American Astronomical Society-Astronomical Society of the Pacific Meeting at Victoria, B.C., June 25–28, 1952," *JRASC* 46 (1952): 129.
2 The fiftieth anniversary of the observatory was marked by further celebrations of Plaskett, which included the AAS again meeting at the DAO. See K.O. Wright, "Fifty Years at the Dominion Astrophysical Observatory," *JRASC* 62 (1968): 269. For events and some papers presented at the seventy-fifth anniversary of the DAO, see *JRASC* 87 (October 1993).
3 Madge G. Adam, "The Changing Face of Astronomy in Oxford (1920–60)," *QJRAS* 37 (1996): 161.
4 L.A. Higgs, "The Plaskett Medal: Our Society's Newest Award," *JRASC* 82 (1988): 312-16. This number of the *Journal* also included an abbreviated version of Thomas Plaskett's family history, Helen Hogg's memories of the Plaskett era at the DAO, and the text of a speech given by the first recipient of the medal, Richard Gray. For information on Plaskett fellowships, see http://www.nrc-cnrc.gc.ca/eng/careers/programs/plaskett_fellowship/index.html on formal terms of the Plaskett Medal, see http://casca.ca/?page_id=524.
5 J.E. Hesser, e-mail to the author, 8 June 2015.
6 Donald C. Morton, "The Naming of Two Rocky Mountains after Canadian Astronomers," *JRASC* 105 (2011): 161–2, and "New Ascents in the Northern Rockies," *Canadian Alpine Journal* 51 (1968): 182. The astronomers on the expedition were George Wallerstein from the University of Washington, Bob O'Dell and Tom Grenfell from the Yerkes Observatory of the University of Chicago, and Lyman Spitzer and Donald Morton from the Princeton University Observatory. Morton, the only Canadian, was later director of the Herzberg Institute of Astrophysics.
7 The coordinates come from the Natural Resources Canada website, http://www4.rncan.gc.ca/search-place-names/search?lang=en.

8 "New Names on the Moon," *Spaceflight* 9 (February 1967): 38-39.
9 "Canadian Name on the Moon" *Victoria Colonist*, 14 March 1967, 21.
10 P.M. Millman, "Canadian Craters on the Moon," *JRASC* 64 (1970): 375–6.
11 The names, sizes, and coordinates of all the lunar features can be found on the gazetteer on the U.S. Geological Survey website. The Canadians, in addition to Plaskett and Chant, include astronomers Beals, Frank Hogg, McKellar, Newcomb, and Petrie, physicians Avery and Banting, geologist Daly, and physicist Foster.
12 Jeffrey A. Foust, "The Moon Rediscovered," *S&T* (1998): 32–4, bases his article on data reported by W.C. Feldman et al., "Fluxes of Fast and Epithermal Neutrons from Lunar Prospector: Evidence for Water Ice at the Lunar Poles" *Science* 281 (1998): 1496–1500. In recent years, the Moon's south pole has been found to hold more promise as a source of water.
13 For details of all named features in the solar system, see https://planetarynames.wr.usgs.gov/Page/MOON/target.
14 Ian Halliday, "Two More Minor Planets Named for Canadian Astronomers," *JRASC* 79 (1985): 26–8.
15 R.A Jarrell, "The Birth of Canadian Astrophysics: J.S. Plaskett at the Dominion Observatory," *JRASC* 71 (1977): 221–33. Many of the ideas in this paragraph are found in Jarrell's chapter, "Honorary American Astronomers: Canada and the American Astronomical Society" in David H. DeVorkin, ed., *The American Astronomical Society's First Century*, 61–73.
16 Ibid., 72.
17 For membership figures, see http://casca.ca/?page_id=61 and https://aas.org/about/what-aas.
18 Jarrell, "Honorary American Astronomers," 71.
19 J.A. Hesser, e-mail to the author, 8 June 2015.
20 R.A. Jarrell, "J.S. Plaskett and the Modern Large Reflecting Telescope," *JHA* 30 (1999): 385.
21 Kathleen Underwood, "He Clocked the Cosmic Whirl," *Maclean's*, 15 May 1937, 48–50; C.A. Chant, "Worcester Reed Warner," *JRASC* 24 (1930): 4.
22 Dennis Overbye, "Canada Looks Up, Way Up," *New York Times*, 24 May 2005, F1; See also *Maclean's*, 5 September 2005 and *Ottawa Citizen*, 17 August 2003, C6.
23 CASCA, *Unveiling the Cosmos: A Vision for Canadian Astronomy: Report of the Long Range Plan 2010 Panel*, http://casca.ca/wp-content/uploads/2016/03/MTR2016nocover.pdf. For funding figures, see p. 116. The quote is found on p. 87, though the facts to back up these claims are found throughout the report.

24 J.A. Hesser, e-mail to the author, 1 November 2015; also *Victoria Times Colonist*, 26 May 2015.

25 K.O. Wright, "The Dominion Astrophysical Observatory, 1918–1975," *JRASC* 69 (1975): 205–11.

26 See http://www.pc.gc.ca/APPS/CP-NR/release_e.asp?bgid=1398&andor1=bg; many other telescopes of great age are still in use as research instruments, including the recently renovated Clark telescope at the Lowell Observatory.

27 Statistics supplied by David Bohlender, manager of the Plaskett Telescope. The numbers could be doubled or even tripled if astronomical circulars and conference papers were included.

28 J.A. Hesser, e-mail to the author, 1 November 2015.

29 Under a 1970 amendment by Parliament to the National Research Council Act, responsibility "to administer and operate" the observatories of the government of Canada was transferred to the NRC, which established the Herzberg Institute of Astrophysics (renamed in 2014 Herzberg Astronomy and Astrophysics Programs).

30 W.H. McCrea, *Biogr. Mems. Fell. R. Soc.* 27 (1981): 446.

31 Ibid.; P.M. Millman, "Plaskett, Harry Hemley, 1893–1980," *JRASC* 74 (1980): 234–6.

32 E.A. Milne, quoted by M.G. Adam and Roger Hutchins in their article on Harry Plaskett in the *Oxford Dictionary of National Biography* at http://www.oxforddnb.com/. He was also described as "extremely modest" in an unsigned obituary in the *Times* (London), 28 January 1980, 14.

33 A number of papers by astronomers at the DAO were written to celebrate the fiftieth anniversary of the observatory and were published in a special number of *JRASC* 62 (December 1968). Those that show the evolution in areas of research begun by Plaskett include Anne B. Underhill, "The Interpretation of Stellar Spectra," Alan H. Batten, "Understanding Spectroscopic Binaries," and G.A.H. Walker, "Studies of the Interstellar Medium at the Dominion Astrophysical Observatory."

34 Judith A. Irwin, "From Wysiswyg to Haloes and Fountains: The Discovery of Galactic Structure," *JRASC* 87 (1993): 368–78, highlights advances in the understanding of the Milky Way from about 1920 to 1993.

35 Underwood, "He Clocked the Cosmic Whirl," 50.

36 J.H. Hodgson, *The Heavens Above* 1: 149, quotes from Beals' notes for an after-dinner talk.

37 B.J. Bok, "John Stanley Plaskett, 1865–1941," *Science* 94 (1941): 453-55; C.S. Beals, "John Stanley Plaskett," *JRASC* 35 (1941): 402.

Selected Bibliography

This bibliography contains only books that are cited more than once in the notes. Nearly all the relevant scientific periodical literature cited in the notes can be located through adsabs.harvard.edu/abstract_service.html. Reports of the Auditor General can be conveniently accessed at http://library.queensu.ca/gov/sessional-papers. Some newspapers have also been digitized: the *Toronto Star* and the *Globe and Mail* are accessible at the Toronto Public Library; several other newspapers, especially U.S. ones, are available through the historic newspaper section of www.ancestry.com; digital copies of Woodstock newspapers were accessed at the Woodstock Public Library. An online edition of the *Victoria Daily Colonist* (up to 1951) is at www.britishcolonist.ca. Unfortunately there is not yet any searchable digitized version of Ottawa newspapers for the relevant period.

Many of Plaskett's friends and associates appear in this book in supporting roles. For further information about them, there are a couple of general reference works that are especially useful: Thomas Hockey, et al., eds., *Biographical Encyclopedia of Astronomers* (New York: Springer, 2014) (also available online to subscribers) and the multi-volume *Dictionary of Canadian Biography*, which is available in print and online at www.biographi.ca, though at present it includes only people who died before 1940.

Aitken, Robert Grant, ed. *The Adolfo Stahl Lectures in Astronomy*. San Francisco: Stanford University Press, 1919.

– *The Binary Stars*. 1935; reprint, New York: Dover, 1964.

Bartusiak, Marcia. *The Day We Found the Universe*. New York: Pantheon Books, 2009.

Becker, Barbara J. *Unravelling Starlight: William and Margaret Huggins and the Rise of New Astronomy*. Cambridge: Cambridge University Press, 2011.

Brashear, John A. *John A. Brashear: The Autobiography of a Man Who Loved the Stars*. Boston: Houghton Mifflin, 1925.

Broughton, R. Peter. *Looking Up: A History of the Royal Astronomical Society of Canada*. Toronto: Dundurn Press, 1994.

Brown, Robert Craig, and Ramsay Cook, *Canada, 1896–1921*. Toronto: McClelland and Stewart, 1974.

Chant, Clarence A. *Astronomy in the University of Toronto: The David Dunlap Observatory*. Toronto: University of Toronto Press, 1954.

Christianson, Gale E. *Edwin Hubble: Mariner of the Nebulae*. New York: Farrar, Straus, Giroux, 1995.

Connor, Ralph [Charles W. Gordon]. *Glengarry School Days*. Toronto: McClelland & Stewart, 1993.

Crelinsten, Jeffrey. *Einstein's Jury: The Race to Test Relativity*. Princeton: Princeton University Press, 2006.

Crichlow, Neville J., and C. Digby Turney. *Stained Glass Windows in the Church of St. Mary the Virgin, Sapperton*. New Westminster, BC: Privately published, 2005.

DeVorkin, David H., ed. *The American Astronomical Society's First Century*. New York: AAS, 1999.

– *Henry Norris Russell: Dean of American Astronomers*. Princeton, NJ: Princeton University Press, 2000.

Evans, David S., and J. Derral Mulholland. *Big and Bright: A History of the McDonald Observatory*. Austin: University of Texas Press, 1986.

Friedland, Martin L. *The University of Toronto: A History*. Toronto: University of Toronto Press, 2002.

Gingerich, Owen. *The Physical Sciences in the Twentieth Century*. New York: Scribner's, 1989.

Green, P. David. *Visionary Veterinarian: The Remarkable Exploits of Dr. Duncan McNab McEachran*. Victoria, BC: Anconalces, 2012.

Hearnshaw, J.B. *The Analysis of Starlight: One Hundred and Fifty Years of Astronomical Spectroscopy*. Cambridge: Cambridge University Press, 2014.

– *Astronomical Spectrographs and Their History*. Cambridge: Cambridge University Press, 2009.

Herzberg, Paul A. *Luise Herzberg, Astrophysicist: A Memoir*. Toronto: York University Bookstore, 2010.

Hodgson, John H. *The Heavens Above and the Earth Beneath: A History of the Dominion Observatories*. Vol. 1. Ottawa: Energy, Mines and Resources Canada, 1989.

Illustrated Historical Atlas of Oxford County, Ontario. Toronto 1876; reprinted Belleville 1972. Available online at http://digital.library.mcgill.ca/CountyAtlas/default.htm.

Jarrell, Richard A. *The Cold Light of Dawn: A History of Canadian Astronomy*. Toronto: University of Toronto Press, 1988.

Jones, Bessie Z., and Lyle G. Boyd. *The Harvard College Observatory*. Cambridge, MA: Bellknap Press, 1971.

Koltun, Lilly, ed. *Private Realms of Light: Amateur Photography in Canada, 1839–1940*. Markham, ON: Fitzhenry & Whiteside, 1984.

Langton, Hugh H. *Sir John Cunningham McLennan*. Toronto: University of Toronto Press, 1939.

Lankford, John. *American Astronomy: Community, Careers and Power, 1859–1940*. Chicago: University of Chicago Press, 1997.

Levy, David. *The Man Who Sold the Milky Way: A Biography of Bart Bok*. Tuscon: University of Arizona Press, 1993.

Longair, Malcolm S. *The Cosmic Century: A History of Astrophysics and Cosmology*. Cambridge: Cambridge University Press, 2006.

Loudon, William J. *Studies of Student Life*. Volumes 4 and 5. Toronto: Macmillan, 1923.

Osterbrock, Donald E. *Pauper and Prince: Ritchey, Hale and Big American Telescopes*. Tuscon: University of Arizona Press, 1993.

– *Yerkes Observatory, 1892–1950*. Chicago: University of Chicago Press, 1997.

Payne-Gaposhkin, Cecilia. *An Autobiography and Other Recollections*. Edited by Katherine Haramundis. Cambridge: Cambridge University Press, 1984.

Pherrill, C.O. *150th Anniversary, Christ Church (Anglican), Huntingford, Ontario, 1839–1989*. (N.p., 1954).

[Plaskett, Harry H. ed.] *Five Halley Lectures*. Oxford: Oxford University Press, 1936.

Plaskett, Thomas Stanley. *Plaskett-Stanley Family History*, 1939. Transcribed for the Oxford Historical Society, Woodstock, by E. Gardhouse, 1993. A portion of this history was published in *JRASC* 82 (1988): 319–27.

Royal Society of Canada. *Fifty Years Retrospect*. Toronto: Ryerson Press, 1932.

Sandage, Allan. *The Mount Wilson Observatory*. Volume 1 of the *Centennial History of the Carnegie Institution of Washington*. Cambridge: Cambridge University Press, 2004.

Smith, Robert W. *The Expanding Universe: Astronomy's "Great Debate," 1900–1931*. Cambridge: Cambridge University Press, 1982.

Stratton, Frederick J.M. *Astronomical Physics*. London: Methuen, 1925.

Tayler, Roger J., ed. *History of the Royal Astronomical Society, 1920–1980*. Volume 2. Oxford: Blackwell, 1987.

Taylor, John H. *Ottawa: An Illustrated History*. Toronto: Lorimer, 1986.

Thistle, Mel W. *The Inner Ring: The Early History of the National Research Council of Canada*. Toronto: University of Toronto Press, 1966.

Thomas, Morley. *The Beginnings of Canadian Meteorology*. Toronto: ECW Press, 1991.

Thomson, Malcolm M. *The Beginning of the Long Dash: A History of Timekeeping in Canada*. Toronto: University of Toronto Press, 1978.

Tuck, Judy. *A History of Harriston: A Commemorative Book for the Harriston Centennial*. Mildmay, ON: Town Crier, 1978.

Van der Kruit, P.C. and K. van Berkel, eds. *The Legacy of J. C. Kapteyn*. Dordrecht: Kluwer Academic, 2000.

Wilson, Margaret. *Ninth Astronomer Royal: The Life of Frank Watson Dyson*. Cambridge: W. Heffer, 1951.

Index

Illustrations found throughout the text are in bold; those in the illustration section are referenced by their photograph number, also in bold.